Mathematical
Perspectives

Professor Dr. Kurt-Reinhard Biermann

Mathematical Perspectives

Essays on Mathematics and Its Historical Development

Edited by

JOSEPH W. DAUBEN

Department of History
Herbert H. Lehman College
City University of New York
Bronx, New York

and

The Graduate Center
City University of New York
New York, New York

Presented to Professor Dr. Kurt-Reinhard Biermann on the Occasion of His 60th Birthday

1981

ACADEMIC PRESS
A Subsidiary of Harcourt Brace Jovanovich, Publishers
New York London Toronto Sydney San Francisco

ACADEMIC PRESS, INC.
111 Fifth Avenue, New York, New York 10003

United Kingdom Edition published by
ACADEMIC PRESS, INC. (LONDON) LTD.
24/28 Oval Road, London NW1 7DX

Library of Congress Cataloging in Publication Data
Main entry under title:

Mathematical perspectives.

A collection of papers in honor of Kurt-R. Biermann.
English, French, or German.
Includes bibliographies and index.
1. Mathematics--History--Addresses, essays, lectures.
I. Dauben, Joseph Warren, Date. II. Biermann, Kurt-R.
QA21.M36 510'.9 80-1781
ISBN 0-12-204050-3

PRINTED IN THE UNITED STATES OF AMERICA

81 82 83 84 9 8 7 6 5 4 3 2 1

Contents

List of Contributors

Numbers in parentheses indicate the pages on which the authors' contributions begin.

Paul P. Bockstaele (1), Department of Mathematics, Katholieke Universiteit Leuven, B-3030 Heverlee, Belgium

Pierre Dugac (13), Université Pierre et Marie Curie, Paris, France

Wolfgang Eccarius (37), Amrastrasse 107, 59 Eisenach, Deutsche Demokratische Republik

E. A. Fellmann (47), Arnold Böcklinstrasse 37, CH-4051 Basel, Switzerland

*Menso Folkerts** (53), Universität Oldenburg, D-2900 Oldenburg, Bundesrepublik Deutschland

I. Grattan-Guinness (95), Middlesex Polytechnic at Enfield, Middlesex EN3 4SF, England

A. P. Juschkewitsch (247), Institute of the History of Science and Technology, Moscow K-12, USSR

*Present address: Institut für Geschichte der Naturwissenschaften der Universität München, Deutsches Museum, D-8000 München 26, Bundesrepublik Deutschland.

Eberhard Knobloch (139), Institut für Philosophie, Wissenschaftstheorie, Wissenschafts- und Technikgeschichte, Technische Universität Berlin, D-1000 Berlin 10, Bundesrepublik Deutschland

Uta C. Merzbach (167), The Smithsonian Institution,Washington, D.C. 20560

Olaf Neumann (179), Sektion Mathematik, Friedrich-Schiller-Universität, DDR-69 Jena, Deutsche Demokratische Republik

Ivo Schneider (201), Institut für Geschichte der Naturwissenschaften der Universität München, Deutsches Museum, D-8000 München 26, Bundesrepublik Deutschland

Christoph J. Scriba (221), Institut für Geschichte der Naturwissenschaften, Mathematik und Technik, Universität Hamburg, D-2000 Hamburg 13, Bundesrepublik Deutschland

Kurt Vogel (237), Isoldenstrasse 14, D-8000 München 40, Bundesrepublik Deutschland

Preface

The 13 essays in this collection were originally meant to comprise an issue of the journal *Historia Mathematica* in honor of Professor Kurt-R. Biermann's 60th birthday. In planning the volume, allowance was made for the possibility that everyone agreeing initially to contribute a paper might not manage to meet the final deadline. This, however, did not prove to be the case, and when all contributions were assembled, it appeared that the entire collection had the makings of a book. In offering these essays as a supplement to *Historia Mathematica*, Academic Press has acknowledged the original intention of the journal to honor Professor Biermann, a member of its Editorial Board, for his many contributions to the history of mathematics. But it is also an indication of the growing interest in the subject that this collection of essays reflects a range of scholarship that is substantial in its own right.

Professor Biermann's many contributions to the history of science are described in the biographical essay which begins this volume. Moreover, the range of his scholarly interests and the extent of his writing may be judged by the selected list of his publications at the end of this book. The 13 articles which follow the biographical essay are a measure of the esteem in which he is held by his colleagues everywhere. Moreover, the editors of *Historia Mathematica* are pleased that these authors could join forces in a supplement to the journal's usual quarterly appearance. We dedicate this book to Professor Biermann, with great appreciation for his scholarly contributions to the discipline and his professional contributions to the promotion of the history of mathematics throughout the world. Among these contributions are the strong support he has always given to *Historia Mathematica*, and the service he continues to render as a member of the Executive Committee of the International Commission on the History of Mathematics.

Kurt-Reinhard Biermann

Kurt-R. Biermann was born on December 5, 1919, in Bernburg. He left school with the *Abitur* from the Lessing-Gymnasium in Berlin in 1938, and thereafter continued his studies from 1940 until 1943 at the Technische Hochschule in Charlottenburg and Stuttgart. Following the interruption of his education by the Second World War, he began a series of correspondence courses in 1952 through the Technische Hochschule in Dresden, which he continued until 1956. The following year he transferred to the Humboldt University in Berlin, German Democratic Republic, where he received his Ph.D. (Dr. rer. nat.) in 1964, summa cum laude. In 1968 he was habilitated to teach the history of mathematics.

Biermann has been active at the Academy of Sciences of the German Democratic Republic since 1952; from 1956 until 1966 he served as Secretary of the Euler Commission. As Secretary of the Commission, Biermann came in contact with A. P. Yushkevich [born 1907; see *Historia Mathematica 3,* 259–278 (1976)], with whom Biermann has been amicably associated ever since. He has also been a scientific member of the Alexander-von-Humboldt Research Institut since 1958, and has served as its director since 1969.

Encouraged by Joseph E. Hofmann [1900–1973; see *Historia Mathematica 2,* 137–152 (1975)], Biermann began to study the manuscripts of G. W. Leibniz on combinatorics and probability theory in 1954. Later he extended these studies to include periods prior to and following the era of Leibniz. Similarly, his activities in connection with the Humboldt Research Institute also occasioned a further extension of his scholarly interests, and he was led to combine his Humbolt studies with his mathematical–historical concerns to investigate thoroughly the connections Humboldt had with German and French mathematicians. This resulted in a large number of biographical articles on mathematicians who had corresponded with Humboldt, who were encouraged and aided by him, or who had various other contacts with him. Included in the list of such mathematicians are N. H. Abel, T. Clausen, A. L. Crelle, C. G. J. Jacobi, F. Minding, J. Steiner, and F. Woepcke. Above all, Biermann concerned himself intensively with P. G. Lejeune Dirichlet and G. Eisenstein. Through the exploitation of previously unnoticed surviving documents, he has done much to illuminate the essential details of the biographies of the latter two mathematicians. Among Biermann's publications on these subjects, mention is made here only of *Dirichlet. Dokumente für sein Leben und Wirken* (Berlin: Akademie-Verlag, 1959) and the contribution to Eisenstein's *Mathematische Werke, 2* (New York: Chelsea Publishing Company, 1975), 919–929. As for the edition of Eisenstein's works, Biermann exercised considerable influence upon the individual items to be included, as well as its general organization.

Biermann has also studied intensively the relations between Humboldt and C. F. Gauss, whose life and work have been of special interest to him since 1959. The publication of a great number of papers was followed in 1977 by his edition of the correspondence between Humboldt and Gauss (Berlin: Akademie Verlag), which appeared as Volume 4 in the series *Beiträge zur Alexander-von-Humboldt-Forschung*, of which Biermann is the editor. Of special interest are a number of articles in which Biermann has also deciphered and interpreted encoded notes by Gauss.

Over the years, Biermann has broadened his research to include nearly all German mathematicians who have worked in Berlin, especially Weierstrass, a subject that provided yet another focus for Biermann's investigations [see, for example, his article in the *Journal für die reine und angewandte Mathematik, 223*, 191–220 (1966)]. All of these detailed studies were incorporated into Biermann's major study of mathematics and mathematicians at the University of Berlin, namely: *Die Mathematik und ihre Dozenten an der Berliner Universität. 1810–1920* (Berlin: Akademie-Verlag, 1973). This work has met with great interest and has been universally acclaimed by critics the world over. For example, Ivor Grattan-Guinness called it a "magnificent book, which will not only forever be authoritative for its particular subject-matter but also stand as a model of institutional history within a scientific discipline" [*Annals of Science, 32*, 404 (1975)]. Biermann went on to publish a series of documents related to the history of the position of mathematics and mathematicians in the Berlin Academy. For the first time, he drew upon a previously untapped source for the history of mathematics: the scholarly recommendations (*Laudationes*) that were officially considered in the election of new members. They convey an impression of the appraisal of many well-known mathematicians by competent contemporary specialists. Ever since Biermann's first paper on this subject appeared (*Vorschläge zur Wahl von Mathematikern in die Berliner Akademie* (Berlin: Akademie-Verlag, 1960)), it has been an example followed by other authors.

The selected bibliography of Biermann's publications listed at the end of this volume includes eight monographs and 179 scientific papers (among them biographies for the *Dictionary of Scientific Biography*) which have appeared in 14 different countries. The list does not include numerous short biographies, popular science articles, or reviews. In all of his work, exactness and reliability have been goals to which he has aspired. Hallmarks of his work include the mining of unpublished archival sources and the critical use of the existing literature as well. Biermann has succeeded in generalizing the experience he has gained, in particular with the dating of manuscripts, in order to make these results accessible to other historians of science.

Biermann's contributions have been recognized in numerous ways; in 1966 he was elected a corresponding member of the Académie internationale d'histoire des sciences, and in 1971 he was made a *Membre effectif* of the Academy; since 1972 he has been a member of the Deutsche Akademie der Naturforscher Leopoldina. Beginning in 1968, Biermann has assumed various functions for the International Union of the History and Philosophy of Science/Division of the History of Science. From its inception in 1971, he has also served as an active member of the Executive Committee of the International Commission on the History of Mathematics of the IUHPS/DHS.

Since 1952 Biermann has been married to Dr. med. dent. Elisabeth Biermann-Appuhn. They are the parents of two sons, Rainer and Jörg. Currently, Biermann lives in Buch, a suburb of Berlin.

This supplement to *Historia Mathematica* is dedicated to Professor Dr. Kurt-R. Biermann on the occasion of his 60th birthday, which was celebrated on December 5, 1979. It is with sincere best wishes for the future from the contributors to this volume, and from Professor Biermann's many colleagues and friends in all parts of the world, that this collection of papers is published as an expression of the esteem in which he is held by historians and mathematicians everywhere.

JOSEPH W. DAUBEN

Adrianus Romanus and Giovanni Camillo Glorioso on Isoperimetric Figures*

PAUL P. BOCKSTAELE

This paper begins with a consideration of studies and commentaries on the problem of "figures of equal perimeter" by various authors in antiquity, including Zenodorus, Theon, and Pappus. Early in the 17th century, new attempts were made to investigate such figures by Joannes Broscius, Adrianus Romanus, and Giovanni Camillo Glorioso, starting from the work of Petrus Ramus. In his *Geometria* of 1569, Ramus discussed the isoperimetric problem, including the theorem that "Among isoperimetric homogeneous figures, the most regular one has the greatest area." The first to criticize Ramus' audacious reformulation of the Greek theory on isoperimetric figures was the Polish mathematician Jan Broscius. In correspondence with Adrianus Romanus in 1610, a counterexample to Ramus' theorem was discussed, leading to publication of their letters on the subject in 1615. Later, while studying medicine at Padua, Broscius became friends with Glorioso, who used experimental methods to investigate the same questions Broscius had sent to Romanus.

The problem of the relations between the perimeter of a plane figure and its area, and between the area of a solid figure and its volume, had been raised early by several Greek authors. The first one to treat this problem mathematically seems to have been Zenodorus (probably 2nd cent. B.C.). His work is lost, but Theon of Alexandria (end of the 4th cent. A.D.) reported on it in his commentary on the *Almagest* of Ptolemy. According to Simplicius

* For K.-R. Biermann on his 60th birthday.

ISBN 0-12-204050-3

(first half of the 6th cent. A.D.) Archimedes too is said to have written a book "on figures of equal perimeter." It is not clear, however, whether Simplicius refers to a separate work, now lost, or to some propositions from *On the Sphere and Cylinder* or *Measurement of the Circle*. Using Zenodorus' work, Pappus (end of the 3rd cent. A.D.) treats the isoperimetric problem in Book V of his *Mathematical Collection* [1]. In its original form, this problem was to find among all closed plane curves with a given perimeter the one that bounds maximum area, and its spatial analogue. Pappus proves among other things the following theorems on isoperimetric plane surfaces:

1. Among regular polygons of equal perimeter, the one having more sides has greater area (Book V, Prop. 1).
2. The circle has greater area than a regular polygon of the same perimeter (Book V, Prop. 2).
3. Among n-sided polygons of the same perimeter, the regular one has greatest area (Book V, Prop. 10).
4. Among triangles with the same base and equal perimeter, the isosceles has the greatest area, and the more a triangle approximates the isosceles, the greater its area (Book V, Prop. 5). This theorem is equivalent to: Among two triangles with the same base and equal perimeter, the one wherein the difference between the two other sides is smaller, has greater area.

Until the discovery of the calculus at the end of the 17th century, virtually nothing was added to the works of Zenodorus, Theon, and Pappus on isoperimetric figures. However, the interest in this problem never disappeared completely. Considerations of isoperimetric figures are found in astronomical writings wherein the form of the celestial bodies is discussed. The fact that, among all bodies with equal surface, the sphere is the one with the greatest volume, has often been used as an argument in favor of the spherical form of earth and heavens. Still at the end of the 16th century, Christoph Clavius published, in view of this argument, a lengthy treatise on isoperimetric figures in his Commentary on Sacrobosco's *Sphaera* [Clavius 1570] and later once again in his *Geometria practica* [Clavius 1604]. In essence though, he does not go any farther than what can already be found in Zenodorus.

In this paper, I want to deal with an attempt, made at the beginning of the 17th century, to throw some new light on the comparison of the areas of polygons with equal perimeter. The actors in our story are Jan Brożek or Broscius (1585–1652) [2], Adriaan van Roomen or Romanus (1561–1615) [3], and Giovanni Camillo Glorioso (1572–1643). Our point of departure is the mathematical work of the French university professor and educator Pierre de la Ramée or Ramus (1515–1572) [4]. In March 1544, a verdict forbade him to continue his philosophy teaching, because of his anti-Aristotelism.

A consequence was his growing interest in mathematics. Initially, his attention was mainly drawn to arithmetic, but from 1558 onward, he became more intensively engaged in geometrical problems. He studied Euclid's *Elements*, the *Spherics* of Theodosius, and several works of Archimedes. He equally took on Apollonius, Serenus, and Pappus. His interest in isoperimetric figures dates from this period. He knew the Greek contribution to this problem from Pappus' *Mathematical Collection*, and was aware that, according to Simplicius, Archimedes was to have written a work on "figures with equal perimeters." Ramus often tried to purchase this work. In March 1564 he wrote to Roger Ascham (1516–1568) in Cambridge that he had heard about a physician in Ascham's circle who was said to own Archimedes' work. In his letter, he asked for a copy, in exchange for which he offered texts of Pappus, Apollonius, and Serenus [5]. Shortly afterward, Ramus was told that John Dee's library contained a number of mathematical manuscripts, among which was Archimedes' "on isoperimetric figures" [6]. In a letter dating from 19 December 1565, he asked Dee to provide him with a copy of this manuscript [7]. We do not know whether Dee ever answered Ramus' letter. In any case, Ramus never seems to have gotten hold of Archimedes' work.

Ramus' interest in isoperimetric figures left its traces in his 1569 *Geometria* [Ramus 1569]. In this work, the isoperimetric problem is discussed in two places: incidentally in Book 19 and more fundamentally in Book 4.

Book 19 is devoted to practical geometry, mainly to the calculation of areas. Paragraph 6 starts off with the following theorem: "Among heterogeneous plane figures of equal perimeter, the circle has the greatest area." In support of this proposition, Ramus compares the area of an isosceles triangle, a square, and a circle, all with perimeter 24. Then he quotes Quintilian's *Institutio oratoria*, Book I, 10, where it is stated that many historiographers were mistaken in holding the view that figures with an equal perimeter have an equal area. By means of a number of isoperimetric rectangles and triangles, Ramus illustrates the incorrectness of this view. To show that the area of isoperimetric regular polygons increases with the number of the sides, he calculates the area of a number of regular polygons with perimeter 360. The proof of this statement, given by Ramus, is, however, quite incomplete. He concludes with the theorem that the circle has greater area than any of the other isoperimetric plane figures.

A more fundamental discussion of the isoperimetric problem is presented by Ramus in Book 4 of his *Geometria*. As a consequence of his rather particular ideas on logic and mathematics, it differs considerably from the classic approach. In order to understand better Ramus' basic ideas on the place of isoperimetrics in geometry, something must first be said about his logical conceptions. According to Ramus, the basis of each of the seven liberal arts consists of a number of principles, which bring out a substantial property of the things under consideration, and therefore need no demonstration, but only some explanation through examples. In defining these principles and in establishing the several arts, one has to respect a number of logical laws, especially the three laws of method: the *lex de omni* or *lex* κατὰ παντός (the law of universal application), the *lex per se* or *lex* καθ' αὑτό (the law of essential application), and the *lex de universali* or *lex* καθ' ὅλου (the law of total application). W. Ong interprets them as follows in his study on Ramus' method [Ong, 260]:

> As applied to an art, the first law is said to mean that in any art a statement is to be taken in its full extension, as admitting no restriction or exception. The second law is said to mean that in an art all statements must "join" things necessarily related—cause and effect, subject and proper adjunct, and so on. The third law means that all statements in an art admit of simple logical conversion—a position which equivalently says that all statements in an art are statements of definitions.

The last law implies that a quality of a genus cannot be predicated of a species, and inversely. Consequently, Ramus had to avoid in geometry all theorems formulating special cases of general assertions. And yet he was led to make generalizations, often very audacious ones. A typical example is his theory on isoperimetrics in Book 4 of his *Geometria*. Compiling the Greek contributions to the isoperimetric problem, he noticed some theorems in which it was proved that the area of a figure is greater, the more it approximates regularity. An example is the above Theorem 4. He studied a similar situation in Book 19 of his *Geometria*, wherein he compared isoperimetric rectangles. Ramus considered these theorems to be special cases of a fundamental property of isoperimetric figures: the more regular a figure, the greater its area. This principle, which, according to Ramus, is "the first in its genus, and therefore needs no demonstration," is formulated as follows:

> Ex isoperimetris homogeneis ordinatius est maius; ex heterogeneis ordinatis terminatius.
>
> Among isoperimetric homogeneous figures (polygons), the most regular one has the greatest area; among heterogeneous but regular figures, the one with more sides has greater area.

Ramus provides no further explanation of this postulate. He just remarks that "more regular" means the same as "less irregular," and adds some examples:

> Ordinatius vero in theoremate intelligatur etiam pro minus ordinato. Sic triangulum aequilaterum erit maius isoperimetro inaequilatero, et aequicrurum vario. Sic in

quadrangulis quadratum maius non quadrato. Sic oblongum ordinatius est maius minus ordinato oblongo. Sic ex heterogeneis ordinatis quadratum maius triangulo, et circulus quadrato, quia quadratum est πολυπλευρώτερον, et pluribus terminis constat quam triangulum, et circulus quam quadratum.

The term "more regular" in this theorem is also to be understood in the sense of "less irregular." Thus an equilateral triangle has greater area than an isoperimetric one with unequal sides, and an isosceles is greater than a scalene. Thus, a square is greater than any other quadrangle, and an oblong quadrangle will be greater than any other quadrangle, more oblong and more irregular. Among heterogeneous regular polygons, the square is greater than the triangle and the circle greater than the square, since the circle has more sides than the square and the square more than the triangle.

The last part of Ramus' proposition is nothing else than the above Theorem 1. The first part goes far beyond the Greek contribution to the problem. Translated into our language, it states that the set of n-sided polygons with equal perimeter can be ordered by the relation "more regular," and that for this ordering, the area is an increasing function on this set.

During several years, this audacious reformulation of the Greek theory on isoperimetric figures remained unnoticed. The first to formulate some criticism, in 1610, was the young Polish mathematician Jan Brożek. Comparing an isosceles triangle with sides 10, 20, 20 with an isoperimetric scalene triangle with sides 19, 20, 11, he discovered that the latter has greater area. This is in contradiction with Ramus' predictions. In a letter from Krakow, dated 1 October 1610, Brożek submitted the problem to Adriaan van Roomen, who at that time was introducing the young nobleman Thomas Zamojski (1594–1638) into the mathematical sciences at Zamość in Poland. Van Roomen sent his answer a few months later. In full admiration for the answer he received, Brożek decided to publish it, together with his own initial letter. The booklet containing this correspondence appeared in Krakow in 1615 under the title *Epistolae ad naturam ordinatarum figurarum plenius intelligendam pertinentes* (*Letters Leading to the Acquisition of a Deeper Insight into the Nature of Regular Polygons*). In his dedication to the reader, Brożek justifies his publication as follows: "Occurit quaedam difficultas in nobilissimo Geometrarum theoremate de figuris ordinatis: eiusque nodi solutionem petii et impetravi ab Adriano Romano Mathematico celeberrimo. Quae vero ab illo percepi Lector, ea te diutius celare nolui" ("A difficulty arose concerning the famous geometrical theorem on regular polygons. I asked for the solution and received it from the famous mathematician Adrianus Romanus. I don't want to keep away from you any longer, dear reader, what he sent to me.") [8].

Brożek's letter to Van Roomen opens with a quotation of Ramus' theorem and its application on isoperimetric triangles: "Among isoperimetric triangles, the equilateral one has greater area than the one with unequal sides and an isosceles has greater area than a scalene." When Brożek checked this last statement, he was forced to conclude that something was wrong.

> Atqui hoc posterius non est καθα παντος: nam si proponantur duo triangula, quorum unum aequicrurum habeat latera 10. 20. 20. alterum vero varium 19. 20. 11. manifestum est per Geodaesiam trianguli hoc maiorem capacitatem habere, nimirum $102.\frac{96}{205}$ quam illud licet sit aequicrurum; habet enim tantum $96.\frac{159}{193}$. Addendum est itaque aliquid theoremati, ut sit καθ' ὅλουπρῶτον.
>
> This last statement, however, cannot be universally applied. If one takes e.g. two triangles, the one an isosceles with sides 10, 20, 20, the other a scalene with sides 19, 20, 11, than it is clear that the latter has greater area, viz., $102\frac{96}{205}$, than the former, which has only $96\frac{159}{193}$, although it is an isosceles. Something has to be added to the theorem, in order to guarantee its universal applicability.

Brożek is clearly referring to Ramus' laws of method, which are evidently not respected in the theorem mentioned above. This is the reason why he would like to know Van Roomen's opinion in this matter.

Van Roomen's extensive answer probably reached Brożek in the early part of 1611. According to Van Roomen's judgment, the theorem that "among isoperimetric homogeneous figures, the most regular one has greater area" is right, but Ramus was wrong in applying it on isosceles and scalene isoperimetric triangles. What is lacking is a clear definition of the meaning of "more regular": "Quid igitur per ordinatius aut minus inordinatum sit intelligendum, id apertius definiendum erat Ramo." Van Roomen will try to fill in the gap by formulating a clear mathematical definition of the terms "more regular" and "less irregular." He proposes two criteria to compare the degree of regularity of two polygons:

> 1. Figuram illa est magis ordinata, cuius maximum latus ad laterum differentiam maximam, rationem habet maiorem. Item, maximus angulus ad differentiam angulorum maximam, rationem habet maiorem, adeo, ut si rationem habeat infinitam, ea sit absolute ordinata.
> 2. Figuram illa est magis ordinata, cuius perimeter circulo circumscripta ad radium circuli, minorem habet rationem. Haec conditio vera quidem est, sed non ita universalis; neque enim omnis figura est circulo circumscriptibilis. Si tamen circumscriptionem lato modo accipere velimus, ita ut non sit necessarium omnia latera tangere, iudicavero statui posse universalem.
>
> 1. A polygon is more regular, in as far as the ratio of its greatest side to the greatest difference of two sides is greater; and in as far as the ratio of its greatest angle to the greatest difference of two of its angles is greater. When the ratio is infinite, then the figure is absolutely regular.
> 2. Among polygons, the more regular one is the one for which the ratio of its perimeter to the radius of the inscribed circle is the smallest. This condition is true indeed, but not universally applicable, because not every polygon can be circumscribed to a circle. When we accept however circumscription in the sense that not each side has to be tangent to the circle, I think that we may take the condition to be general.

Taking the first criterion as a definition of "more regular," Van Roomen reassumes Ramus' theorem:

> Ex duobus Isoperimetris homogeneis id quod est ordinatius (hoc est cuius maximum latus ad maximam laterum differentiam, maximusque angulus ad maximam angulorum differentiam rationem habet maiorem) id etiam esse maius.

> Among two isoperimetric homogeneous polygons, the most regular one (i.e., the one of which the greatest side has a greater ratio to the greatest difference of two sides, and the greatest angle to the greatest difference of two angles) has the greater area.

No demonstration, however, is attempted. In its general form, the first criterion is indeed inadequate, because it only takes into account the polygon's greatest and smallest sides and angles. Even in the case of a triangle, the most regular one in the sense of Criterion 1 does not necessarily have the greatest area [9]. But it is applicable to some specific categories of polygons, e.g., to equiangular polygons with alternating equal sides or to equilateral polygons with alternating equal angles. Van Roomen chooses exactly this type of polygon to illustrate the adequacy of his criterion. As a first example, he takes five equiangular hexagons with alternating equal sides and perimeter 6. The ratio of two successive sides is consecutively

1 to 1 5 to 4 4 to 3 3 to 2 2 to 1.

The ratio of the greatest side to the greatest difference of two sides is respectively

infinite 5 to 1 4 to 1 3 to 1 2 to 1.

The five polygons thus are placed according to their decreasing regularity, the first being "absolute ordinatum," the last one "valde inordinatum"; the regularity of the other three decreases gradually. Van Roomen concludes that the area will decrease accordingly. It is consecutively

$$2\tfrac{5980}{10000} \quad 2\tfrac{5873}{10000} \quad 2\tfrac{5803}{10000} \quad 2\tfrac{5634}{10000} \quad 2\tfrac{5020}{10000}.$$

As a second example, Van Roomen presents four equilateral hexagons with perimeter 6 and alternating equal angles. The unequal angles relate to each other as

1 to 1 7 to 5 [10] 2 to 1 11 to 4.

The ratio of the greatest angle to the greatest difference of two angles is

infinite 7 to 2 2 to 1 11 to 7.

The area of the polygons is

$2\tfrac{5980}{10000}$ $2\tfrac{4936}{10000}$ $2\tfrac{1928}{10000}$ [11] $1\tfrac{8345}{10000}$ [12].

Van Roomen still remarks that if the equal angles or sides are not alternating, the area will decrease even more intensively. To illustrate this (see the figure), he takes an equilateral hexagon *ABCDEF* with perimeter 6, its four obtuse angles *A*, *B*, *D*, and *E* being equal, just as are its two acute angles *C* and *F*. The ratio of an obtuse angle to an acute is as 19 to 2. He does not pursue this example any further. He ends his letter with the words: "Sed iam lusum est satis. Ex hisce veritatem propositionis colliges satis." ("But enough of this game. The correctness of the theorem has been amply illustrated.")

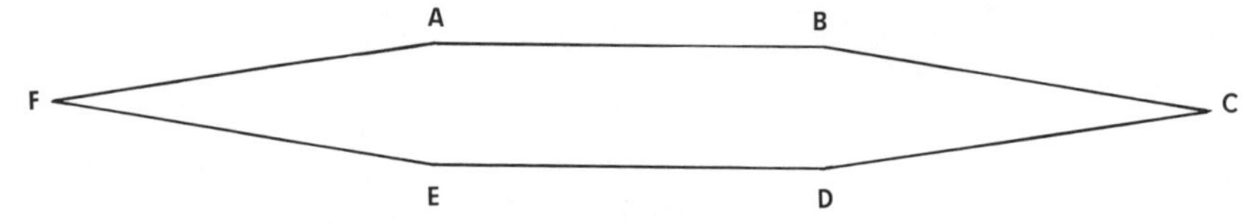

Van Roomen's second criterion is fully applicable to polygons in which a circle can be inscribed (*strictu sensu*); it is useless for the other ones, among other reasons because it is not clear which inscribed circle is referred to.

We already pointed out how greatly Brożek admired Van Roomen's solution of the problem at hand. This becomes even more clear when he publishes it once again unaltered in his *Apologia pro Aristotele* [Brożek 1652, 43–45]. Even the faults we indicated remained uncorrected. Apparently, Brożek never read Van Roomen's answer with a critical eye. It would have been easy to undermine it, in the same way he did with Ramus' theorem, by simply giving counterexamples [9].

During the years 1620–1624, while studying medicine in Padua, Brożek became friends with Giovanni Camillo Glorioso, who since 1613 had been professor of mathematics at the university. He submitted to him the same problem that he had submitted to Van Roomen. Glorioso, who did not know Van Roomen's answer, published his comment in his *Exercitationum mathematicarum decas prima* [Exercitatio decima, pp. 97–107]. Unlike Van Roomen, his method was experimental. He compared the area of all kinds of isoperimetric polygons, from which he deduced the five following theorems:

Theorem 1. Among isoperimetric n-sided polygons, the regular one has greater area.

Comparing rectangles and trapezoids with the same perimeter, he came to the conclusion that the area of a rectangle can be greater, equal or smaller than the area of the trapezoid. Taking for granted that a rectangle is more regular than a trapezoid, he summarized the results in

Theorem 2. Among n-sided irregular polygons of the same perimeter, the most irregular one can have greater, equal or smaller area.

Theorem 3. Among regular polygons of equal perimeter, the one having more sides has greater area.

The comparison of isoperimetric polygons with an unequal number of sides brought him to

Theorem 4. Among heterogeneous nonregular polygons, the one with the greatest number of sides can have greater, equal or smaller area.

Theorems 1 and 3 are synthesized in

Theorem 5. Among isoperimetric heterogeneous polygons, partly regular and partly irregular, the regular one with the greatest number of sides has the greatest area.

Glorioso notes the fact that Theorems 1 and 3 can be found in Theon and Pappus, and that Theorems 2 and 4 have been demonstrated by Clavius in his *Geometria practica*. Discussing Theorems 2 and 4, Glorioso concludes that nobody has yet been able to formulate a general rule on the comparison of the areas of nonregular isoperimetric figures. Considering, however, the fact that among isoperimetric n-sided polygons the regular one "may be taken as rule and norm" and has the greatest area, he restates Ramus' theorem as follows:

> Ex isoperimetris homogeneis universis ordinatum est maius, ex inordinatis ordinatius, hoc est quod magis accedit ad ordinatum.
>
> Of all isoperimetric homogeneous polygons, the regular one has the greatest area. Among isoperimetric nonregular n-sided polygons, the most regular one, i.e., the one that most approximates the regular one, has greatest area.

Glorioso confesses that he is not completely certain about this theorem, but that it seems reasonable to accept it. He ends with the wish to know Van Roomen's opinion about this subject.

If one is to compare the solutions proposed both by Van Roomen and Glorioso, one comes to the conclusion that they are to a great extent analogous. The two mathematicians accept Ramus' theorem as right, but point out that the term "more regular" has to be defined more accurately. Both propose to call a polygon "more regular," the more it approximates the regular polygon with the same number of sides. It remains Van Roomen's merit that he tried to formulate a mathematical formula for this approximation. Surprisingly, he only includes the greatest and the smallest sides and angles. A few experiments might have taught him that all differences between two sides and between two angles have to be taken into consideration.

In 1652, Brożek took up the whole question again in his *Apologia pro Aristotele* [Brożek 1652, Chapters XIV and XVIII, 40–61]. He quotes with high praise the answer of Van Roomen and also mentions Glorioso's work. He adds some further criticism on Ramus, using nonconvex polygons, but he does not make any substantial contribution to the problem.

NOTES

1. For the history of the isoperimetric problem in antiquity, see Müller, 1953.
2. B. Knaster, "Brożek, Jan" in *Dictionary of Scientific Biography*, vol. 2, 526–527.
3. On Van Roomen, see Bockstaele, 1966.
4. On Ramus as a mathematician, see Verdonk.
5. "Alterum est de libro Archimedis περὶ ἰσομέτρων, quem audivi penes quendam eruditum vestrae aulae medicum esse: si facultas ulla sit describendi, habeo rariora quaedam in hoc genere, et Pappi et Apollonii et Sereni, quae perlubenter vicissim cum eo communicabo." [Ascham, 97]
6. In the catalogue of the manuscripts in Dee's library, compiled by Dee himself in 1583, a work of Archimedes "on figures with equal perimeters" is mentioned twice: under nr. 100 as "Archimedes de figuris isoperimetris," and under nr. 109 as "Archimede de figuris ysoperimetrorum" [Halliwell, 77–78].
7. "Fac, amabo te, ut elenchus tuorum mathematicorum nobis communicetur, ut si quid divitiis tuis nobis opus erit, a te opem expetamus, et quidem tam liberaliter, quam ingenue nostra tibi communia fieri velim: ἰσοπερίμετρα vero Archimedea quae sint et quanta vehementer aveo scire. Habeo enim Pappi quaedam ejusdem argumenti paulo ampliora, quae suspicor eadem illa esse." [Ramus and Talaeus, 174]
8. The correspondence between Brożek and Van Roomen is also to be found in Bockstaele 1976, letters 39 and 40, pp. 292–298.
9. For instance, of the two isoperimetric triangles 41, 23, 36 and 45, 26, 29, the latter has the smaller area, although it is the more regular one in the sense of Van Roomen's first criterion.
10. In the *Epistolae*, there is a 3 instead of the correct 5.
11. The *Epistolae* wrongly says: 7928.
12. In the *Epistolae*, there is the wrong $2\frac{8345}{10000}$ instead of the correct $1\frac{8345}{10000}$.

REFERENCES

Ascham, Roger. 1864. *The whole Works of Roger Ascham, now first collected and revised with the Life of the Author by J. A. Giles* (Vol. 2, London: John Russell Smith).

Bockstaele, Paul P. 1966. "Roomen, Adriaan van," *Nationaal Biografisch Woordenboek* (Vol. 2, Brussels: Koninklijke Vlaamse Academiën van België), 751–767.

Bockstaele, Paul P. 1976. "The correspondence of Adriaan van Roomen," *Lias*, *3*, 85–129, 249–299.

Brożek, Jan. 1615. *Epistolae ad naturam ordinatarum figurarum pleniùs intelligendam pertinentes* (Krakow: Andreas Petrocovius).

Brożek, Jan. 1652. *Apologia pro Aristotele et Euclide, contra Petrum Ramum, et alios. Additae sunt duae Disceptationes de Numeris Perfectis* (Danzig: Georg Förster).

Clavius, Christoph. 1570. *In Sphaeram Ioannis de Sacro Bosco Commentarius* (Rome: Victorius Helianus). There were several other editions.

Clavius, Christoph. 1604. *Geometria practica* (Rome: Aloisius Zannetti; Mainz, 1606).
Glorioso, Joannes Camillus. 1627. *Exercitationum mathematicarum Decas prima. In qua continentur varia et theoremata et problemata, tum ei ad solvandum proposita, tum ab eo inter legendum animadversa* (Naples: Lazarus Scorigius).
Halliwell, J. O. ed. 1892. *The private Diary of Dr. John Dee, and the Catalogue of his Library of Manuscripts, from the original manuscripts in the Ashmolean Museum at Oxford, and Trinity College Library, Cambridge* (London: Camden Society).
Müller, Wilhelm. 1953. "Das isoperimetrische Problem im Altertum," *Sudhoffs Archiv, 37,* 39–71.
Ong, Walter J. 1958. *Ramus. Method, and the Decay of Dialogue, From the Art of Discourse to the Art of Reason* (Cambridge, Mass.: Harvard University Press).
Ramus, Petrus. 1569. *Arithmeticae libri duo, Geometriae septem et viginti* (Basel: E. Episcopius; other editions: Basel, 1580; Frankfurt, 1599 and 1627).
Ramus, Petrus, and Audomarus Talaeus. 1599. *Collectaneae Praefationes, Epistolae, Orationes* (Marburg: Paulus Egenolphus; reprinted, with an introduction by W. J. Ong, Hildesheim: Georg Olms Verlagsbuchhandlung, 1969).
Verdonk, J. J. 1966. *Petrus Ramus en de Wiskunde* (Assen: Van Gorcum).

Department of Mathematics
Katholieke Universiteit Leuven
Heverlee, Belgium

Des fonctions comme expressions analytiques aux fonctions représentables analytiquement*

PIERRE DUGAC

The concept of function constitutes the central core of the foundations of analysis in the 19th century: from its conceptualization as an analytic expression, as it was defined by Euler and was still current at the beginning of the 19th century, until the formulation of function representable analytically, culminating through numerous studies of the 19th century in the work of René Baire, research which contributed to the founding of a new analysis. Such is at least the general sense which emerges from recent research by the author [Dugac 1976, 1978, and 1979]. In these publications, the very rich publications of K.-R. Biermann have been used, in particular *Die Mathematik und ihre Dozenten an der Berliner Universität. 1810–1920* [1973], one of the most important of all books in the history of mathematics to appear in the last ten years.

LA NOTION EULÉRIENNE DE FONCTION

C'est en 1748, dans son *Introduction à l'analyse infinitésimale* [Euler 1835, 2], qu'Euler donne sa définition d'une fonction: "Une fonction de quantité variable est une expression analytique composée, de quelque manière que ce soit, de cette même quantité et de nombres, ou de quantités constantes."

* Kurt-R. Biermann freundschaftlichst gewidmet.

ISBN 0-12-204050-3

Il s'agit donc d'une "expression analytique" formée à partir des fonctions usuelles de l'analyse en utilisant les opérations élémentaires de l'arithmétique, la composition des fonctions, les séries et les produits infinis, l'intégration et la dérivation. Ainsi était formé le champ des fonctions "arbitraires," sans que les mathématiciens du XVIIIe siècle sentent un besoin urgent de définir avec précision les opérations arithmétiques répétées une infinité de fois, ni celles d'intégration et de dérivation.

De plus, les deux théorèmes clés de l'analyse classique se rapportant aux fonctions réelles de la variable réelle, à savoir le théorème des valeurs intermédiaires et le théorème des accroissements finis, n'étaient pas démontrés rigoureusement. La démonstration du premier consistait à vérifier son évidence géométrique, quant au second, il n'avait pas encore émergé sous forme analytique, sa forme géométrique reposant sur le fait également évident pour les mathématiciens de ce siècle qu'une fonction arbitraire était monotone par morceaux et que son graphe admettait une tangente parallèle à la corde.

A la fin du XVIIIe siècle, en 1797, Lagrange publie sa *Théorie des fonctions analytiques*, la plus vaste tentative, avant le début du XIXe siècle, pour donner à la science mathématique des bases rigoureuses. Son livre a deux objectifs principaux: le premier était de "débarasser le calcul différentiel des considérations métaphysiques d'infiniment petits"—d'éviter l'emploi de la "méthode des fluxions" qui introduit en analyse la notion "étrangère" de mouvement et l'usage de la notion de limite car elle "n'est pas assez claire pour servir de principe à une science dont la certitude doit être fondée sur l'évidence"—et le second objectif était de "rattacher le calcul au reste de l'algèbre de manière à ne faire du tout qu'une seule méthode."

Lagrange reprend dans son livre [Lagrange 1797, 1] la définition de fonction d'Euler et élabore une théorie des fonctions basée sur le développement des fonctions en série de Taylor, codifiant ainsi l'idée des mathématiciens du XVIIIe siècle que les fonctions qui interviennent en analyse—les fonctions "analytiques"—sont en général localement développables en séries entières. Cauchy, entre autres, critiquera les méthodes de Lagrange et ses critiques porteront sur les questions de convergence des séries et sur la validité du théorème de Taylor. Mais il ne faut pas perdre de vue que la tentative de Lagrange couronne une époque mathématique où le développement des fonctions en séries régnait en maître en analyse. Sa volonté de donner une théorie générale à cette nouvelle analyse, basée sur un principe simple, même si cette théorie est fondée sur un formalisme de calcul, aura une grande portée.

Il faut remarquer aussi que ce fut Lagrange qui donna la première démonstration, certes lacunaire, du théorème des accroissements finis dans ce traité sur les fonctions analytiques [Lagrange 1797, 45–49], "théorème nouveau et remarquable par sa simplicité et par sa généralité," ouvrant la voie au rôle essentiel que va jouer la fonction dérivée en analyse.

Ce fut Gauss qui établit le premier rigoureusement, en 1813 et dans le cas de la série hypergéométrique, quand une série représente effectivement une fonction. Il montre [Gauss 1813, 126] que la série considérée est convergente pour $|x| < 1$; elle est divergente pour $|x| > 1$ et "il ne peut être question de sa somme." En effet, puisque "notre fonction est définie comme la somme de la série, la recherche sur sa nature doit être restreinte aux cas où elle converge". Pour $x = 1$, Gauss fait ensuite une étude très soigneuse de la convergence de la série, ouvrant ainsi l'ère de la rigueur en analyse.

LA FONCTION CONTINUE AU SENS DE BOLZANO

C'est avec B. Bolzano que se produit la première extension de la notion de fonction, plus particulièrement de fonction continue, mais il ne faut pas oublier que pour les mathématiciens du début du XIXe siècle une fonction "arbitraire" était, entre autres, dérivable sauf en un nombre fini de points. En fait, Bolzano exprime, dans son mémoire de 1817 sur le théorème des valeurs intermédiaires, les préoccupations plus ou moins clairement manifestées de ses contemporains; et, en donnant un statut mathématique bien déterminé aux fonctions continues, ce que fera aussi Cauchy indépendamment en 1821, il posera la question de savoir ce qu'est une fonction et quelles sont les propriétés minimums qui la caractérisent, question à laquelle Dirichlet donnera une réponse d'une très grande généralité.

Bolzano avait déjà critiqué avec vigueur, dans son livre sur le *Théorème du binôme* paru en 1816, l'usage que faisaient les mathématiciens du XVIIIe siècle des séries infinies, et, dans les remarques finales de son livre [Bolzano 1816, 143–144], il notait que la notion de nombre irrationnel n'était pas encore sérieusement fondée.

Son mémoire de 1817 sur le théorème des valeurs intermédiaires est à notre avis capital, au point de vue où nous nous sommes placés dans ce travail, pour trois raisons:

1° Bolzano ne définit pas une fonction "arbitraire" continue par une "expression analytique," mais par des inégalités, ce qui va engendrer deux familles de fonctions continues et qui posera deux problèmes: Quelles sont les "expressions analytiques" qui sont continues au sens de Bolzano; et: Toute fonction continue au sens de Bolzano a-t-elle une expression analytique?

2° La démonstration de Bolzano du théorème des valeurs intermédiaires montre qu'il a eu le premier l'immense mérite de comprendre le processus logique de théorèmes en cascades qui conduit du théorème à démontrer, en passant par le théorème sur la borne supérieure d'un ensemble de nombres réels majoré, au critère "de Cauchy"—que Bolzano énonce dans ce mémoire avant Cauchy—critère dont la démonstration par Bolzano de la condition suffisante, démonstration tout à fait incomplète, laisse entrevoir, certes vaguement, la nécessité de "construire" l'ensemble des nombres réels.

3° Le théorème de Bolzano sur la borne supérieure d'un ensemble majoré est un des grands moments d'histoire des mathématiques: d'une part par sa démonstration "topologique" basée sur la dichotomie et, d'autre part, par l'influence que son énoncé et sa démonstration ont exercé sur Weierstrass.

Bolzano donne la définition suivante d'une fonction continue: dire qu'une fonction réelle f de la variable réelle x est continue, pour toutes les valeurs de x appartenant à un intervalle donné, ne signifie "rien d'autre que ceci: si x est une telle valeur quelconque, la différence $f(x-\omega)-f(x)$ peut être rendue plus petite que toute grandeur donnée, si l'on peut toujours prendre ω aussi petit que l'on voudra."

Il énonce ensuite, pour la première fois en histoire des mathématiques, le critère "de Cauchy" [Bolzano 1817, 150], ou plutôt la condition suffisante de ce critère:

> Si une suite de grandeurs $F_1(x)$, $F_2(x)$, $F_3(x)$, . . . , $F_n(x)$, . . . , $F_{n+r}(x)$ est telle que la différence entre son $n^{\text{ième}}$ terme $F_n(x)$ et tout terme ultérieur $F_{n+r}(x)$, aussi éloigné soit-il du $n^{\text{ième}}$, reste plus petite que toute grandeur donnée, si l'on a pris n suffisamment grand, alors il existe une certaine *grandeur constante*, et *une* seule, dont s'approchent toujours davantage les termes de cette suite, et dont ils peuvent s'approcher d'aussi près que l'on voudra, lorsqu'on prolonge la suite suffisamment loin.

Bolzano ne réussit pas à démontrer que son critère est une condition suffisante de convergence de la suite $(F_n(x))$, à savoir qu'il existe X tel que $\lim_{n\to+\infty} F_n(x)=X$. Il affirme seulement que l'on peut déterminer X "avec la précision que l'on voudra," et sa démonstration ne prouve rien d'autre que, si X existe—ce qui est à démontrer—alors, quel que soit $\varepsilon>0$, il existe un entier positif N tel que $n\geq N$ entraîne $|F_n(x)-X|\leq\varepsilon$. D'ailleurs, il ne pouvait pas faire une démonstration rigoureuse de son critère tant que l'ensemble des nombres réel n'était pas construit. Mais une relecture attentive de cette démonstration, à la lumière de son ouvrage inédit *Théorie des nombres pure*, publié en 1976, écrit vers 1835, indique que l'insistance de Bolzano sur le fait que l'on peut déterminer X, "avec la précision que l'on voudra," était le premier pas, certes encore intuitif en 1817, vers la démonstration correcte; ce pas sera précisé dans la preuve, encore lacunaire, de son critère qu'on trouve dans cet ouvrage de 1835.

Ce critère admis, Bolzano démontre ensuite rigoureusement le théorème sur la borne supérieure d'un ensemble majoré [Bolzano 1817, 153]: "Si une propriété M n'appartient pas à *toutes* les valeurs d'une grandeur variable x, mais appartient à *toutes* celles qui sont *plus petites* qu'un certain u, alors il existe toujours une grandeur U qui est la plus grande de celles dont on peut affirmer que toutes les valeurs inférieures x possèdent la propriété M."

La démonstration est basée sur la dichotomie d'un intervalle et l'application du critère de Bolzano à la suite de terme général $1/2^n$. Il en déduit ensuite [Bolzano 1817, 159–162], en utilisant la définition de la continuité d'une fonction et le théorème sur la borne supérieure d'un ensemble majoré, le théorème des valeurs intermédiaires.

CAUCHY ET LA CONSERVATION DE LA CONTINUITÉ PAR DES OPÉRATIONS DE L'ANALYSE

Nous allons encore voir avec Cauchy comment la transformation de l'analyse qui s'opère au début du XIXe siècle tourne essentiellement autour de l'axe formé par les fonctions "arbitraires" les plus générales, à savoir les fonctions continues, plus précisément continues par morceaux. Cournot a été bien conscient de ce changement en analyse lorsqu'il indiquait le pivot autour duquel allait s'articuler son traité de calcul infinitésimal [Cournot 1841, VIII]: "J'ai cherché à faire comprendre comment, par le progrès de l'abstraction mathématique, on est amené à concevoir l'existence d'une théorie qui a pour objet les propriétés générales des fonctions continues."

Mais cette mutation de l'analyse demandait la renovation de ses fondements, car, comme l'a bien vu Cauchy dans l'introduction à son *Cours d'analyse* de 1821, on tendait "à faire attribuer aux formules algébriques une étendue indéfinie, tandis que, dans la réalité, la plupart de ces formules subsistent uniquement sous certaines conditions et pour certaines valeurs des quantités qu'elles renferment." Pour déterminer ces conditions et ces valeurs, Cauchy sent le besoin de fixer "d'une manière précise le sens des notations" qu'il utilise, plus précisément des notions, ce qui fera que le concept de fonction, élément central constitutif des fondements de l'analyse, va à la fois se clarifier et se généraliser.

Cauchy redonne dans son *Cours d'analyse* la définition de la continuité d'une fonction et énonce le critère "de Cauchy" (dont il ne démontre pas lui non plus la condition suffisante), notions que l'on trouve déjà chez Bolzano dans son mémoire de 1817; mais il est nécessaire de remarquer que ces concepts vont être repris par d'autres mathématiciens grâce à l'extraordinaire influence de ce traité de Cauchy. Il donne également [Cauchy 1821, 17], pour la première fois, une définition, précise et rigoureuse, d'un infiniment petit: c'est une suite qui a "zéro pour limite." Il définit aussi un nombre irrationnel (une fois introduits les nombres rationnels à partir de la mesure des grandeurs) comme "la limite des diverses fractions qui en fournissent des valeurs de plus en plus approchées." Bien qu'on ait ici une première idée qu'un nombre irrationnel peut être défini par des suites de nombres rationnels, il est clair que, pour parler de la limite des nombres rationnels, il est nécessaire qu'au préalable l'ensemble des nombres réels soit déjà défini, ce dont Cauchy, comme d'ailleurs tous les mathématiciens de ce temps, n'avait pas pris conscience.

Ce qui est particulièrement intéressant dans ce cours, c'est la notation spéciale qu'utilise Cauchy pour ce qu'on nomme actuellement les valeurs d'adhérence d'une suite, et qu'il ne définit pas avec beaucoup de précision, mais contribue toutefois à généraliser la notion de limite. Il note seulement que si une suite de nombres réels (a_n) converge vers une "limite fixe," et si f est une fonction réelle de la variable réelle, alors la suite $(f(a_n))$ peut converger "à la fois vers plusieurs limites différentes"; il propose dans ce cas la notation (en prenant l'exemple donné par Cauchy, avec a_n tendant vers l'infini et $f = \sin$): $\lim((\sin a_n))$. C'est cet exemple qui aurait pu donner à Dirichlet l'idée d'une fonction continue et qui n'est pas monotone par morceaux (fonctions dont parlera Dirichlet en 1829): $f(O) = 0$ et $f(x) = x \sin(1/x)$ si $x \neq 0$.

Dans le paragraphe 1 du premier chapitre de son livre [Cauchy 1821, 32], *Considérations générales sur les fonctions*, Cauchy donne sa définition de la notion de fonction qui dérive de celle d'Euler, mais ce qui est important c'est sa caractérisation, qui contient implicitement la future définition de Dirichlet: "Pour qu'une fonction d'une seule variable soit complètement déterminée, il est nécessaire et il suffit que de chaque valeur particulière attribuée à la variable on puisse déduire la valeur correspondante de la fonction."

Mais le chapitre le plus important de ce cours de Cauchy est celui consacré aux séries convergentes et divergentes, où il définit [Cauchy 1821, 114], la première fois dans l'histoire des mathématiques, la notion de série convergente, élément essentiel de l'analyse en ce début du XIXe siécle. De plus, contrairement à tous les mathématiciens de son temps pour qui cela allait de soi, il essaie de montrer [Cauchy 1821, 120] que la propriété de continuité est conservée par le passage à la limite définissant la convergence d'une série: la somme d'une série de fonctions continues convergente dans un intervalle représente une fonction continue dans cet intervalle. Il s'agit là d'un théorème inexact, concernant une des plus importantes propriétés à cette époque en analyse, et qui va poser des questions aux chercheurs dont les réponses toucheront toutes les parties de l'analyse.

Un autre théorème inexact, également "évident" pour ses contemporains, énoncé dans ce *Cours d'analyse* de Cauchy, est intéressant pour notre propos [Cauchy 1821, 45–46]: Si une fonction réelle de deux variables réelles est séparément continue par rapport à x et y, alors elle est continue par rapport à (x, y).Ce théorème aura aussi une longue postérité qui culminera avec les travaux de Baire sur les fonctions représentables analytiquement, car il ramènera l'étude des propriétés d'une fonction de deux variables f séparément continue par rapport à x et y à celle d'une fonction d'une variable définie, en particulier, par $g(x) = f(x, x)$.

On trouve une autre erreur, d'où surgira par un long cheminement une notion importante, dans le *Résumé des lecons sur le calcul infinitésimal* de Cauchy, paru en 1823. Notons d'abord que Cauchy y définit, pour la première fois, la notion de dérivée [Cauchy 1823, 22]: c'est la limite, "lorsqu'elle existe," du rapport $[f(x+i) - f(x)]/i$ quand i tend vers zéro; définition qui va permettre de répondre à la question: Toute fonction continue est-elle dérivable?

Cauchy donne dans ce livre sa définition de l'intégrale pour une fonction continue sur un intervalle fermé et borné, et c'est là qu'on trouve l'origine de la théorie moderne de l'intégration, théorie qui sera un des éléments essentiels d'approfondissement de la notion de fonction. Mais, au cours de sa démonstration [Cauchy 1823, 123–125], Cauchy suppose implicitement que la fonction considérée est uniformément continue sur l'intervalle $[a, b]$. A force d'essayer de démontrer la possibilité de définir l'intégrale pour les fonctions continues, Cauchy se heurte, sans en prendre clairement conscience, à une de leurs propriétés associée à celle d'un intervalle fermé et borné. Lorsqu'on définit la continuité d'une fonction à l'aide de (ε, η), où $\eta = \eta(\varepsilon, x)$, $x \in [a\ b]$, définition que Cauchy maîtrise parfaitement dès 1821 [Cauchy 1821, 59–60], la question surgit de savoir si $\inf_{x \in [a,b]} \eta(\varepsilon, x)$ est strictement positif ou nul, c'est-à-dire si on peut ou non recouvrir, pour chaque $\varepsilon > 0$, l'intervalle $[a, b]$ par un nombre fini de sous-intervalles où la propriété considérée est vérifiée. Le premier pas dans cette direction sera fait par Dirichlet, vers les années 1850, lorsqu'il énoncera [Dirichlet 1904, 4] dans ses cours sur l'intégrale définie le théorème sur la continuité uniforme d'une fonction continue sur $[a, b]$.

Le coup de grâce à la théorie lagrangienne de fonctions sera donné par le contre-exemple fameux de Cauchy [Cauchy 1823, 229–230] montrant que si la série de Maclaurin d'une fonction f converge en un point x, elle n'est pas nécessairement égale à $f(x)$. Ainsi c'est l'ensemble des fonctions continues qui devient, vers les années 1820, le champ d'activité des mathématiciens, les fonctions continues étant dans leur esprit monotones et dérivables par morceaux.

C'est Abel qui montrera le premier, en donnant un contre-exemple [Abel 1826, 223–224], que le théorème de Cauchy sur la continuité de la somme d'une série convergente de fonctions continues est inexact. Ce qui est curieux, c'est que cette remarque d'Abel figure en note d'un théorème où il commet aussi une erreur de même nature que celle de Cauchy. En effet, il considère la série de terme général $v_n(x)\alpha^n$, les fonctions $x \mapsto v_n(x)$ étant continues dans l'intervalle $[a, b]$ et la série convergente pour $|\alpha| = \delta > 0$, et il veut montrer que la série est convergente pour $|\alpha| < \delta$ (ce qui est exact) et que sa somme est une fonction continue dans $[a, b]$. L'intérêt de sa preuve lacunaire est qu'il introduit l'expression

$$\theta(x) = \sup_{p \in \mathbf{N}}(|v_m(x)\delta^m|, |v_m(x)\delta^m + v_{m+1}(x)\delta^{m+1}|, \ldots, |v_m(x)\delta^m + v_{m+1}(x)\delta^{m+1} + \cdots + v_{m+p}(x)\delta^{m+p}|, \ldots),$$

et qu'il suppose implicitement que $\sup_{x \in [a,b]} \theta(x)$ est fini, introduisant ainsi une notion qui conduira directement Weierstrass, vers les années 1840, à la notion de convergence uniforme, et d'abord de convergence qu'on appelle depuis Baire normale.

Ainsi, l'infini pose de nouvelles questions aux mathématiciens et il leur en posera de plus en plus. Par ailleurs, il apparaît nécessaire, si l'on veut conserver les propriétés des fonctions dans les opérations où intervient l'infini, qu'elles soient "uniformes."

DIRICHLET ET LES "FONCTIONS QUI NE SONT ASSUJETTIES Á AUCUNE LOI ANALYTIQUE"

Dans son mémoire de 1829, *Sur la convergence des séries trigonométriques qui servent à représenter une fonction arbitraire*, Dirichlet démontre que si une fonction réelle f est continue et monotone par morceaux dans $[-\pi, \pi]$, alors sa série de Fourier converge dans $[-\pi, \pi]$ vers $[f(x+) + f(x-)]/2$. Ainsi donc les fonctions continues et monotones par moreceaux sont représentables analytiquement à l'aide des séries trigonométriques.

A la fin de son mémoire [Dirichlet 1829, 168–169], Dirichlet examine les cas où le nombre de discontinuités, ainsi que celui des maximums et minimums, de la fonction dans l'intervalle considéré est infini. Il était convaincu, en ce qui concerne le second cas, qu'il peut "être ramené à celui qu'il venait de considérer." Donc, en particulier, Dirichlet pensait que la série de Fourier d'une fonction continue f converge vers f (ce qui n'est pas exact comme le montrera en 1873 du Bois-Reymond). Quant au cas d'une infinité de discontinuités—problème qui est posé pour la première fois de façon aussi claire—il mettait en question la notion d'intégrale de Cauchy, et Dirichlet écrit à ce propos: "Il est nécessaire qu'alors la fonction $\varphi(x)$ soit telle que, si l'on désigne par a et b deux quantités quelconques comprises entre $-\pi$ et π, on puisse toujours placer entre a et b d'autres quantités r et s assez rapprochées pour que la fonction reste continue dans l'intervalle de r à s."

Dirichlet impose donc à la fonction que ses points de discontinuité forment un ensemble rare. Avec cette condition, portant sur les singularités de la fonction qui forment un ensemble infini, condition de nature "topologique," s'ouvre une ère nouvelle en analyse, dont un des aboutissements sera l'oeuvre de Baire. Notons que lorsque R. Lipschitz poursuivra [Lipschitz 1864, 284] les travaux de Dirichlet sur les séries trigonométriques, il considérera que cette condition de Dirichlet correspond au fait que cet ensemble rare est tel que son dérivé—c'est-à-dire l'ensemble de ses points d'accumulation—est fini. H. Hankel utilisera en 1870 cette condition de Dirichlet dans son mémoire consacré aux *Recherches sur les fonctions indéfiniment oscillantes et discontinues* pour définir les fonctions ponctuellement discontinues.

La portée de cette condition "topologique" sera, à notre avis, immense, d'autant plus que Dirichlet donne ensuite le premier exemple d'une fonction vraiment "arbitraire" définie sur **R** et qui ne vérifie pas cette condition, à savoir: $f(x) = c$ si x est rationnel et $f(x) = d \neq c$ si x est irrationnel. Cette fonction va fournir à l'analyse un très important contre-exemple (en particulier, elle ne sera pas intégrable au sens de Riemann). De plus, en prenant $c = 0$ et $d = 1$, on obtient ainsi la fonction caractéristique de l'ensemble des irrationnels, c'est-à-dire que c'est le début du transfert des propriétés de la fonction, de ses singularités, à son ensemble de définition, donc un pas de plus vers la "topologie générale."

Cette fonction "totalement discontinue" était une étape vers la définition générale par Dirichlet de fonction "arbitraire" qu'il donne en 1837: f est une fonction si elle fait correspondre à toute valeur de x une valeur bien déterminée $f(x)$. Il définit aussi [Dirichlet 1837, 38] les "fonctions arbitraires" comme "des fonctions qui ne sont assujetties à aucune loi analytique." La netteté de cette définition donne raison à Dini [Dini 1880, 2] lorsqu'il écrit que la définition d'une fonction "arbitraire" que donnaient d'autres mathématiciens avant Dirichlet, entre autres Poinsot en 1814 et Fourier en 1822, "exprimait ou sous-entendait toujours le concept d'existence d'une expression analytique donnée pour la fonction même."

Ainsi est posé, dans toute sa généralité, le problème des fonctions "arbitraires," c'est-à-dire des fonctions discontinues les plus générales et de leur "expression analytique," sans oublier qu'on n'avait pas encore à cette époque de représentation analytique d'une fonction continue quelconque.

De quelles fonctions discontinues disposaient à cette époque les analystes? Essentiellement des fractions rationnelles et des fonctions usuelles composées avec des fractions rationnelles, ainsi que des fonctions continues par morceaux. Ils avaient de plus maintenant la fonction de Dirichlet, définie de façon "ensembliste," dont les singularités forment un sous-ensemble infini de la droite. Mais il y a de plus les singularités des fonctions représentées par la somme d'une série convergente de fonctions continues.

C'est ce dernier mode de génération des fonctions discontinues qui a attiré l'attention de Seidel qui dans son mémoire *Note sur une propriété des séries qui représentent des fonctions discontinues*, après avoir cité le théorème inexact de Cauchy sur la continuité de la somme d'une série convergente de fonctions continues, affirme [Seidel 1847, 35]:

> Il en résulterait que les séries de cette forme ne sont pas appropriées à représenter encore les fonctions *discontinues* au voisinage de leurs sauts; autrement dit: il est impossible de représenter, à l'aide d'un ensemble de grandeurs continues, le discontinu, même lorsqu'on fait appel à la forme de l'infini; ainsi ce dernier ne constituerait pas un passage du rationnel à l'irrationnel, ni même ne permettrait de jeter un pont entre les quantités continues et discontinues.

Ce passage contient en germe certaines des idées que développeront plus tard G. Cantor et R. Baire, mais il contient essentiellement une idée nouvelle: atteindre le discontinu à l'aide du continu en faisant intervenir l'infini. Mais qu'est-ce que l'infini, ou plutôt, et c'est sous cette forme qu'il se présente surtout en mathématiques, que sont et quelle est la nature des ensembles infinis?

Dans cette première moitié du XIXe siècle, les mathématiciens commencent à avoir une meilleure connaissance des nombres irrationnels grâce, d'une part, à la démonstration de l'impossibilité de résoudre, en général, par radicaux les équations algébriques et, d'autre part, à la démonstration, faite par Liouville en 1844, que l'ensemble des nombres transcendants est infini.

Mais c'est encore Bolzano, dans ses *Paradoxes de l'infini*, parus en 1851, qui a posé, avec le plus de vigueur et malgré de nombreux raisonnements qui laissent à désirer, la question fondamentale de l'infini. Il écrit, en effet, au début de son livre [Bolzano 1920, 1]: "Il est certain que la plupart des énoncés paradoxaux que l'on rencontre dans le domaine des mathématiques sont des théorèmes qui contiennent le concept de l'infini, soit directement, soit qu'ils s'appuyent sur lui, au moins d'une certaine façon, lorsqu'on cherche à les démontrer."

On y remarque [Bolzano 1920, 13–14] la proposition, reprise plus tard par Dedekind, sur l'existence d'un ensemble infini. Pour cela, Bolzano "démontre" que "l'ensemble des propositions et vérités en soi" est infini, mais il revient en réalité à la notion classique de l'infini et à la construction intuitive de la suite des entiers; sa "démonstration" présuppose clairement que l'ensemble des entiers positifs est infini.

Ce qui est aussi important dans cet ouvrage, c'est que Bolzano y explicite particulièrement bien [Bolzano 1920, 28–29] la possibilité de mettre en "bijection" deux sous-ensembles infinis différents de la droite, premier pas vers la notion d'équipotence des ensembles.

C'est pourtant Riemann, poursuivant les travaux de Dirichlet sur les fondements de l'analyse, qui accomplira un pas décisif vers la clarification de la notion de fonction et l'étude de ses singularités. Il avait présenté, en 1854, comme thèse à l'Université de Göttingen, publiée seulement en 1867, le mémoire *Sur la possibilité de représenter une fonction par une série trigonométrique*. Il est intéressant de noter que l'adjectif "arbitraire," qui figurait dans les titres des mémoires sur le même sujet de Fourier et de Dirlichet, accolé au mot fonction, a disparu du titre du mémoire de Riemann.

En vue d'élargir la classe des fonctions représentables par des séries trigonométriques, Riemann élabore une nouvelle définition de l'intégrale définie, permettant d'intégrer les fonctions bornées ayant une infinité dénombrable de discontinuités. Il donne une condition d'intégrabilité qui va être une étape vers la théorie moderne d'intégration. En effet, aprés avoir défini l'oscillation de la fonction réelle f dans l'intervalle $[x_i, x_{i+1}]$ par

$$D_i = \sup_{x,y \in [x_i, x_{i+1}]} |f(x) - f(y)|,$$

Riemann énonce la condition suivante: "Pour qu'une fonction bornée soit intégrable dans $[a, b]$, il faut et il suffit qu'on puisse diviser $[a, b]$ en intervalles partiels tels que la somme des longueurs de ceux de ces intervalles dans lesquels l'oscillation est plus grande que ε, quel que soit $\varepsilon > 0$, soit aussi petite que l'on veut."

Cette théorie d'intégration de Riemann permet de mesurer des ensembles de points qui ne sont pas nécessairement des intervalles. Riemann donne également [Riemann 1867, 243–244] un exemple d'une fonction bornée, ayant une infinité dénombrable de discontinuités et intégrable, exemple d'une grande portée. Cet exemple montre qu'une fonction ayant un ensemble de points de discontinuité partout dense peut être intégrable. H. J. Smith donnera, en 1875, un exemple d'une fonction qui n'est pas intégrable au sens de Riemann et dont l'ensemble des points de discontinuité est rare. C'est à partir de ce genre d'exemples qu'on commencera à faire la distinction entre les ensembles de mesure nulle et les ensembles rares. Et enfin, avec ce mémoire de Riemann la séparation sera consommée entre les fonctions continues par morceaux et les fonctions dérivables.

WEIERSTRASS ET LES "EXPRESSIONS ANALYTIQUES" DES FONCTIONS CONTINUES

Bien qu'une partie essentielle de l'oeuvre weierstrassienne tourne autour de la possibilité de développer localement en série entière les fonctions "usuelles" de son temps, le développement et le raffinement nécessaires pour rendre rigoureuse cette théorie ont contribué puissamment à l'épanouissement de la notion de fonction.

Une époque importante dans l'activité mathématique de Weierstrass, au point de vue où nous nous sommes placés dans notre article, se situe entre 1865 et 1870.

Il élabore d'abord une théorie correcte des nombres réels, base de l'analyse, comme en témoigne, entre autres, la lettre du 4 septembre 1867 de H. Hankel à H. Grassmann [Grassmann 1911], dans laquelle Hankel écrit que Weierstrass, dans ses cours, s'est "attaché à une construction rigoureuse et soigneuse" des nombres réels.

On trouve ensuite dans son cours inédit de 1865–1866 sur les *Principes de la théorie des fonctions analytiques*, rédigé par M. Pasch, l'énoncé du théorème qu'un ensemble borné infini de points du plan admet au moins un point d'accumulation, théorème dont la formulation est très voisine de celle du théorème de Bolzano de 1817 sur la borne supérieure d'un ensemble majoré: "Si on a à l'intérieur d'une partie bornée du plan une infinité de points possédant une certaine propriété, alors il existe dans son intérieur ou sur sa frontière au moins un point tel que dans tout voisinage de ce point il y a une infinité de points ayant cette propriété."

D'ailleurs, il n'y a pas de doute que Weierstrass connaissait bien avant 1870 le mémoire de Bolzano sur le théorème des valeurs intermédiaires, comme en témoigne aussi la lettre de H. A. Schwarz à G. Cantor du 1er avril 1870 et où écrit: "Je suis aussi d'accord, comme toi, avec l'opinion, soutenue par Monsieur Weierstrass dans ses leçons, que sans les conclusions qui ont été développées par Monsieur Weierstrass à partir des principes de Bolzano on n'aurait pas pu réussir dans de nombreuses recherches."

C'est également à cette époque que Weierstrass démontre rigoureusement les théorèmes essentiels de l'analyse, base de la théorie des fonctions, sur la borne supérieure d'un ensemble majoré et sur la borne supérieure d'une fonction continue sur un intervalle fermé et borné comme en temoignent, en particulier, G. Cantor [Cantor 1870, 141] et U. Dini [Dini 1878, IV, 43, 51].

Un bon témoignage sur l'acquis weierstrassien de ces années est le cours inédit de Weierstrass de 1874 sur l'*Introduction à la théorie des fonctions analytiques*, rédigé par G. Hettner.

Le début de ce cours est consacré à l'étude de la notion de fonction. Après avoir donné la définition d'une fonction, telle que nous la connaissons aujourd'hui, l'attribuant à Fourier, Cauchy et Dirichlet, Weierstrass fait des réserves sur sa trop grande généralité. Pour lui, si l'on ne sait rien d'autre sur une fonction que ce que cette définition en exprime, on ne pourra en tirer aucune conclusion sur ses propriétés. Cette réflexion sur la notion de fonction est cependant plus profonde qu'elle ne paraît peut-être à première vue, et un de ses aboutissements se trouvera dans l'oeuvre de Baire. Weierstrass lui-même s'assigne comme tâche—une des tâches qui transparaît lorsqu'on jette un regard synthétique sur son oeuvre—de déterminer des classes de fonctions les plus larges possibles dont on puisse donner une représentation analytique et qui répondent de façon aussi complète que possible aux besoins de l'analyse. De plus, il faut que cette représentation analytique soit utilisable de manière aussi commode que générale.

Mais comment arriver à cette notion de fonction réellement utilisable et rigoureusement définie? Weierstrass va partir des nombres entiers, supposés donnés, et aboutir, dans la première partie de son cours, à la "construction" des nombres réels. Il étudie ensuite les nombres complexes, les polynômes et les fractions rationnelles. Le but que Weierstrass poursuit dans ce cours est de développer la notion de fonction analytique; il lui en associe un deuxième: introduire dans les enchaînements analytiques la même clarté que l'on trouve dans les enchaînements algébriques. Weierstrass reprend ainsi à son compte le projet de Lagrange, mais il utilise les outils nouveaux qui ont été forgés depuis le début du XIXe siècle: les notions de limite, de borne supérieure et inférieure, de point d'accumulation, de convergence uniforme, et des premiers éléments de topologie générale.

Pour réaliser son but, Weierstrass aborde l'étude des séries entières. Dans cette partie de son cours, il a besoin du théorème sur la borne supérieure d'un ensemble majoré—où il fait une démonstration inspirée par celle de Bolzano. On y trouve également le théorème affirmant qu'un ensemble borné infini de nombres réels admet au moins un point d'accumulation, et la démonstration de ce théorème est apparentée à la précédente. Ce qui est important de noter c'est qu'on trouve dans ce cours un renvoi au mémoire de Bolzano de 1817 sur le théorème des valeurs intermédiaires. Weierstrass utilise ensuite ce théorème sur les points d'accumulation pour prouver qu'une fonction continue sur un intervalle fermé et borné atteint sa borne supérieure et sa borne inférieure.

Pour étudier les propriétés des fonctions analytiques, Weierstrass introduit dans son cours un certain nombre de notions de topologie générale: ensemble borné, ensemble ouvert, voisinage d'un point, point extérieur, point frontière, et ensemble connexe.

Après avoir exposé sa théorie des fonctions, Weierstrass précise le sens de ses recherches et le but de la théorie des fonctions, montrant ainsi qu'il avait bien compris le sens dans lequel évoluait cette théorie. En effet, pour lui, le sens de ces recherches est que les propriétés caractéristiques des fonctions reposent sur leurs points singuliers; quant au "but idéal" de la théorie des fonctions, il est de "représenter analytiquement" les fonctions définies par ailleurs de façons diverses; et il est fondamental "d'obtenir *a priori* la forme et les conditions d'une telle représentation."

Weirstrass présente en 1872 à l'Académie des Sciences de Berlin le premier exemple d'une fonction continue sur la droite et qui n'est dérivable en aucun point, et sépare ainsi définitivement les notions de fonction continue et de fonction dérivable.

Un autre pas important dans le développement et la clarification du concept de fonction est accompli par Weierstrass en 1885 lorsqu'il démontre que toute fonction réelle continue peut être représentée par une série convergente de polynômes, faisant ainsi rentrer toutes les fonctions continues dans la classe de fonctions représentables analytiquement.

DEDEKIND ET L'APPLICATION COMME FONDEMENT DES MATHÉMATIQUES

Pour avancer dans la connaissance des fonctions "arbitraires," il était donc nécessaire de connaître mieux les ensembles infinis, lieux de leurs singularités. Mais il était non moins nécessaire de connaître la nature de ces deux êtres mathématiques—ensemble et fonction et leurs relations—qui commençaient à apparaître comme les éléments premiers constitutifs des mathématiques. La contribution de R. Dedekind à leur étude fut décisive.

C'est entre 1869 et 1870 que Dedekind réussit à vaincre les dernières difficultés "pour fonder une théorie des idéaux, rigoureuse et sans exceptions," théorie qu'il va publier dans le Xe supplément de la deuxième édition des *Leçons sur la théorie des nombres* de Dirichlet, parue en 1871. Nous pensons que c'est aussi là, dans ce X^e^ supplément de Dedekind, qu'il faut chercher le "lieu de naissance" de la théorie des ensembles. Cantor lui-même soulignera en 1877 l'importance de ces recherches ensemblistes de Dedekind. Dedekind y introduit les notions de corps et d'idéal et définit les opérations sur les ensembles—considérés pour la première fois, avec tant de netteté, de nécessité mathématique et de généralité, comme des objets, tels des points, soumis aux lois mathématiques—"traduction" des opérations algébriques sur les nombres.

Dedekind publie en 1872, dans son livre *Continuité et nombres irrationnels*, sa théorie des nombres réels qu'il avait élaboré dès 1858. L'intérêt de cette théorie, au point de vue qui nous intéresse ici, est qu'un nombre réel est défini de façon "ensembliste": une "coupure" qui le définit est une certaine partition de l'ensemble des nombres rationnels en deux sous-ensembles, et dans cette définition intervient l'idée de la totalité des nombres rationnels, c'est-à-dire la totalité des éléments d'un ensemble infini. Ce fait a été souligné par Cantor en 1883. A la fin de son livre, Dedekind se pose la question de savoir si "les opérations arithmétiques elles-mêmes possèdent une certaine continuité." On a ici un embryon de la future théorie des algèbres topologiques et il énonce même le théorème très général suivant:

> Si le nombre x est le résultat d'un calcul faisant intervenir les nombres $a, b, c, \ldots$ et si x appartient à l'intervalle X, alors il existe des intervalles A, B, C, . . . contenant les nombres $a, b, c, \ldots$ tels que le résultat du même calcul, dans lequel les nombres $a, b, c, \ldots$ peuvent être remplacés par des nombres quelconques des intervalles A, B, $C, \ldots$, sera toujours un nombre appartenant à l'intervalle X.

Dedekind commence à rédiger son livre *Que sont et que représentent les nombres?* dès 1872, livre qui paraîtra en 1888, où il donnera les premiers éléments de la future théorie "naïve" des ensembles. La première rédaction de ce livre est de 1872–1878 et la deuxième de juin–juillet 1887, et Dedekind avait discuté de ses idées fondamentales sur le fini et l'infini, entre 1877 et 1882, avec H. Weber, H. A. Schwarz, et G. Cantor.

Le but de son livre est de "construire" l'ensemble des nombres entiers positifs et il rappelle dans son introduction les mots de Dirichlet: tout théorème d'algèbre et d'analyse "peut s'énoncer comme un théorème sur les nombres entiers."

Le paragraphe 1, *Ensembles d'éléments*, débute par la définition d'une "chose": c'est "tout objet de notre pensée." Si on la désigne par a, alors on écrira que $a = b$ si "tout ce qu'on peut penser de a on peut le penser de b" et réciproquement. Si $a = b$ et $b = c$, alors on a $a = c$. Si les différentes choses $a, b, c, \ldots$ sont considérées comme rassemblées sous un même point de vue, alors on dira qu'elles forment un "ensemble" S et on appellera $a, b, c, \ldots$ les "éléments" de S. Un tel élément peut être considéré, à son tour, en tant qu'un objet de notre pensée, également comme une chose. Ainsi Dedekind indique clairement qu'un ensemble peut être envisagé comme un élément d'un autre ensemble. Il écrit ensuite: "Un ensemble S est complètement déterminé lorsqu'on a déterminé pour toute chose si elle est élément de S ou non."

Puis il définit l'égalité de deux ensembles: deux ensembles S et T sont égaux, et on écrit $S = T$, si tout élément de S est un élément de T et réciproquement. Un ensemble A sera une "partie" d'un *ensemble* S "si tout élément de A est aussi un élément de S," et si A est une partie de B et B une partie de A, alors on a $A = B$. Si A est une partie de S et si $A \neq S$, dans ce cas on dira que A est une "partie propre" de S. Après avoir indiqué la propriété de transitivité de l'inclusion des ensembles, Dedekind définit l'ensemble "réunion"—et il précise qu'il faut distinguer l'ensemble réunion de A,

B, C, . . . de l'ensemble dont A, B, C, . . . sont des éléments—ainsi que l'ensemble "intersection" des ensembles A, B, C, ...

Le paragraphe 2, *Application d'un ensemble*, est un des plus importants de ce livre, car non seulement c'est autour de l'idée centrale d'application qu'il bâtira sa théorie des nombres entiers, mais surtout parce que Dedekind montre, pour la première fois, l'utilité et l'importance de considérer les applications les plus générales.

Dedekind donne d'abord la définition d'une application due à Dirichlet: l'application f est une "loi" qui à tout élément déterminé s d'un ensemble S fait correspondre un élément bien déterminé $f(s)$, "image" de s. Si T est contenu dans S, alors l'application f "contient une application déterminée de T," c'est-à-dire que Dedekind définit ici la restriction de f à T. Il montre ensuite comment une application f agit sur les opérations définies sur les ensembles, et introduit la notion d'application "composée." Puis il démontre que la composition des applications est associative.

Le titre du paragraphe 3 pourrait être traduit, en utilisant les termes d'aujourd'hui: *Bijectivité d'une application. Ensembles équipotents.* Alors f admet une "application inverse," et l'image par f de l'intersection des ensembles A, B, C, . . . est l'intersection de $f(A)$, $f(B)$, $f(C)$, ... Deux ensembles S et R seront dits "équipotents" s'il existe une application "bijective" de S sur R. Cette relation d'équipotence est transitive. Il en résulte qu'on peut "partager tous les ensembles en classes" de façon qu'à une classe déterminée appartiennent tous les ensembles équipotents à un ensemble donné R, "le représentant de la classe." Dedekind ne soupçonne évidemment pas qu'on se trouve ici en présence d'un paradoxe dû à la notion d'ensemble de tous les ensembles sur lequel on définit la relation d'équivalence considérée.

Le paragraphe 5, intitulé *Le fini et l'infini*, aura une grande résonnance dans les discussions sur les fondements de mathématiques qui vont s'ouvrir bientôt. Dedekind commence par donner une définition d'un ensemble infini, la première définition que l'on rencontre dans l'histoire des mathématiques et qui est basée sur la notion d'application "bijective": un ensemble S est infini s'il est équipotent à une de ses parties propres. C'est dans ce paragraphe que se trouve le seul "théorème" de Dedekind dont la démonstration ne cadre pas avec sa pensée mathématique: "Il existe des ensembles infinis."

Ce théorème ne figure pas dans la première rédaction de *Que sont et que représentent les nombres?*, écrite entre 1872 et 1878; la deuxième rédaction est de 1887. Or, Dedekind avait reçu en 1882, envoyé par Cantor, le livre de Bolzano *Paradoxes de l'infini.* C'est en s'inspirant de la démonstration de Bolzano du théorème sur l'existence d'un ensemble infini qu'il a élaboré la sienne, basée sur sa définition d'un ensemble infini.

Il nous semble enfin intéressant de souligner que les trois concepts essentiels de la mathématique dedekindienne, ceux d'idéal, de coupure et de chaîne (une application d'un ensemble dans lui-même et qui joue un rôle primordial dans sa théorie des nombres entiers), sont fondés sur les notions premières d'ensemble et d'application.

CANTOR, LA TOPOLOGIE GÉNÉRALE, ET LE TRANSFINI

Les contributions essentielles de Georg Cantor au développement de la notion de fonction sont dues à ses recherches sur les ensembles infinis, leurs propriétés topologiques et la possibilité de les "bien ordonner."

Tandis que Dedekind introduit les notions ensemblistes pour résoudre des problèmes d'algèbre et fonder la notion de nombre entier, les concepts ensemblistes de Cantor proviennent de ses recherches sur les séries trigonométriques issues du mémoire de Riemann sur le même sujet.

Il est à remarquer que Cantor avait fait ses études à Berlin, entre 1863 et 1866, au moment où Weierstrass y exposait ses premiers éléments de topologie générale et sa théorie des nombres réels, sous-tendue par des notions ensemblistes. Sur l'influence de Weierstrass sur Cantor et sur le développement de la théorie cantorienne des ensembles, un témoignage intéressant nous est fourni par la lettre de G. Mittag-Leffler à P. Jourdain du 11 avril 1919:

> L'influence directe de Weierstrass sur Cantor était en réalité beaucoup plus grande qu'on ne le sait en général. Je les ai connus tous les deux très intimement pendant des années. Ne croyez pas que je veuille discuter la priorité de Cantor sur bien des choses. Rien n'était plus loin de la pensée de Weierstrass. *La théorie des ensembles* comme nouvelle théorie mathématique est certainement la création de Cantor. Mais Cantor était en réalité plus impressionné par Weierstrass qu'il ne le savait lui-même.

Le début de la théorie cantorienne des ensembles peut être trouvé dans son mémoire de 1871 [Cantor 1871, 85], dans lequel il considère une série trigonométrique telle que $f(x) = 0$ quel que soit x: il en résulte alors que $a_n = b_n = 0$, quel que soit l'entier positif n. Cantor va maintenant affaiblir les hypothèses de son théorème d'unicité. Il va supposer que, dans un intervalle borné, il y a un nombre fini de points en lesquels ou bien la série trigonométrique n'est pas convergente, ou bien $f(x) \neq 0$. Il démontre que son théorème d'unicité reste valable dans ce cas et ajoute que "cette généralisation n'est nullement la dernière," car il est arrivé à une "extension qui va plus loin."

Cette généralisation, Cantor l'expose dans son mémoire sur les séries trigonométriques de 1872. Il y introduit la notion de point d'accumulation, à la suite de Weierstrass, et considère l'ensemble P' des points d'accumulation de P, appelé "le premier dérivé" de P. Si P' n'est pas composé d'un nombre fini d'éléments, on peut de même considérer l'ensemble P'' de ses points d'accumulation, ou "le second ensemble dérivé" de P; et ainsi de suite. S'il arrive, au cours de cette itération, que l'ensemble $P^{(n)}$ soit composé d'un nombre fini d'éléments, alors $P^{(n+1)}$ est vide et on dira que "P est un ensemble de $n^{\text{ème}}$ espèce." Cantor ajoute alors: "Ainsi, dans cette théorie, le domaine de tous les ensembles d'espèce déterminée sera considéré comme un genre particulier du domaine de tous les ensembles concevables, et les ensembles dits de $n^{\text{ème}}$ espèce forment une espèce particulière dans ce genre."

On retrouve ici aussi la notion d'ensemble de tous les ensembles et, il semble, l'influence de Bolzano dont Cantor a connu l'oeuvre déjà en 1870.

Grâce à la notion d'ensemble de $n^{\text{ème}}$ espèce, il généralise son théorème sur l'unicité du développement en séries trigonométriques: Si $f(x) = 0$ dans $]0, 2\pi[$, sauf aux points d'un ensemble de $n^{\text{ème}}$ espèce, alors on a $a_n = b_n = 0$ quel que soit l'entier n.

Le pas décisif que Cantor va faire faire à la théorie des ensembles est inséparable de la rencontre qu'il a faite, "par hasard," de Dedekind en 1872 en Suisse. Le 29 novembre 1873 il écrit à Dedekind qu'il voudrait lui "soumettre une question qui a pour moi un certain intérêt théorique, mais à laquelle je ne puis répondre." Il s'agissait de savoir s'il existe une bijection entre l'ensemble des nombres entiers positifs et l'ensemble des nombres réels. Dedekind répond, "par retour de courrier," en formulant et en démontrant complètement la proposition que "l'ensemble de tous les nombres algébriques" peut être mis en correspondance bijective avec l'ensemble des nombres entiers. La lettre de Cantor du 7 décembre 1873 contient la première démonstration, qu'il venait d'achever, de la non-existence d'une bijection entre l'ensemble des nombres entiers et $[0, 1]$.

A peine son théorème envoyé au *Journal de Crelle*, Cantor écrit à Dedekind le 5 janvier 1874 pour lui poser un nouveau problème, une de ces questions qui allaient changer l'essence d'une partie des mathématiques et qui correspondent à une mutation profonde de leur caractère. En effet, Cantor écrit:

> A propos des questions qui m'ont occupé ces derniers temps, je m'apercois que, dans cet ordre d'idées, se présente aussi la suivante: Est-ce qu'une surface (par exemple un carré, frontière comprise) peut être mise en relation univoque avec une courbe (par exemple un segment de droite, extrémités comprises), de telle sorte qu'à tout point de la surface corresponde un point de la courbe, et réciproquement à tout point de la courbe un point de la surface?

C'est seulement plus de trois ans plus tard, en 1877, que Cantor démontrera qu'on peut mettre en bijection une droite avec un plan. Il publie ce théorème dans le mémoire *Une contribution à la théorie des ensembles*, paru en 1878, mémoire qui commence par une introduction où il définit la notion de puissance: deux ensembles M et N ont même puissance s'il existe une application bijective de M sur N, et on dit alors que M et N sont équivalents. Il nomme de première classe, ou de première puissance, les ensembles équivalents à l'ensemble des nombres entiers positifs. C'est dans ce mémoire que Cantor pose le problème de la classification, du point de vue de la puissance, de tous les sous-ensembles infinis de la droite.

Dans une série d'articles, *Sur les ensembles infinis linéaires de points*, parus entre 1879 et 1884, Cantor donne un exposé général de sa théorie des sous-ensembles infinis de la droite, munie de sa topologie naturelle, en les groupant suivants leurs caractères communs et en utilisant, comme notion primitive, le concept d'ensemble dérivé. Si, pour un entier strictement positif n, $P^{(n+1)}$ est vide, on dit que P est du "premier genre," et si $P^{(n)}$ n'est vide quel que soit l'entier positif n, dans ce cas on dira que P est du "second genre." Pour étudier les sous-ensembles de la droite, Cantor introduit la notion d'ensemble partout dense, ainsi que celle d'ensemble parfait P, tel que $P = P'$. Après l'introduction de l'ensemble vide, Cantor définit l'ensemble dérivé de P d'ordre ω, c'est l'intersection de tous les $P^{(n)}$, n étant un entier strictement positif quelconque; les ensembles du premier genre sont caractérisés par la propriété que cet ensemble est vide. On peut alors définir, en général, l'ensemble dérivé $P^{(\omega+1)}$ de $P^{(\omega)}$, ainsi que l'ensemble dérivé de $P^{(\omega)}$ d'ordre ω désigné par $P^{(2\omega)}$, et ainsi de suite, pour former l'ensemble $P^{(\omega^2)}$, et finalement on construit les ensembles $P^{(\omega^\omega)}$, etc.

La construction précédente devait amener Cantor, tout naturellement, à l'idée d'un "ensemble bien ordonné." Les ordinaux finis forment la classe I des ordinaux, tandis que les ordinaux des ensembles de première classe (les ensembles dénombrables) forment la classe II des ordinaux: ω, $\omega + 1$, $\ldots$, 2ω, $\ldots$, ω^2, $\ldots$, ω^ω, $\ldots$; les ordinaux des ensembles infinis non dénombrables formant la classe III. Un nombre de la classe II ou de la classe III est appelé un nombre transfini.

Ces recherches de Cantor conduisent Bendixson [Bendixson 1883] à prouver le théorème suivant caractérisant les ensembles fermés rares, permettant de mieux connaître les sous-ensembles de la droite: Si P est un ensemble fermé rare, alors ou bien P est dénombrable ou bien P est la réunion d'un ensemble parfait et d'un ensemble dénombrable.

BAIRE ET LES FONCTIONS REPRÉSENTABLES ANALYTIQUEMENT

C'est Dini qui pose le premier, avec toute la netteté désirable, le problème de savoir si une fonction au sens de Dirichlet est représentable analytiquement. En effet, il écrit dans son traité *Fondements de la théorie des fonctions de variables réelles* (où il démontre aussi pour la première fois de manière rigoureuse le théorème des accroissements finis) [Dini 1878, 37]:

> Avec une telle définition se pose naturellement la question: "si, en conservant toute la généralité contenue dans la définition, une fonction y de x, donnée dans un certain intervalle, pourrait toujours s'exprimer analytiquement ou non, pour toutes les valeurs de la variable du même intervalle, par une ou plusieurs séries finies ou infinies d'opérations du calcul à faire sur la variable"; et à cette question, dans l'état actuel de la science, on ne peut pas répondre de façon pleinement satisfaisante, puisque, bien que l'on sache maintenant que, pour des classes de fonctions les plus étendues et aussi pour des fonctions qui présentent des singularités les plus grandes, on peut donner une représentation analytique, il subsiste pourtant encore le doute que, en ne faisant aucune restriction, il puisse aussi exister des fonctions pour lesquelles toute expression analytique, au moins avec les signes actuels de l'analyse, soit tout à fait impossible.

C'est R. Baire qui tentera de répondre dans ses travaux à cette question de Dini.

C'est vers le 25 novembre 1896 que Baire redécouvre qu'une fonction de deux variables, continue séparément par rapport à x et par rapport à y, peut être discontinue par rapport à (x, y), et il introduit les notions de semi-continuité inférieure et supérieure. En effet, il constate alors qu'une des premières propriétés de ces fonctions, en considérant une fonction réelle de deux variables f, définie dans $[0,1] \times [0,1]$ et continue séparément par rapport à x et y, est que la fonction $M(x) = \sup_{y \in [0,1]} f(x, y)$ vérifie la condition suivante: Quel que soit $\varepsilon > 0$, il existe $\alpha > 0$ tel que, si $|x - x_0| \leq \alpha$, alors $M(x_0) - M(x) \leq \varepsilon$. Le contre-exemple de la fonction définie par $f(x, y) = 2xy/(x^2 + y^2)$ si $(x, y) \neq (0, 0)$ et $f(0, 0) = 0$, et telle que $M(x) = 1$ si $x \neq 0$ et $M(0) = 0$, lui montre immédiatement que la fonction M n'est pas, en général, continue et ne possède pas la deuxième propriété de continuité, à savoir $M(x) \leq M(x_0) + \varepsilon$. En janvier 1897, il découvre que les fonctions semi-continues jouissent de certaines propriétés des fonctions continues. Ainsi, si la borne inférieure des valeurs prises par une fonction semi-continue inférieurement sur un compact est nulle, alors elle prend cette valeur en au moins un point de l'intervalle considéré.

Ces recherches conduisent Baire à relier les questions de continuité d'une fonction réelle de deux variables réelles à celles de continuité d'une fonction réelle d'une variable réelle. En particulier, il s'agit de savoir à quelles conditions est assujettie la fonction $x \mapsto f(x, x)$, si la fonction $(x, y) \mapsto f(x, y)$ est continue séparément par rapport à x et à y. Ainsi, par exemple, la fonction définie par $f(x, y) = 2xy/(x^2 + y^2)$ si $(x, y) \neq (0, 0)$, $f(0, 0) = 0$, et $(x, y) \in [0, 1] \times [0, 1]$, est discontinue sur la droite $x = y$, puisque $f(x, x) = 1$ si $x \neq 0$ et $f(0, 0) = 0$. Cette question ramène Baire à l'étude des fonctions réelles de la variable réelle discontinues. Il s'aperçoit, vers le 25 janvier, qu'une partie des résultats qu'il venait de trouver ont été déjà obtenus par Dini. Il découvre aussi, en particulier, qu'une fonction semi-continue inférieurement (resp. supérieurement) sur un intervalle est ponctuellement discontinue sur cet intervalle.

Baire est donc amené à étudier les fonctions ponctuellement discontinues et il utilise pour étudier leurs ensembles des points de discontinuité les théorèmes de Cantor sur les ensembles dérivés et les ensembles parfaits.

Au mois de mars 1898 Baire prouve d'abord le théorème suivant: "Si une série, dont les termes sont des fonctions continues en x, est convergente pour chaque valeur de x, elle représente une fonction qui est ponctuellement discontinue relativement à tout ensemble parfait."

Quelques jours plus tard, il démontre que cette condition nécessaire est aussi suffisante, généralisant ainsi aux fonctions ponctuellement discontinues le théorème de Weierstrass sur les fonctions continues. Par son théorème, Baire élargit la classe des fonctions représentables analytiquement en y incluant les fonctions ponctuellement discontinues.

Pour démontrer cette réciproque il utilise les travaux de Cantor, y compris ceux sur le transfini.

Dans sa Note aux Comptes Rendus de l'Académie des Sciences de Paris du 6 juin 1898, Baire introduit sa fameuse classification des fonctions. Les fonctions continues forment la classe 0, et les fonctions discontinues limites de fonctions continues forment la classe 1. Soit maintenant une suite de fonctions appartenant aux classes 0 et 1, et possédant une limite n'appartenant à aucune de ces deux classes. On dira alors que cette fonction appartient à la classe 2. Ainsi, la fonction de Dirichlet égale à 0 sur les rationnels et à 1 sur les irrationnels appartient à la classe 2. On définit ainsi par récurrence les fonctions de classe n: une fonction sera de classe n si elle est la limite d'une suite de fonctions appartenant aux classes $0, 1, 2, \ldots, n-1$, et si elle n'appartient pas elle-même à l'une de ces classes. Baire utilise ensuite la notion de nombre transfini pour définir la classe α, α étant un nombre transfini quelconque de la deuxième classe.

Il énonce ensuite le théorème de stabilité suivant:

> Considérons l'ensemble E de toutes les fonctions, continues et discontinues, qui viennent d'être définies, c'est-à-dire l'ensemble des fonctions appartenant aux classes marquées par un nombre de la première ou de la deuxième classe de nombres. Si une suite de fonctions appartenant à l'ensemble E a une fonction limite, cette fonction limite appartient aussi à l'ensemble E.

L'ensemble E est donc stable par passage à la limite, et Baire en tire la conclusion suivante:

> L'ensemble E a la puissance du continu tandis que l'ensemble de toutes les fonctions discontinues a une puissance supérieure; de sorte que l'ensemble E, tout en étant beaucoup plus général que l'ensemble des fonctions continues, ne forme qu'une catégorie très particulière de fonctions par rapport à l'ensemble de toutes les fonctions que l'on peut concevoir.

Ainsi, Baire est conscient qu'à côté des fonctions de E, fonctions représentables analytiquement, et obtenues par un nombre dénombrable de passages à la limite, il y a des fonctions discontinues que l'on ne peut pas obtenir par ce procédé analytique.

A propos de ces problèmes, Baire écrit à Volterra le 25 octobre 1898:

> Il ne me semble pas inutile de réfléchir un peu à ces questions; ne peut-on pas espérer qu'on arrivera de cette manière à préciser dans quelle mesure il nous est permis d'employer la notion de *fonction arbitraire*? Peut-être y a-t-il, par la nature même des choses, une limite dans la conception de fonction arbitraire, il s'agirait de voir quelle est cette limite.

C'est en 1904 que Lebesgue annonça dans sa note *Sur les fonctions représentables analytiquement* qu'il avait réussi à "nommer une fonction qui, n'étant d'aucune classe, échappe à toute définition analytique," c'est-à-dire qu'elle n'est pas représentable analytiquement. Cette découverte, d'après Lusin, "a produit une impression aussi abasourdissante qu'en son temps la découverte de Fourier." Dans son mémoire de 1905, portant le même titre, Lebesgue écrivait qu'il n'était pas évident "qu'il existe de fonctions non représentables analytiquement." Il y démontre aussi l'identité entre l'ensemble des fonctions de Baire, l'ensemble des fonctions représentables analytiquement, et, résultat surprenant, l'ensemble des fonctions mesurables au sens de Borel. Ainsi la fonction "construite" par Lebesgue et qui n'est pas représentable analytiquement n'est pas borélienne.

Dans la conclusion de sa thèse de 1899, *Sur les fonctions de variables réelles*, Baire note, montrant sa prise de conscience de la transformation que l'analyse venait de subir, transformation que nous avons analysée en étudiant le développement de la notion de fonction, des fonctions comme expressions analytiques aux fonctions représentables analytiquement:

> La théorie des ensembles de points joue un rôle important dans ces méthodes; on peut même dire, d'une manière générale, que, dans l'ordre d'idées où nous nous sommes placés, tout problème relatif à la théorie des fonctions conduit à certaines questions relatives à la théorie des ensembles; et c'est dans la mesure où ces dernières questions sont avancées ou peuvent l'être qu'il est possible de résoudre plus ou moins complètement le problème donné.

BIBLIOGRAPHIE

Abel, N. H. 1826. "Untersuchungen über die Reihe $1 + \frac{m}{1}x + \frac{m(m-1)}{1.2}x^2 + \frac{m(m-1)(m-2)}{1.2.3}x^3 + \cdots$," *Journal für die reine und angewandte Mathematik*, *1*, 311–339 = "Recherches sur la série $1 + \frac{m}{1}x + \frac{m(m-1)}{1.2}x^2 + \frac{m(m-1)(m-2)}{1.2.3}x^3 + \cdots$," *Oeuvres complètes* (Kristiania: Grøndahl, 1881), volume 1, 219–250.

Baire, René. 1897. "Sur la théorie générale des fonctions de variables réelles," *Comptes Rendus de l'Académie des Sciences de Paris, 125*, 691–694.

Baire, René. 1898. "Sur les fonctions discontinues développables en séries de fonctions continues," *Comptes Rendus de l'Académie des Sciences de Paris, 126*, 884–887.

Baire, René. 1899. "Sur les fonctions de variables réelles," *Annali di matematica pura ed applicata*, (3), *3*, 1–123.

Bendixson, Ivar. 1883. "Quelques théorèmes de la théorie des ensembles," *Acta mathematica, 2*, 415–429.

Biermann, Kurt-R. 1973. *Die Mathematik und ihre Dozenten an der Berliner Universität 1810–1920*. (Berlin: Akademie-Verlag).

Bolzano, Bernard. 1816. *Der binomische Lehrsatz, und als Folgerung aus ihm der polynomische, und die Reihen, die zur Berechnung der Logarithmen und Exponentialgrössen dienen, genauer als bisher erwiesen* (Prag: Enders).

Bolzano, Bernard. 1817. "Rein analytischer Beweis des Lehrsatzes, dass zwischen je zwey Werthen, die ein entgegengesetztes Resultat gewähren, wenigstens eine reelle Wurzel der Gleichung liege," *Abhandlungen der K. Böhm. Gesellschaft der Wissenschaften*, (3) *5*, 1–60 = (Berlin: Mayer und Müller, 1894) = "Démonstration purement analytique du théorème: entre deux valeurs quelconques qui donnent deux résultats de signes opposés se trouve au moins une racine réelle de l'équation, traduit en français par J. Sebestik," *Revue d'Histoire des Sciences, 17*, 136–164.

Bolzano, Bernard. 1851. *Paradoxien des Unendlichen* (Leipzig: Reclam) = (Leipzig: Meiner, 1920).

Bolzano, Bernard. 1976. *Reine Zahlenlehre* (Stuttgart-Bad Cannstatt: Friedrich Frommann).

Cantor, Georg. 1870. "Beweis, dass eine für jeden reellen Werth von x durch eine trigonometrische Reihe gebende Function $f(x)$ sich nur auf eine einzige Weise in dieser Form darstellen lässt," *Journal für die reine und angewandte Mathematik, 72*, 139–142.

Cantor, Georg. 1871. "Notiz zu dem Aufsatze: Beweis dass eine für jeden reellen Werth von x durch eine trigonometrische Reihe gegebene Function $f(x)$ sich nur auf eine einzige Weise in dieser Form darstellen lässt," *Journal für die reine und angewandte Mathematik, 73*, 294–296.

Cantor, Georg. 1872. "Ueber die Ausdehnung eines Satzes aus der Theorie der trigonometrischen Reihen," *Mathematische Annalen, 5*, 123–132 = "Extension d'un théorème de la théorie des séries trigonométriques," *Acta mathematica, 2* (1883), 336–348.

Cantor, Georg. 1874. "Ueber eine Eigenschaft des Inbegriffes aller reellen algebraischen Zahlen," *Journal für die reine und angewandte Mathematik, 77*, 258–262 = "Sur une propriété du système de tous les nombres algébriques," *Acta mathematica, 2* (1883), 305–310.

Cantor, Georg. 1878. "Ein Beitrag zur Mannigfaltigkeitslehre," *Journal für die reine und angewandte Mathematik, 84*, 242–252 = "Une contribution à la théorie des ensembles," *Acta mathematica*, 2 (1883), 311–328.

Cauchy, A. L. 1821. *Cours d'analyse de l'Ecole royale polytechnique* (Paris: Debure) = *Oeuvres complètes*, (2) *3*(Paris: Gauthier-Villars).

Cauchy, A. L. 1823. *Résumé des leçons données à l'Ecole royale polytechnique sur le calcul infinitésimal* (Paris: Debure) = *Oeuvres complètes*, (2) *4*, 7–261 (Paris: Gauthier-Villars).

Cournot, A. A. 1847. *Traité élémentaire de la théorie des fonctions et du calcul infinitésimal, 1* (Paris: Hachette).

Dedekind, Richard. 1872. *Stetigkeit und irrationale Zahlen* (Braunschweig: Vieweg).

Dedekind, Richard. 1888. *Was sind und was sollen die Zahlen*? (Braunschweig: Vieweg).

Dini, Ulisse. 1878. *Fondamenti per la teorica delle funzioni di variabili reali* (Pisa: Nistri).

Dini, Ulisse. 1880. *Serie di Fourier e altre rappresentazioni analitiche delle funzioni di una variabile reale* (Pisa: Nistri).

Dirichlet, Lejeune P. G. 1829. "Sur la convergence des séries trigonométriques qui servent à représenter une fonction arbitraire entre des limites données," *Journal für die reine und angewandte Mathematik*, *4*, 157–169.

Dirichlet, Lejeune P. G. 1837. "Sur les séries dont le terme général dépend de deux angles, et qui servent à exprimer des fonctions arbitraires entre des limites données," *Journal für die reine und angewandte Mathematik*, *17*, 35–56.

Dirichlet, Lejeune P. G. 1871. *Vorlesungen über Zahlentheorie* (zweite Auflage, Braunschweig: Vieweg).

Dirichlet, Lejeune P. G. 1904. *Vorlesungen über die Lehre von den einfachen und mehrfachen bestimmten Integralen* (Braunschweig: Vieweg).

Dugac, Pierre. 1976. *Richard Dedekind et les fondements des mathématiques* (Paris: Vrin).

Dugac, Pierre. 1978. *Sur les fondements de l'analyse de Cauchy à Baire* (Paris: Université Pierre et Marie Curie).

Dugac, Pierre. 1979. *Histoire du théorème des accroissements finis* (Paris: Université Pierre et Marie Curie).

Euler, Leonhard. 1748. *Introductio in analysin infinitorum* (Lausanne) = *Introduction à l'analyse infinitésimale*, *1*, traduit du latin par J. B. Labey (Paris: Bachelier).

Gauss, C. F. 1813. "Disquisitiones generales circa seriem infinitam $1 + \frac{\alpha, \beta}{1.\gamma} x + \frac{\alpha(\alpha+1)\beta(\beta+1)}{1.2.\gamma(\gamma+1)} xx + \frac{\alpha(\alpha+1)(\alpha+2)\beta(\beta+1)(\beta+2)}{1.2.3.\gamma(\gamma+1)(\gamma+2)} x^3 + \text{etc.}$," *Comm. soc. reg. Gott. rec.*, *2* = *Werke*, *3* (Göttingen: Königliche Gesellschaft der Wissenschaften, 1886), 123–162.

Grassmann, Hermann. 1911. *Gesammelte mathematische une physikalische Werke 3*(2) (Leipzig: Teubner).

Lebesgue, Henri. 1904. "Sur les fonctions représentables analytiquement," *Comptes Rendus de l'Académie des Sciences de Paris*, *139*, 29–31.

Lebesgue, Henri. 1905. "Sur les fonctions représentables analytiquement," *Journal des Mathématiques pures et appliquées*, (6) *1*, 139–216.

Lipschitz, Rudolf. 1864. "De explicatione per series trigonometricas instituenda functionum unius variabilis arbitrariarum, et praecipue earum, quae per variabilis spatium finitum valorum maximorum et minimorum numerum habent infinitum, disquisitio, "*Journal für die reine une angewandte Mathematik*, *63*, 296–308 = "Recherches sur le développement en séries trigonométriques des fonctions arbitraires d'une variable et principalement de celles qui, dans un intervalle fini, admettent une infinité de maxima et de minima, traduit par P. Montel," *Acta mathematica*, *36*(1913), 281–295.

Riemann, Bernhard. 1867. "Ueber die Darstellbarkeit einer Function durch eine trigonometrische Reihe," *Abh. K. Gesell. Wiss. Gött. Math. Classe*, *13*, 87–132 = "Sur la possibilité de représenter une fonction par une série trigonométrique," *Oeuvres mathématiques* (Paris: Blanchard, 1968), 225–279.

Seidel, P. L. 1847. "Note über eine Eigenschaft der Reihen, welche discontinuirliche Functionen darstellen," *München Akad. Wiss. Abh.*, *51*, 379–394 = (Leipzig: Engelmann, 1900).

Weierstrass, Karl. 1885. "Ueber die analytische Darstellbarkeit sogenannter willkürlicher Functionen reeller Argumente," *Sitzungsberichte K. Pr. Akad. Wiss. zu Berlin*, 633–639, 789–805 = "Sur la possibilité d'une représentation analytique des fonctions arbitraires d'une variable réelle, traduit par L. Laugel," *Journal des mathématiques pures et appliquées*, (4) *2*(1886), 105–138.

Weierstrass, Karl. 1895. *Mathematische Werke*, *2* (Berlin: Mayer and Müller).

Université Pierre et Marie Curie
Paris, France

August Leopold Crelle und die Berliner Akademie der Wissenschaften*

WOLFGANG ECCARIUS

August Leopold Crelle has won a permanent place in the history of mathematics for his scientific and organizational ability. He is especially important for the history of the Berlin Academy of Sciences, to which he was elected in 1827 with the strong support of Alexander von Humboldt. Crelle was also instrumental in advancing mathematics as the founding editor (1826) of the *Journal für die reine und angewandte Mathematik*, and as the specialist advisor for mathematics to the Prussian Ministry of Education and Cultural Affairs. This paper is devoted primarily to a study of the recommendations Crelle wrote for prospective members of the Academy. The mathematical papers he read there, as well as the prize-problems he proposed and evaluated for the Academy are also discussed.

Es ist das Verdienst von Kurt-R. Biermann, die Wahlvorschläge von Mathematikern für die Berliner Akademie der Wissenschaften erstmalig der Historiographie der Mathematik erschlossen zu haben [1]. Wenn Adolf P. Juškevič in einer Rezension sagt [2], daß diese Dokumente "richtiggehende mathematikhistorische Skizzen" darstellen, so erhebt

* Herrn Prof. Dr. Kurt-R. Biermann zum 60. Geburtstag gewidmet. Überarbeitete und ergänzte Fassung eines Abschnittes der Dissertation des Verfassers, die von Herrn Prof. Dr. Kurt-R. Biermann als einem der Gutachter betreut wurde.

ISBN 0-12-204050-3

sich ganz naturgemäß die Frage nach den näheren Umständen, denen sie ihr Zustandekommen verdanken. Das ist schon deshalb der Fall, weil die von Biermann veröffentlichten ausführlichen Gutachten erst relativ spät—nämlich seit 1839—allgemein in Gebrauch gekommen sind. Letzten Endes hängt das Entstehen von Wahlvorschlägen in extenso mit der immer stärker werdenden Differenzierung der einzelnen Wissenschaftsdisziplinen zusammen, welche die zunächst unterstellte Urteilsfähigkeit *aller* Akademiemitglieder über die personellen Probleme auch ihnen fremder Fachgebiete zu einer Illusion werden ließ. Da es auf die Dauer auch nicht angehen konnte, daß sich das Wahlgremium dem Sachverstand der Vorschlagenden auf Treu und Glauben unterwarf, wurde schließlich in den ausführlichen "Motivationen" ein für beide Seiten gangbarer Kompromiß gefunden. Für die Berliner Akademie ist dieser Vorgang aufs engste mit dem Namen von August Leopold Crelle verknüpft, der sich besonders durch seine wissenschaftsorganisatorische Tätigkeit einen bleibenden Platz in der Geschichte der Mathematik erworben hat [3].

Schon seine Aufnahme in diese traditionsreiche Gelehrtengesellschaft stand eindeutig unter dem Zeichen seiner wissenschaftsorganisatorischen Aktivitäten. Sie ging übrigens auf die Anregung keines Geringeren als Alexander von Humboldt zurück, in dessen viel breiter angelegten Bestrebungen Crelles Tätigkeit eingebettet war. Humboldt, selber einer der bedeutendsten Naturforscher des 18. und 19. Jahrhunderts, besaß durch seinen langjährigen Aufenthalt in Frankreich einen klaren Blick für die Notwendigkeit der Entwicklung der exakten Wissenschaften auch in Deutschland. Als er 1827 endgültig von Paris nach Berlin übersiedelte, brachte er deshalb ein umfassendes wissenschaftsorganisatorisches Programm mit, welches sich besonders auf Astronomie, Chemie, Biologie und Mathematik erstreckte [4]. Es war nur natürlich, daß sich Humboldt bei der Verwirklichung seiner Pläne nach Verbündeten umsah. Auf mathematischem Gebiet mußte seine Aufmerksamkeit dabei zwangsläufig auf Crelle fallen. In erster Linie ist das der 1826 erfolgten Gründung des "Journals für die reine und angewandte Mathematik" zuzuschreiben, mit der Crelles Name als Gründer und Herausgeber verbunden war [5]. Schon die ersten Bände hatten überall eine ungewöhnliche Aufmerksamkeit erregt, besonders durch die Arbeiten Niels Henrik Abels, Carl Gustav Jacob Jacobis und Jakob Steiners. Das Interesse, welches insbesondere in Frankreich der neuen mathematischen Zeitschrift entgegengebracht wurde, war Humboldt natürlich nicht entgangen. Da Humboldts Pläne sich auch auf das Ausbildungswesen der genannten Wissenschaften bezogen und Crelle ähnliche Intentionen besaß [6], ergaben sich auch hier Berührungspunkte. So kam es, daß Crelle gewissermaßen der Vollstrecker der Humboldtschen Pläne auf mathematischem Gebiet wurde.

Die erste persönliche Bekanntschaft der beiden Männer datiert aus dem Jahre 1826, als sich Humboldt zur Vorbereitung seiner Übersiedlung aus Paris in Berlin aufhielt [7]. Sollten die Pläne, die beide sicher schon damals hegten, mit einiger Aussicht auf Erfolg betrieben werden, so mußte Crelle in eine passende einflußreiche Stellung einrücken. Das geschah auf zwei verschiedene Arten: Einmal wurde Crelle dank Humboldtscher Vermittlung zum Fachreferenten für Mathematik an das preußische Kultusministerium berufen, zum anderen regte Humboldt beim Sekretär der mathematischen Klasse der Berliner Akademie, Johann Franz Encke, die Aufnahme zum ordentlichen Mitglied an [8]. Encke ging bereitwillig auf diesen Vorschlag ein, der auch die Unterstützung der anderen Klassenmitglieder fand [9]. In seinem Wahlvorschlag hob Encke bezeichnenderweise die Bedeutung des Crelleschen Journals auch für die Akademie hervor, weil sich über diese Zeitschrift und ihren Herausgeber eine vorzügliche Verbindung zu ausländischen Mathematikern ergäbe [10]. Die Wahl Crelles zum Akademiemitglied erfolgte dann auch mit überwältigender Mehrheit [11]. Damit war aber eine wesentliche Voraussetzung für die Verwirklichung Humboldtscher und Crellescher Absichten erfüllt worden.

Wenn auch die Bedeutung der Akademien im 19. Jahrhundert ganz allgemein zurückgegangen war—"die Mathematiker lebten nicht mehr in der Umgebung von Königshöfen oder in den Salons der Aristokratie" [12]—so übten sie doch nach wie vor einen erheblichen Einfluß auf das wissenschaftliche Leben aus. Mitglied einer Akademie zu sein bedeutete in den meisten Fällen für die Betroffenen eine Aufwertung der sozialen Stellung und wirkte sich dadurch günstig auf ihren weiteren beruflichen Werdegang aus. Die Akademien nahmen aber auch auf den Gang der Wissenschaftsentwicklung direkten Einfluß, sei es im Gedankenaustausch ihrer Mitglieder untereinander oder durch Förderung des wissenschaftlichen Nachwuchses, zum Beispiel mit Hilfe von Preisaufgaben oder durch andere Einrichtungen. Durch verschiedene wissenschaftliche Publikationsorgane boten sie ferner den Mitgliedern eine günstige Gelegenheit, ihre Forschungsergebnisse allgemein bekanntzumachen. Zurückgeblieben waren freilich ihre Verbindungen zu den verschiedenen Formen der Praxis, besonders wenn man beachtet, daß die Akademien ursprünglich unter dem Aspekt der Nützlichkeit gegründet worden waren. Diese Erscheinung besaß allerdings nur vorübergehenden Charakter und wird verständlich, wenn man berücksichtigt, daß die ursprünglichen Akademien als Institutionen des Feudalstaates für die Verbindung zwischen Wissenschaft und Praxis [13] in einem komplizierten und relativ langwierigen Prozeß ihre Stellung in der bürgerlichen Gesellschaft, und zwar als Einrichtung der Bourgeoisie, neu bestimmen mußten. Gerade in diesem Zusammenhang hat Crelle speziell bei der Berliner Akademie eine nicht geringe Rolle gespielt.

Crellesche Vorschläge zur Wahl von Mathematikern in die Berliner Akademie

Name	Vorschlagsdatum	Aktenzeichen*	Crelle Autor/Mitunterzeichn.
Ordentliche Mitglieder			
Dirichlet	4.11.1830	II: IIIa, Bd. 3, 30	Mitunterz.
Steiner	21.4.1834	II: IIIa, Bd. 3, 139	Autor
Hagen	9.1.1842	II: IIIa, Bd. 5, 16	Autor
Minding	9.1.1842 13.2.1842	II: IIIa, Bd. 5, 27 II: IIIa, Bd. 5, 28	Autor Mitunterz.
Auswärtige Mitglieder			
Poisson	vor dem 13.11.1829	II: IIIb, Bd. 2, 42	Mitunterz.
Jacobi	27.4.1835	II: IIIb, Bd. 3, 12	Mitunterz.
Korrespondierende Mitglieder			
Jacobi	vor dem 13.11.1829	II: IIIb, Bd. 2, 42	Mitunterz.
Möbius	vor dem 13.11.1829	II: IIIb, Bd. 2, 42	Mitunterz.
Poncelet	6.12.1830	II: IIIb, Bd. 2, 79	Mitunterz.
Gergonne	6.12.1830	II: IIIb, Bd. 2, 79	Mitunterz.
Quetelet	6.12.1830	II: IIIb, Bd. 2, 79	Mitunterz.
Libri	6.12.1830	II: IIIb, Bd. 2, 79	Mitunterz.
Sturm	19.1.1835	II: IIIb, Bd. 3, 23	Autor
de Pambour	25.2.1839	II: IIIb, Bd. 3, 122	Autor
Kummer	18.3.1839	II: IIIb, Bd. 3, 112	Mitunterz.
Richelot	8.7.1842 13.7.1842 18.7.1842	II: IIIb, Bd. 4, 39 II: IIIb, Bd. 4, 40 II: IIIb, Bd. 4, 41	Autor Mitunterz. Mitunterz.

* Archiv der Akademie der Wissenschaften der DDR.

Die mathematische Klasse der Akademie, in die Crelle nach seiner Wahl im Jahre 1827 eintrat, bedurfte allerdings einer Auffrischung durch wirklich bedeutende Mathematiker, falls sie den neuen Anforderungen gewachsen sein sollte, denn keiner ihrer derzeitigen Mitglieder war in der Lage, der Wissenschaftsentwicklung neue Impulse zu geben und dadurch an der Verwirklichung der Humboldtschen Ideen mitzuwirken [14]. Es kommt nun in der Hauptsache Crelle das Verdienst zu, hier grundlegenden Wandel geschaffen zu haben. Von dieser Tatsache legt die beigefügte Übersicht über die Crelleschen Wahlvorschläge ein beredtes Zeugnis ab! Unter den von Crelle unterbreiteten Vorschlägen dominieren bei weitem die Namen ersten Ranges [15].

Charakteristischerweise befinden sich unter den vorgeschlagenen Korrespondenten auch zwei Herausgeber mathematischer Fachzeitschriften [16]; auf die große Bedeutung ihrer Tätigkeit wird in den kurzen Wahlvorschlägen ausdrücklich hingewiesen. Im Oberbaurat Gotthilf Hagen und dem französischen Ingenieur Francois-Marie de Pambour haben wir zwei Vertreter der angewandten Mathematik vor uns, durch deren von Crelle inspirierten Wahl gewissermaßen die allmähliche Entwicklung der mathematischen Klasse der Berliner Akademie zur Klasse für Mathematik, Physik und Technik eingeleitet wurde. In beiden Fällen konnte er bei den anderen Klassenmitgliedern nicht auf eine allgemeine Bekanntschaft mit Person und Leistung der Kandidaten rechnen; um sich trotzdem ihre Unterstützung beim Wahlvorgang zu sichern, fügte er seinen Anträgen ausführliche "Motivationen" bei und begründete damit eine Tradition, als sich zuerst Johann Peter Gustav Lejeune-Dirichlet und später auch die anderen führenden Mathematiker der Berliner Akademie bei ihren Wahlvorschlägen seinem Vorgehen anschlossen [17].

An den Crelleschen Wahlvorschlägen ist aber nicht nur diese Tatsache interessant, vielmehr wird aus ihrem Wortlaut auch besonders deutlich, welche Vorstellungen er mit dem Begriffe der Anwendungen der Mathematik verband. Doch lassen wir ihn dazu selbst zu Worte kommen! Im Wahlvorschlag für Hagen führte er aus:

> Eine ersprießliche und naturgemäße Anwendung der Mathematik auf das Bauwesen kann schon nicht anders als von Jemand gemacht werden, der Mathematiker und Practiker, beides in zureichendem Umfange, *zugleich* ist. Aber auch das reicht noch nicht hin, und die Anwendung kann dennoch wenig naturgemäß ausfallen, wenn ein Drittes fehlt, welches beide Arten von Kenntnissen und Einsichten, die theoretischen und die practischen, gleichsam beherrscht und lenkt. Dieses Dritte ist ein eigenthümliches *Talent*, die Gegenstände der Praxis mathematisch zu erfassen und zu durchschauen, und die Mathematik nur so wirken zu lassen, wie sie es nach der Natur complicirter Dinge vermag: in die Erscheinungen nicht bloß mathematische Formeln zu bringen, sondern sie mit mathematischem Geiste zu durchschauen und zu durchdringen. [18]

In Hagen fand er diese Voraussetzungen auf glückliche Art vereint, und daß er mit dieser Meinung nicht alleine stand, beweist der Umstand, daß sich Dirichlet und Steiner seinem Vorschlag anschlossen [19]. Auch Humboldt hat den Crelleschen Vorschlag unterstützt [20].

In Pambour würdigte Crelle einen der Pioniere des Dampfmaschinen- und Lokomotivbaues, dem das Verdienst zukommt, sich als einer der ersten mit theoretischen Problemen dieses ingenieurtechnischen Gebietes befaßt zu haben [21]. So heißt es denn auch im Crelleschen Wahlvorschlag:

> Pambour hat sich um die Theorie der Dampfmaschinen, dieser so unendlich wichtigen Werkzeuge, die in der neueren Zeit zur Föderung des industriellen und selbst intellectuellen Verkehrs, also zur Förderung der gesammten Civilisation, so mächtig beigetragen haben, in der That überaus verdient gemacht; er hat die bessere Theorie dieser wichtigen Maschinen nach meiner Meinung erst geschaffen, denn alles, was bis dahin für Theorie galt, war offenbar unrichtig. Außerdem sind die Ausdauer und die Einsicht, mit welcher er zahlreiche Versuche im Großen über die Wirkung locomotiver Dampfmaschinen auf der Eisenbahn zwischen Liverpool und Manchester angestellt hat, höchst achtens werth. Das . . . Werk des Herrn v. Pambour über Dampfwagen ist nach meiner Meinung so vortrefflich und in so echt mathematischem Geiste verfaßt, daß es in der That in seiner Art classisch genannt zu werden verdient. [22]

Die Hochachtung Crelles vor den Leistungen Pambours war so groß, daß er ihn sogar mehrfach, wenn auch erfolglos, für die Friedensklasse des Ordens "Pour le mérite" vorschlug [23].

Aber nicht nur durch Wahlvorschläge für solche Wissenschaftler, die in erster Linie als Techniker und erst in zweiter als Mathematiker einzuordnen sind, hat Crelle das technische Element in der mathematischen Klasse der Berliner Akademie heimisch gemacht: Seine Vorträge in der Klasse und seine Akademieveröffentlichungen wandten sich neben rein mathematischen Themen immer häufiger solchen der Technik zu. So finden wir neben der Behandlung von Fragen des Eisenbahnbaues Crellesche Auseinandersetzungen mit Problemen der Theorie der Dampfmaschinen, der Statik, der Trinkwasserversorgung, der Abwasserbeseitigung und der barometrischen Höhenmessung [24].

Auf mathematischem Gebiet war Crelle in der Akademie indes nicht weniger aktiv. Da sind zunächst seine mathematischen Vorträge vor der Klasse zu nennen, die auch in den Akademieabhandlungen Aufnahme fanden. Ihre Thematik schloß sich meist eng den besonderen Crelleschen Interessengebieten an: Neben einigen Beiträgen aus dem Gebiete der Analysis—zum Beispiel Konvergenzuntersuchungen bei Reihen—finden wir in der Hauptsache Untersuchungen zur Zahlentheorie [25].

Die Ausarbeitung von Tabellenwerken war bekanntlich ein weiterer wichtiger Schwerpunkt in Crelles mathematischer Tätigkeit [26]. Es war deshalb ganz natürlich, daß er auch in seiner akademischen Wirksamkeit darauf Bezug nahm. Hierbei handelte es sich allerdings weniger um seine Klassenvorträge, als vielmehr um finanzielle Unterstützung von seiten der Akademie für die Ausarbeitung einer Faktorentafel für die natürlichen Zahlen zwischen 4 000 000 und 6 000 000, ein Unternehmen, für welches sich auch Gauß interessierte [27]. Für die Besoldung der bei diesem umfangreichen Vorhaben tätigen Hilfskräfte erhielt Crelle von der Akademie zwischen 1834 und 1835 insgesamt 400 Taler [28]. Später hat er sich nochmals für den gleichen Zweck eingesetzt, als die "lebende Rechenmaschine" Zacharias Dase seine Faktorentafel bis zur 9. Million fortsetzte [29]. Zusammen mit Dirichlet trat er für eine angemessene Besoldung des Kopfrechners Dase ein, als die Berliner Akademie zur Begutachtung dieser Angelegenheit aufgefordert wurde [30].

Einen aufschlußreicheren Einblick in Crelles mathematische Interessengebiete gewährt freilich seine Mitarbeit auf dem Gebiete der akademischen Preisaufgaben, wo er sowohl als Gutachter für die eingegangenen Lösungen tätig war [31], aber auch durch eigene Vorschläge für derartige Aufgaben hervortrat [32]. Wenn nun auch diese Vorschläge bei den tatsächlich gestellten Preisaufgaben keine Berücksichtigung fanden, so erhält man doch gerade aus ihnen den erwähnten Einblick und gleichzeitig einen hochinteressanten Eindruck von den engen Wechselbeziehungen, denen in Crelles Bewußtsein die von Traditionen bestimmten Interessen und die von der zeitgenössischen Forschung neu aufgeworfenen Probleme unterworfen waren, mit denen er durch seine Herausgebertätigkeit ständig konfrontiert wurde. Am deutlichsten wird diese Erscheinung in seinem Vorschlag vom 24.4.1836 [33], der aus diesem Grunde und gewissermaßen als mathematisches Credo Crelles hier zum Abschluß des Berichtes über seine akademische Wirksamkeit vollständig wiedergegeben werden soll. Er lautet:

Zunächst wiederhole ich den Vorschlag folgender in den Jahren 1828 und 1832 zur Auswahl gestellten Aufgaben, weil meines Wissens noch keine derselben von den Mathematikern in Folge eigener Anregung befriedigend gelöset worden ist.

1.

Wenn auch noch keine vollständige Theorie der algebraischen Gleichungen von der Form

$$x^n + Ax^{n-1} + Bx^{n-2} + Cx^{n-3} \dots + N = 0,$$

wo n eine ganze Zahl ist, für Gleichungen von höheren Graden als dem vierten, gegeben werden kann, zu welcher Theorie insbesondere der allgemeine geschlossene Ausdruck von x durch $A, B, C, \dots N$ gehören würde; so kann doch Folgendes verlangt werden:

Erstlich, ein elementarer, den Lehrbüchern und dem Unterrichte zugänglicher, strenger Beweis, daß die Wurzeln nicht allgemein algebraisch durch die Coefficienten ausgedrückt werden können, sobald n größer ist als 4, in so fern solches, nach den Arbeiten von Riccati, Abel etc. zu schließen, wirklich der Fall sein sollte.

Zweitens die Beantwortung der Frage, von welcher *Art* im Allgemeinen die geschlossenen Ausdrücke der Wurzeln durch die Coefficienten sein dürften, welche Ausdrücke sich, wie bewiesen werden kann, immer auf die Form $\alpha \pm \beta\sqrt{-1}$ reduciren lassen, und ob etwa, z.B. von Gleichungen fünften Grades, die Wurzeln durch elliptische Functionen ausgedrückt werden können, ungefähr auf ähnliche Weise, wie die Wurzeln der Gleichungen dritten und vierten Grades durch Kreis-Functionen, welches, nach der Zahl der Elemente in diesen und jenen Functionen zu schließen, nicht ganz unwahrscheinlich ist.

Drittens die Beantwortung der Frage: welche Anwendungen auf die algebraische Auflösung der Gleichungen überhaupt von den elliptischen Functionen gemacht werden können.

Viertens, eine allgemeine Angabe der besonderen Fälle, wo Gleichungen von höheren Graden als dem vierten, *algebraisch* auflösbar sind, und der Bedingungs-Gleichungen zwischen den Coefficienten $A, B, C, \ldots N$, oder auch zwischen den Wurzeln selbst, für diese Fälle.

2.

Eine systematische Theorie der Convergenz der unendlichen Reihen zu geben, wozu insbesondere die Angabe allgemeiner *Kennzeichen* der Convergenz der Reihen, deren Fortschreitungs-Gesetz gegeben ist, gehören würde.

3.

Eine allgemeine Theorie der Polyeder, das heißt, beliebiger, von Ebenen umschlossener, körperlicher Räume zu geben, wenn auch nur der *convexen* Polyeder, oder derjenigen, deren Durchschnitte mit beliebigen Ebenen geradlinige Figuren sind, die keine ausspringenden Winkel haben. Zu dieser Theorie würden zunächst allgemeine Ausdrücke der Bedingungen gehören, welche zwischen der Anzahl der umschließenden Seiten-Ebenen, der Anzahl der Seiten der einzelnen Ebenen etc. Statt finden müssen, wenn die Polyeder *möglich* sein sollen. Ferner die Angabe der Stücke (Seiten und Winkel), welche ein Polyeder *bestimmen*; desgleichen die Berechnung der übrigen Stücke aus den bestimmenden Stücken u.s.w.

4.

In der Variations-Rechnung kommt es bekanntlich auf die Veränderung und Differentiation von Integralen an, die von entwickelt oder unentwickelt gegebenen Differential-Ausdrücken vorausgesetzt werden, weshalb die Variations-Methode auch direct die sogenannten Bedingungen der Integrabilität giebt: also auf die Differentiation unter dem Integralzeichen. Es wäre zu untersuchen, welcher Gebrauch von den Resultaten jener Differentiation, noch außer demjenigen zur Bestimmung der größten und kleinsten Werthe der Integrale zwischen bestimmten Grenzen für die Reihen, für die Integration selbst, oder sonst in der Analysis gemacht werden könne.

5.

Das Gesetz der Kettenbrüche zu untersuchen, welche die Wurzeln einer Gleichung vom dritten Grade ausdrücken; die Verwandlung und Reduction der ganzen rationalen Functionen dritten Grades von zwei Veränderlichen abzuhandeln, und eine directe Methode zur Auflösung in ganzen Zahlen der Gleichungen dritten Grades zwischen zwei Veränderlichen zu geben, wenn die Unbekannten ganze Zahlen sein können.

Sodann füge ich, für die Auswahl, jetzt noch folgende Aufgabe hinzu.

6.

Die Bedeutung der *Facultäten* in der Analysis umfassend nachzuweisen, und zwar der Facultäten für beliebige rationale oder irrationale, mögliche oder unmögliche Exponenten, nicht bloß der Factoriellen, oder Producte äquidifferenter Größen. Die Convergenz der Reihen auf welche die Entwicklung der Facultäten, und zwar auch dieser selbst, nicht bloß ihrer Logarithmen führt, zu untersuchen; die allgemeine Reduction bestimmter hypergeometrischer Reihen auf Facultäten auszuführen; den Zusammenhang der Facultäten mit den Kreis- und hyperbolischen Functionen, und dann mit den elliptischen Functionen, ausführlich nachzuweisen; allgemeine Gleichungen zwischen den Facultäten-Coefficienten, den Differenz-Coefficienten und den Bernoullischen Zahlen zu geben. Zu untersuchen, ob und wie etwa, z.B. durch die Vermittlung bestimmter Integrale, die Facultäten mit den Wurzeln der algebraischen Gleichungen, vielleicht von höheren als dem vierten Grade, zusammenhängen; auch die Anwendungen der Facultäten auf die Theorie der Zahlen zu berücksichtigen.

NOTES

1. Biermann, Kurt-R. 1960. *Vorschläge zur Wahl von Mathematikern in die Berliner Akademie. Ein Beitrag zur Gelehrten- und Mathematikgeschichte des 19. Jahrhunderts. Abhandlungen der Deutschen Akademie der Wissenschaften zu Berlin, Klasse für Mathematik, Physik und Technik*, Nr. 3 (Berlin).
2. Juškevič, Adolf P. (1961). [Besprechung von 1], in *Deutsche Literaturzeitung, 82*, 1031–1033.
3. Siehe hierzu Eccarius, Wolfgang. (1972). "Der Techniker und Mathematiker August Leopold Crelle (1780–1855) und sein Beitrag zur Förderung und Entwicklung der Mathematik im Deutschland des 19. Jahrhunderts," *NTM—Schriftenreihe für Geschichte der Naturwissenschaften, Technik und Medizin, 12*(2), 38–49.
4. Näheres hierzu in Biermann, Kurt-R. (1967). "Alexander von Humboldts wissenschaftsorganisatorisches Programm bei der Übersiedlung nach Berlin," *Monatsberichte der Deutschen Akademie der Wissenschaften zu Berlin, 9*, 216–222.
5. Siehe hierzu Eccarius, Wolfgang. (1976). "August Leopold Crelle als Herausgeber des Crelleschen Journals," *Journal für die reine und angewandte Mathematik, 286/287*, 5–25.
6. Siehe hierzu Eccarius, Wolfgang. (1977). "August Leopold Crelle und der Versuch einer Reorganisation des preußischen Mathematikunterrichts in der Periode der industriellen Revolution," *Mitteilungen der Mathematischen Gesellschaft der DDR, 10*, 90–95.
7. Vergleiche Biermann, Kurt-R. 1973. *Die Mathematik und ihre Dozenten an der Berliner Universität 1810–1920. Stationen auf dem Wege eines mathematischen Zentrums von Weltgeltung* (Berlin), 22.
8. Archiv der Akademie der Wissenschaften der DRR (zukünftig als AAW zitiert), II: IIIa, Bd. 3, Bl. 1. In einem Rundschreiben Enckes an die Mitglieder der Klasse heißt es hier: "Von einer Seite, von woher jede Aeußerung aufmerksame Beachtung verdient, bin ich daran erinnert worden, ob es nicht vielleicht rathsam seyn möchte den Geheimen Oberbaurath Crelle, den Herausgeber des mathematischen Journals . . . zum Mitglied unserer Classe zu wählen." Da Crelle zu jener Zeit keine anderen einflußreichen Gönner besaß, kommt als Anreger praktisch nur Humboldt in Frage, selbst wenn sein Name hier nicht genannt wird.
9. Das waren zu dieser Zeit Enno Heeren Dirksen, Johann Albert Eytelwein, Ernst Gottfried Fischer, Johann Philipp Gruson, Jabbo Oltmanns und Friedrich Theodor Poselger. Siehe dazu auch: AAW, II: IIIa, Bd. 3, Bl. 1–3.

10. *Ebenda*, Bl. 3.
11. *Ebenda*, Bl. 4. Das Abstimmungsverhältnis bei der am 12.7.1827 von der Gesamtakademie vorgenommenen Wahl betrug 22:1.
12. Struik, Dirk J. 1972. *Abriß der Geschichte der Mathematik* (Berlin, 5. Auflage), 148.
13. Vergleiche Wussing, Hans. (1958). "Die École polytechnique—eine Errungenschaft der französischen Revolution," *Pädagogik*, *13*, 646–662.
14. Siehe hierzu Lenz, Max. 1910. *Geschichte der Königlichen Friedrich-Wilhelm-Universität zu Berlin*, *2*(1) (Halle), 374–375. Lenz spricht hier davon, daß es keinem der Berliner Vertreter der Mathematik jener Zeit gelungen sei, "die Zeiten eines Leibniz und Euler zu erneuern."
15. Selbst bei den Wahlvorschlägen, die Crelle nur mitunterzeichnet hat, darf doch angenommen werden, daß sie durch seine Initiative zustandekamen. Dafür sprechen schon seine engen Beziehungen, die er als Zeitschriftenherausgeber zu den Vorgeschlagenen unterhielt. Im Einzelfall lassen sich dazu noch weitere Indizien angeben, auf die hier einzugehen zu weit führen würde.
16. Es waren dies Joseph Duez Gergonne und Adolphe Quetelet.
17. Siehe Biermann, Kurt-R. [Anmerkung 1], 8–9.
18. AAW, II:IIIa, Bd. 5, Bl. 16.
19. Näheres über Hagen findet man in: Ottmann, Ernst. 1934. *Gotthilf Hagen*, *Altmeister der Wasserbaukunst* (Berlin).
20. *Ebenda*, 19.
21. Maedel, Karl-Ernst. 1966. *Die deutschen Dampflokomotiven gestern und heute* (Berlin), 73.
22. AAW, II:IIIb, Bd. 3, Bl. 122. Die von Crelle hier zitierte Abhandlung Pambours erschien in mehreren Fortsetzungen in den Bänden 23 bis 27 seines "Journals für die Baukunst."
23. AAW, II:IIIa, Bd. 1.
24. Eine vollständige Übersicht gewährt das Verzeichnis der Crelleschen Akademiepublikationen in Harnack, Adolf. 1900. *Geschichte der Königlich Preußischen Akademie der Wissenschaften zu Berlin*, *3* (Berlin), 43–44.
25. *Ebenda*.
26. So zum Beispiel seine berühmten "Rechentafeln, welche alles Multipliciren und Dividiren unter Tausend ganz ersparen, bei größeren Zahlen aber die Rechnung erleichtern und sicherer machen", die in 15 Auflagen, zuletzt 1954 (!) erschienen.
27. Brief Gauß' an Schumacher vom 25.1.1842, in Peters, Christian August Friedrich (Hrsgb.). 1862. *Briefwechsel zwischen C. F. Gauß und H. C. Schumacher*, *4*(Altona).
28. Siehe hierzu Harnack, Adolf. [Anmerkung 24], I (2), 775.
29. Näheres hierzu in Biermann, Kurt-R. (1967). "Beurteilung und Verwendung einer 'lebenden Rechenmaschine' durch C. F. Gauß und die Berliner Akademie," *Forschungen und Fortschritte*, *41*, 361–364.
30. AAW, II:VIb, Bd. 13, Bl. 194.
31. AAW, II:VIIIc, Bd. 2, Heft 2.
32. Zu den Preisaufgaben findet man Näheres bei Biermann, Kurt-R. (1964). "Aus der Geschichte Berliner mathematischer Preisaufgaben," *Wissenschaftliche Zeitschrift der Humboldt-Universität Berlin*, *mathematisch-naturwissenschaftliche Reihe*, *13*, 185–198.
33. AAW, II:VIIIc, Bd. 2, Heft 2. Für die Erlaubnis zur Veröffentlichung der Crelle betreffenden Dokumente aus dem Archiv der Akademie der Wissenschaften der DDR spreche ich der Leitung dieser Einrichtung meinen herzlichsten Dank aus.

Eisenach, Deutsche Demokratische Republik

Hermite–Weber–Neumann: Kleine Briefgeschichte eines grossen Irrtums*

E. A. FELLMANN

Irrtümer sind Schattenklippen, die erst
das Profil der Wahrheit erkennen lassen.
M.F.

In 1978 the author published an article in the journal *Humanismus und Technik*, "An unpublished letter from Charles Hermite to Heinrich Weber," (in German). The addressee of this previously unknown four-page letter (dated February 8, 1891) was not named, and no envelope giving the receiver's name survives. Based upon a number of indications within the letter, it seemed that the recipient must have been Heinrich Weber (1842–1913). As soon as the article appeared, the author sent an offprint to Kurt-R. Biermann, who responded with the surprising opinion that the addressee was not H. Weber, but probably Carl Neumann (1832–1925). This article acknowledges that Biermann's suggestion is correct, and describes the errors which led to the original identification of Weber as recipient of Hermite's letter, and the reasons why Neumann must, in fact, have been the actual addressee.

I

In der Zeitschrift *Humanismus und Technik* 22.Band 2.Heft Dezember 1978, pp. 71–83, erschien mein Beitrag "Ein unveröffentlichter Brief von

* Herrn Prof. Dr. Kurt-R. Biermann zum 60. Geburtstag gewidmet.

ISBN 0-12-204050-3

Charles Hermite an Heinrich Weber". Darin präsentierte ich einen bisher unbekannten vierseitigen Originalbrief* Hermites vom 8.2.1891 an einen nicht namentlich genannten Adressaten; ein Umschlag ist nicht erhalten geblieben. Aufgrund einiger mir als schlüssig erscheinender Indizien[1] glaubte ich, im Empfänger Heinrich Weber (1842–1913) sehen zu müssen. Bald nach Erscheinen des Heftes sandte ich ein Separatum an K. R. Biermann, der mir den Empfang mit der für mich zunächst überraschenden Meinungsäusserung quittierte,[2] der Adressat sei nicht H. Weber, sondern Carl Neumann (1832–1925). Nach wiederholter Abwägung aller Gründe, deren wichtigste ich dem Leser—quasi als prächtiges Lehrstück, wie man in die Irre gehen kann—nicht vorenthalten möchte,[3] gebe ich heute K. R. Biermann öffentlich recht und danke ihm bei der sich nunmehr bietenden Gelegenheit herzlich für die Korrektur eines mathematikhistorischen Irrtums aus unseren Tagen.

II

Da Hermites Brief sich um die Forscherpersönlichkeiten von Riemann, Bessel, Kovalevskaja, Appell, Picard und Goursat dreht und von problemgeschichtlichem Interesse ist, möge seine Transkription hier wiedergegeben werden:

Transkription

Paris 8 Février 1890

1 Monsieur,
2 J'ai recu avec la plus grande reconnaissance les
3 ouvrages de physique mathématique que vous avez
4 eu la bonté de m'envoyer; permettez moi de vous
5 offrir mes remerciements bien sincères et de saisir
6 cette occasion pour vous exprimer les sentiments de
7 la plus haute estime que m'ont inspirés depuis
8 longtemps vos travaux. Je suis resté en dehors
9 de ces belles et importantes recherches sur l'électricité,
10 le magnétisme, l'hydrodynamique auxquelles vous
11 avez consacré vos efforts, mais vous avez fait aussi
12 une part de votre activité scientifique aux
13 questions d'Analyse qui m'ont le plus interessé

14 à la théorie des fonctions abéliennes. C'est
15 en étudiant la première édition de votre ouvrage
16 que la lumière s'est faite dans mon esprit

* Das Original befindet sich im Privatbesitz des Herausgebers in Basel. Es ist ein Geschenk von Herrn Prof. Dr. István Szabó (†21.1.1980), das auch an dieser Stelle herzlich verdankt sei.

17 sur les théories restées jusq'alors inaccessibles
18 de Riemann. Vous avez rendu Monsieur, un
19 incomparable service en exposant avec tant de
20 clarté et de facilité, des conceptions entièrement
21 profondes, d'une puissance admirable, mais bien
22 difficiles à comprendre dans les mémoires
23 du grand géomètre. Sans vous l'oeuvre du
24 merveilleux génie de Riemann n'aurait pas
25 porté tous ses fruits pour la marche en avant
26 de la science, vous l'avez fait connaître, vous
27 avez donné à Mr. Appell la trémie de
28 son beau mémoire sur les fonctions à

29 multiplicateurs. Bien souvent je m'entretiens de vous
30 avec l'éminent analyste et c'est sa reconnaissance que
31 je vous exprime en même temps que la mienne[.]
32 Les richesses analytiques dont il a fait un si heureux emploi
33 seront mises à la disposition des lecteurs français qui ignorent
34 votre langue, dans un travail destiné aux Annales de
35 l'Ecole Normale Supérieure, qu'il rédige en ce moment
36 en collaboration avec Mr. Goursat. Les méthodes exposées
37 dans votre ouvrage se trouveront ainsi répandues parmi
38 nous et entreront dans le courant de notre enseignement.
39 Vous me permettez encore à propos des fonctions abéliennes
40 de rappeler votre excellent mémoire du Ch. [Cahier] 56 du Journal
41 de Crelle, où une si remarquable application de ces fonctions
42 est faite à une question mécanique, comme celle que
43 devait découvrir longtemps après Madame de Kovalevski.
44 Enfin Monsieur, je suis comme vous un ami

45 des fonctions de Bessel, et je me dédommage avec vos
46 mémoires d'analyse de ne pouvoir suivre les travaux [où]
47 à l'exemple de votre illustre et vénéré père, vous appliquez
48 le calcul aux questions physiques qui occupent le plus
49 l'attention de notre époque. Je confierai les ouvrages
50 dont vous avez bien voulu me faire don, à un lecteur
51 très compétent, à mon gendre Mr. Picard, dont le
52 domaine mathématique est plus étendue que le mien
53 et c'est par lui que je compte ne pas rester entièrement étranger
54 à cette partie de vos recherches qui est dehors de mon travail
55 habituel.
56 Je suis Monsieur, profondément reconnaissant envers la Société Royale
57 des Sciences de Saxe, pour le témoignage de sympathie dont elle m'a
58 honoré à l'occasion du soixante dixième anniversaire de ma naissance.
59 En pensant que peut-être vous n'y êtes pas étranger, j' oserai vous demander
60 d'être auprès de vos eminents confrères, l'interprète de ma respectueuse
61 et bien vive gratitude, et je saisirai cette occasion pour vous offrir avec
62 l'expression de ma plus haute estime, celle de ma profonde sympathie
63 et de mes sentiments bien sincèrement dévoués.

Ch. Hermite

III

Kernstück meines Irrtums ist zweifellos ein Lesefehler, der zu vermeiden gewesen wäre. In der hier original abgebildeten Briefzeile 40 (cf. die Transkription) verweist Hermite auf eine Arbeit des Adressaten im Crelle-Journal "X6". Hinsichtlich der enthusiastischen Ausdrucksweise Hermites in den Zeilen 23–26 drängt sich nun der Name Heinrich Webers unmittelbar auf, wenn man an dessen Bücher von 1869 und 1876 denkt.[4] Die sehr merkwürdig geschriebene Ziffer X (cf. Abbildung) lässt sich nur entweder als 7 oder als 5 interpretieren. Im Band 56 des Crelle-Journals findet sich nun keine Abhandlung H. Webers, hingegen im Band 76, die sich einigermassen in den Kontext einfügen liesse.[5] Hält man sich an die Leseart "76", so scheidet Neumann als Adressat aus, da er in diesem Band keine Arbeit publiziert hat—es sei denn, Hermite habe sich in der Band-Nr. geirrt.

Auf H. Weber völlig unpassend sind Hermites Briefzeilen 47–49, denn Webers Vater war nicht Physiker, sondern Historiker. Ich versuchte in "Humanismus und Technik," diese Unpässlichkeit mit der Hypothese zu überbrücken, dass Hermite vielleicht angenommen habe, Heinrich Webers Vater sei der berühmte Physiker und Freund von Gauss, Wilhelm Weber (1804–1891), der allerdings Junggeselle war (!). Exakt an dieser Stelle setzte Biermann mit seiner Kritik ein: ". . . Hermite kannte die mathematische "Szene" in Deutschland sehr gut, und ich halte es für ganz ausgeschlossen, dass er der Meinung war, der (unverheiratete) Wilhelm Weber sei der Vater Heinrich Webers".[6] Ferner ". . . wäre es übrigens erklärlicher, dass jemand eine 5 und eine 7 verwechselt, als dass er einem ihm bekannten Junggesellen einen Sohn zuspricht".[7] Das war ein gewichtiges Argument, doch der Schlusspunkt musste durch ein textkritisches gesetzt werden. Der vorliegende Autograph enthält keine andern Ziffern 5 oder 7, und die einzige (damals sofort zugängliche) Quelle war die Briefreproduktion in den gedruckten "Oeuvres" von Hermite. Dort allerdings kommt eine "5 oder 7" nur ein einzigesmal vor, und der Sinn ist zweideutig! Einem versierten Kenner von Hermites Handschrift (der ich nicht bin) wäre es wahrscheinlich sofort klar gewesen, dass die "Problemziffer", die wie ein kleines griechisches *phi* geschrieben ist, eine Hermitesche 5 ist, doch ich wurde davon erst im Juni 79 in Leningrad überzeugt, als ich die Originalbriefe Hermites an Markov einsehen konnte.[8] Durch die erhärtete Leseart "Crelle 56" rückt nun alles in ein ganz klares Licht. Das "Vaterproblem" löst sich bestens auf [denn Carl Neumanns Vater war der hochbedeutende Physiker Franz Ernst Neumann (1798–1895)], und im Crelle-Journal 56 findet sich eine gut in den Briefkontext passende Arbeit Carl Neumanns.[9] Schliesslich fügt sich auch

Hermites Hinweis auf die Sächsische Akademie (Z. 56/57), zu welcher Carl Neumann in direkter Beziehung stand, gut ein. Ob die Briefzeilen 23–26 ihre historische Berechtigung beanspruchen können im Hinblick auf die entsprechenden Verdienste H. Webers um die Kolportage der Ideen Riemanns in Frankreich, (cf. N 4), wäre zweifellos einer speziellen Untersuchung wert.[10] Biermann ist geneigt, diese fast überschwenglichen Zeilen einem "Schuss *flatterie*" zuzuschreiben. Wie dem auch sein mag: das hier vorliegende "Werkstattbeispiel" zeigt wieder einmal mehr, dass die Geschichte der Wissenschaften—wenn sie selbst eine Wissenschaft sein will—eine äusserst diffizile Sache ist.

NOTES

1. Ich habe sie seinerzeit am 18. Internationalen Kolloquium für Geschichte der Mathematik in Oberwolfach (1974) vorgetragen.
2. Brief K. R. Biermanns vom 22.4.1979 an mich.
3. Diese Darstellung soll nicht als Apologie verstanden werden. "Man schützt sich vor dem Irrtum und zieht sogar Nutzen aus demselben, indem man die Motive, welche verführend gewirkt haben, aufdeckt," from Mach, Ernst. 1906. *Erkenntnis und Irrtum* (Leipzig, 2. Auflage), 119.
4. Weber, H. 1869. *Die partiellen Differentialgleichungen der mathematischen Physik. Nach Riemanns Vorlesungen bearbeitet.* (Braunschweig, 1. Auflage), Sowie *B. Riemanns Gesammelte mathematische Werke und wissenschaftlicher Nachlass*, herausgegeben von R. Dedekind und H. Weber (Leipzig, 1876); Französische Ausgabe von L. Laugel, mit einem Vorwort von Ch. Hermite und einem Essay von Felix Klein, *Oeuvres mathématiques* (Paris, 1898).
5. Weber, H. 1876. "Zur Theorie der Transformation algebraischer Functionen," *Journal für die reine und angewandte Mathematik*, *76*, 345–348.
6. Cf. N.2.
7. Brief K. R. Biermanns vom 11.7.79 an mich.
8. Archiv der Akademia Nauk USSR, Leningrad, F.173, o.1, N 38. Die Korrespondenz ist ediert von Ošigova, H. P. 1967. Revue d'Histoire des Sciences et de leurs Applications *20*, 2–32.
9. Neumann, C. 1869. "De problemate quodam mechanico, quod ad primam integralium ultraellipticorum classem revocatur," *Journal für die reine und angewandte Mathematik*, *56*, 46–64.
10. Neumann, C. 1865. *Vorlesungen über Riemanns Theorie der Abelschen Integrale* (Leipzig).

Basel, Switzerland

Mittelalterliche mathematische Handschriften in westlichen Sprachen in der Berliner Staatsbibliothek. Ein vorläufiges Verzeichnis*

MENSO FOLKERTS

In numerous works Professor K.-R. Biermann has utilized previously unknown materials from Berlin archives which shed considerable information on the history of modern mathematics. This paper also concerns documents to be found in Berlin which provide an overview of those mathematical texts which were written in Western languages in the Middle Ages and are now preserved in the Staatsbibliothek in Berlin. It is generally known that this library possesses many manuscripts, but for numerous codexes relatively little is known about their contents. Unfortunately, the printed catalogues do not give sufficient information about these sources, especially for the important group of Latin manuscripts, numbering nearly 2500, of which three-fourths are described only in the library's own hand-written catalog. This paper intends to remedy this state of affairs by identifying all those manuscripts containing mathematical texts. Manuscripts up to the year 1500 have been included. In all, 72 manuscripts have been found containing mathematical texts, many with several treatises.

In vielen Arbeiten hat Herr Professor Biermann bisher unbekannte Materialien aus Berliner Archiven herangezogen, die über die Entwicklung der Mathematik in neuerer Zeit informieren. Auch die vorliegende Arbeit

* Herrn Prof. Dr. Kurt-R. Biermann zum 60. Geburtstag gewidmet.

ISBN 0-12-204050-3

befaßt sich mit kaum ausgewerteten Dokumenten, die sich in Berlin befinden: Sie soll einen Überblick geben über diejenigen mathematischen Texte, die im Mittelalter in westlichen Sprachen geschrieben wurden und in Handschriften der Berliner Staatsbibliothek vorhanden sind. Es ist allgemein bekannt, daß diese Bibliothek viele Handschriften besitzt, doch weiß man über den Inhalt zahlreicher Codices relativ wenig. Diese erstaunliche Tatsache hängt mit der Entwicklung der Bibliothek zusammen. Da es über die Geschichte der Staatsbibliothek im allgemeinen und der Handschriftenabteilung speziell gedruckte Arbeiten gibt,[1] genügt es hier, das Wichtigste kurz zusammenzufassen.

Die spätere Preußische Staatsbibliothek wurde 1661 vom Kurfürsten Friedrich Wilhelm gegründet. Indem er vor allem den Bücherbestand in der Kurmark und in den neuerworbenen Landesteilen heranzog, erreichte er, daß die Bibliothek bei seinem Tode (1688) schon 1618 Handschriften besaß. Unter den späteren Herrschern wurde der Bestand nur unsystematisch vermehrt; 1735 sollen ca. 2000 Bände Handschriften vorhanden gewesen sein. Zum Teil in Verbindung mit der Gründung der Berliner Universität begann man seit Beginn des 19.Jahrhunderts, den Bücher- und Handschriftenbestand konsequent zu vermehren: Unter dem Oberbibliothekar Friedrich Wilken (1817–1840) wurden bisher verstreute Handschriftenbestände, z.T. aus rheinischen Klöstern, der Stiftsbibliothek Kamin, dem Domstift Havelberg und der Ritterakademie Brandenburg, in Berlin konzentriert, und nach dem Tod von Heinrich Friedrich Diez (1817) gelangte erstmals eine große Privatsammlung (836 Handschriften) in die Bibliothek. Kurz vor Wilkens Tod (1839) gab es schon etwa 6000 Handschriften. Diese konsequente Erwerbungspolitik wurde unter Wilkens Nachfolgern Georg Heinrich Pertz (1842–1873) und Richard Lepsius (1873–1884) fortgesetzt. Durch systematische Ankäufe versuchte man, der Bibliothek eine Bedeutung zu verschaffen, die mit der Pariser und Londoner vergleichbar war. Vor allem Friedrich Wilhelm IV. förderte die Erwerbungen durch finanzielle Zuwendungen. Unter dem ersten Generaldirektor August Wilmanns (1886–1905) wurde 1886 die Handschriftenabteilung als Sondersammlung begründet, deren erster Abteilungsdirektor Valentin Rose († 1916) sich um die wissenschaftliche Erfassung der Handschriften verdient machte.

Seit 1880 wurden zahlreiche größere Handschriftensammlungen erworben. Die bedeutendsten waren die Hamilton-Sammlung (1882; 506 Handschriften an die Staatsbibliothek); diejenigen Handschriften von Sir Thomas Phillipps, die er 1824 von Gerard Meerman gekauft hatte—sie stammen aus dem Collège de Clermont bei Paris—(1887; 190 lateinische Handschriften); die Handschriften von Joseph Goerres (1902; 64 Handschriften) und aus der Königlichen Bibliothek Erfurt (1909; 203 Handschriften). Unter Wilmanns' Nachfolger Adolf von Harnack (1905–1921) wurden insgesamt 1522 abendländische Handschriften erworben. Nach 1920 konnten nur noch relativ wenig Handschriften gekauft werden: Die Zahl der westlichen Handschriften stieg von 12091 im Jahre 1910 über 13138 (1926), 14600 (1939) auf 14745 (1942); im Jahre 1942 gab es insgesamt 71892 Handschriften.

Die meisten westlichen Handschriften überstanden den 2. Weltkrieg unbeschädigt. Diejenigen Codices, die im Krieg in der späteren französischen und amerikanischen Besatzungszone ausgelagert worden waren, wurden zunächst im Tübinger bzw. Marburger Depot der Staatsbibliothek gesammelt und später nach Berlin (West) gebracht; gegenwärtig befinden sie sich in der Staatsbibliothek Preußischer Kulturbesitz (in Zukunft abgekürzt: SBPK). Zu ihnen gehören die meisten Handschriften der Signaturgruppen Ms.germ., Ms.lat., Ms.theol.lat. Die übrigen erhaltenen Handschriften, darunter die mit den Signaturen Ms.Diez B und C, Ms.Hamilton und Ms.Phillipps, befinden sich wieder an ihrem alten Aufbewahrungsort, der Deutschen Staatsbibliothek, Berlin (DDR) (in Zukunft abgekürzt: DSB).

Die vorliegende Arbeit befaßt sich nur mit abendländischen Handschriften der Staatsbibliothek, wobei auch die griechischen Codices nicht einbezogen sind. Die westlichen Handschriften stehen heute im wesentlichen unter folgenden Signaturgruppen: Ms.lat., Ms.theol.lat., Ms.germ., Ms.gall., Ms.ital., Ms.hisp., Ms.slav., Ms.boruss., Ms.catal.A + B, Ms.geneal., Alba amicorum, Ms.Diez B + C, Ms.Hamilton, Ms.Phill., Savigny, Fragm. Dazu kommen die Ms.Magdeb., die nach dem 2.Weltkrieg in die DSB gelangten. Von diesen Abteilungen sind für meine Untersuchungen die nach Sprachen geordneten Handschriftengruppen die wichtigsten. Im folgenden möchte ich angeben, für welche Handschriften es gedruckte Kataloge gibt.

Die wissenschaftliche Katalogisierung der abendländischen Berliner Handschriften beginnt erst mit der Gründung der Handschriftenabteilung (1886). In zwanzigjähriger Arbeit beschrieb V. Rose die lateinischen Phillipps-Handschriften aus dem Besitz von G. Meerman[2] und die sogenannten “Codices electorales”, sofern sie lateinische Texte enthalten: dies sind die in kurfürstlicher Zeit erworbenen Handschriften einschließlich derjenigen Codices aus kurfürstlichen Landen, die später in die Berliner Bibliothek gelangten.[3] Roses Beschreibungen zeichnen sich durch große Akribie und wissenschaftliche Zuverlässigkeit aus, sind jedoch recht unübersichtlich und bieten deshalb für den Benutzer Probleme; außerdem verwirrt die eigenwillige Numerierung Roses. Als dritten Band der lateinischen Handschriften publizierte F. Schillmann 1919 das Verzeichnis der Görres-Handschriften. Die Kataloge von Rose und Schillmann beschreiben die lateinischen Phillipps-Handschriften aus Meermans Besitz vollständig, von der Signaturgruppe Ms.theol.lat. etwa die Hälfte und von den Ms.lat. etwa ein Viertel; für die restlichen theologischen und lateinischen Handschriften gibt es nur ein handgeschriebenes Inventar.

1925–1933 gab H. Degering ein kurzes Verzeichnis der germanischen Handschriften heraus,[4] das im Gegensatz zu Roses Katalog den Inhalt der Codices nur ganz kurz angibt. Schon 1918 war ein entsprechendes Verzeichnis der romanischen Handschriften erschienen.[5] Sie geben einen ersten Überblick über den Inhalt der Handschriften Ms.germ., gall., ital., hisp. und lus., die zum Zeitpunkt der Publikation vorhanden waren. 1926 begann man eine neue Reihe über die Berliner Miniaturen-Handschriften. Von ihr erschienen nur zwei Bände zu den Phillipps-Handschriften und den Ms.germ., die naturgemäß keine wesentlichen Informationen zum Inhalt der Texte in den Handschriften geben.[6] Mit den Beschreibungen der lateinischen Hamilton-Handschriften durch H. Boese (1966)[7] wurde ein vorläufiger Schlußpunkt erreicht. Gegenwärtig arbeitet man an Katalogen der Ms.Diez (DSB) und der in der SBPK befindlichen Ms.theol.lat.

Dieser Überblick zeigt, daß die gedruckten Handschriftenkataloge uns noch keine ausreichende Kenntnis der westlichen Handschriften vermitteln. Dies gilt insbesondere für die besonders wichtige Gruppe der Manuscripti latini, die fast 2500 Nummern umfaßt, von denen drei Viertel nur im handgeschriebenen Bandkatalog der Staatsbibliothek beschrieben sind.

Die vorliegende Arbeit möchte diesem Mangel auf einem kleinen Gebiet abhelfen: den mittelalterlichen mathematischen Handschriften. Ich habe versucht, mit Hilfe der gedruckten Kataloge und der in Berlin existierenden handschriftlichen Inventare möglichst alle Codices zu ermitteln, die mathematische Texte enthalten, und diese dann bei Besuchen in der DSB und SBPK oder mit Hilfe von Mikrofilmen zu identifizieren. Bei dieser Gelegenheit möchte ich nicht versäumen, den Mitarbeitern in den Handschriftenabteilungen der DSB und der SBPK für bereitwillige Auskunft und für die Zusendung von Mikrofilmen zu danken. Es ist bei dem geschilderten Katalogisierungszustand zu vermuten, daß ich dabei Texte übersehen habe, so daß es sich um ein vorläufiges Verzeichnis handelt, doch dürfte der überwiegende Teil der Texte, speziell die längeren, erfaßt sein.

Die Auswahl der Texte bedarf einer Rechtfertigung: Nach dem mittelalterlichen Verständnis des Begriffes Mathematik hätten auch astronomische, speziell komputistische, und physikalische Traktate aufgenommen werden müssen. Aus pragmatischen Gründen habe ich mich entschlossen, dies nicht zu tun, sondern den Begriff Mathematik möglichst eng zu fassen. So fehlen Traktate zur Logik, Bewegungslehre, Zeitrechnung; von astronomisch-mathematischen Instrumenten wurde nur der Quadrant aufgenommen, da er auch für die Vermessung auf der Erde wesentlich benutzt wurde. Traktate "de ponderibus et mensuris" wurden nicht berücksichtigt, wenn sie zu medizinischen oder pharmazeutischen Texten gehörten. Ebenfalls fehlen mathematische Texte, wenn sie Teile größerer, nicht speziell mathematischer

Werke sind; dagegen sind sie vorhanden, wenn sie als eigener Traktat überliefert werden. So habe ich Bedas Abhandlung über die Fingerzahlen, die Kapitel 1 von De temporum ratione bildet, aufgenommen, wenn sie in einer Handschrift einen selbständigen Traktat bildet, nicht aber De temporum ratione als Ganzes.

Als zeitliche Begrenzung wurde etwa das Jahr 1500 gewählt. Bis zu diesem Zeitpunkt geschriebene Texte habe ich berücksichtigt, wenn sie in Latein oder in einer westlichen Nationalsprache verfaßt sind; außer den orientalischen Sprachen wurden also auch griechische Handschriften nicht einbezogen. Es zeigte sich, daß es in Berlin außer lateinischen Texten nur noch solche in deutscher oder italienischer Sprache gibt, nicht aber französische oder englische, die vor 1500 entstanden.

Die Handschriften wurden nach ihren Signaturen alphabetisch geordnet und, um Rückverweise zu erleichtern, durchnumeriert. Die Zahl nach der laufenden Nummer bezeichnet die Nummer des Traktats in der betreffenden Handschrift. Es war nicht möglich und auch nicht beabsichtigt, kodikologisch exakte Beschreibungen zu geben. Ich habe mich vielmehr darauf beschränkt, nach der Signatur die Entstehungszeit, Blattzahl und den gegenwärtigen Aufbewahrungsort (DSB bzw. SBPK) anzugeben. Es folgen Informationen über Vorbesitzer oder Entstehungsort der Handschrift, sofern mir aus der Literatur oder aus Bemerkungen in der Handschrift selbst etwas darüber bekannt war. Unter der Überschrift "Inhalt" habe ich ganz pauschal die Art der Texte im Codex bezeichnet und dann speziell die mathematischen Traktate angegeben, jeweils mit Blattzahl, Incipit und Explicit. Dabei habe ich Überschriften auch dann aufgenommen, wenn sie von späterer Hand nachgetragen wurden. Bei der Wiedergabe von Incipits und Explicits wurden fehlende Initialen ergänzt, Abkürzungen, wenn möglich, aufgelöst, u und v an die übliche Schreibweise angepaßt, e caudata mit ae wiedergegeben und Überschriften auch dann in Minuskeln gedruckt, wenn sie im Codex in Majuskeln geschrieben waren. Die Identifikation der Texte erfolgte im allgemeinen mit Hilfe des Incipits. Wo mir neuere Editionen des Textes bekannt waren, habe ich diese vermerkt und nach Möglichkeit überprüft, ob die Handschrift mit dem Druck übereinstimmt. Falls derselbe Text an einer späteren Stelle erneut auftaucht, habe ich auf die erste Erwähnung verwiesen. Alle Textanfänge, die im Katalog von Thorndike-Kibre verzeichnet sind, habe ich angegeben. Wegen der großen Anzahl der Texte war es für mich nicht möglich, in jedem Fall zu einer sicheren Identifikation zu kommen, so daß eine nähere Beschäftigung mit den Texten an vielen Stellen zu neuen Ergebnissen führen dürfte; dies gilt speziell für manche Quadrant-Traktate, für italienische Texte und für Aufgabensammlungen und -bücher. Auch in dieser Hinsicht ist das Verzeichnis also vorläufig. Zum

Abschluß jeder Nummer habe ich Literaturstellen erwähnt, an denen der Inhalt der betreffenden Handschrift beschrieben (sie also nicht nur genannt) wird; für einen Großteil der Handschriften habe ich keine Literaturangaben gefunden.

Legt man die oben genannte Definition zugrunde, so gibt es—jedenfalls nach meinen Untersuchungen—in der Staatsbibliothek 72 Handschriften mit mathematischen Texten; viele von ihnen enthalten mehrere Traktate. Somit ist die Berliner Staatsbibliothek für die Erforschung der mittelalterlichen Mathematik Westeuropas außerordentlich bedeutend. Eine Durchsicht des Verzeichnisses läßt erkennen, daß fast das ganze Spektrum mittelalterlicher Mathematik vertreten ist: Aus der griechischen Antike gibt es Übersetzungen von Euklids Elementen in den Versionen Adelhard II (35.1)[8] und Campanus (23.6, 36.), von Euklids Data (35.10), von Archimedes' Kreismessung (35.5) und von Theodosios' Sphärik (21., 35.6). Boethius' Arithmetik ist dreimal vertreten (18., 39., 45.1); auch Gerberts Scholion dazu fehlt nicht (45.2). Aus dem frühen Mittelalter finden wir Isidor-Exzerpte über Maße und Gewichte (55.10, 68.), den Calculus des Victorius mit Abbos Kommentar (66.1, 66.2), Bedas Traktat über die Fingerzahlen gleich sechsmal, z.T. mit sehr sauberen zeichnerischen Darstellungen der Fingerhaltungen (11.6, 14., 62., 69.2, 69.3, 70.), ferner das Beda zugeschriebene Schriftchen De arithmeticis propositionibus (65.) und Fortolfs Rithmimachie (12.13). Gerberts Geometrie (13.2) und ihre anonyme Fortsetzung (3.4, 13.1, 55.6) sind ebenso vorhanden wie seine Schrift zum Rechnen auf dem Abakus (64.) und die Traktate anderer Abazisten (Bernelinus, Gerlandus, Herigerus, anonymer Autor: alles in 55.). Die Schriften, die nach der Übersetzungstätigkeit aus dem Arabischen entstanden, nehmen erwartungsgemäß einen noch größeren Raum ein. Die Algebra Alchwarizmis in der Übersetzung Gerhards von Cremona ist zweimal vertreten (9.3, 40.2); vorhanden ist auch der Traktat De curvis superficiebus (35.6). Abhandlungen zum Algorismus belegen zahlenmäßig den ersten Rang. Spitzenreiter ist—nicht überraschend—die Abhandlung des Johannes de Sacrobosco, die in 21 Exemplaren vorhanden ist, darunter einer deutschen (4.) und einer niederdeutschen Übersetzung (44.1), die meines Wissens bisher unbekannt waren. Viele dieser Handschriften enthalten zum Teil umfangreiche Glossen und Kommentare, die noch ausgewertet werden müssen. Neben der Algorismus-Abhandlung von Alexander de Villa Dei (3.1, 32.1, 38.1, 42.2) gibt es noch andere Anweisungen über das Rechnen mit ganzen Zahlen (z.B. 26.1, 26.2, 46.8, 48.1, 51.4) oder mit Brüchen (38.3, 46.9, 46.11, 46.18; in deutscher Sprache: 2.2); der bekannte Algorismus de minuciis des Johannes de Lineriis fehlt natürlich auch nicht (12.1, 17., 46.13). Von Jordanus Nemorarius sind die Arithmetik (52.1), die Demonstratio de algorismo (35.3) und de minuciis (35.4) ebenso vorhanden wie seine Abhandlung über die Proportionen (35.9) und diejenige des Campanus (40.6). Von Bradwardine finden wir die Geometria speculativa (41.1) und seinen Tractatus de proportionibus, der zusammen mit zahlreichen Abhandlungen über Formlatituden und Bewegungstraktaten in der noch genauer zu analysierenden Handschrift 23. steht. Die Arithmetica speculativa des Johannes de Muris ist in 27. erhalten, geometrische und trigonometrische Schriften des Johannes de Lineriis in 12.9, 12.10, 12.12. Bei den Handschriften des 15. Jahrhunderts sei auf zahlreiche italienische Abhandlungen zur Algebra (9.1, 40.8), Geometrie (9.1, 40.11, 40.12, 43.1, 43.2) und kaufmännischen Mathematik verwiesen (40.9, 46.12, 46.20, 51.3). Rechenbücher (2.1, 6.) und praktische Aufgabensammlungen in deutscher Sprache (2.3) sind ebenso vorhanden wie eine lateinische Algebra (40.7), Abhandlungen über das Rechnen auf den Linien (44.6, 46.19) und Aufgabensammlungen der Unterhaltungsmathematik, die oft ältere Probleme wiedergeben (32.2, 40.4, 42.1, 42.4, 46.2, 46.15, 48.2, 57., 59.2).

Die genannten Schriften vermitteln eine Vorstellung von der Vielfalt mathematischer Texte, die in den Handschriften der Staatsbibliothek sich vorfinden. Sieht man die nachfolgenden Beschreibungen durch, so wird das Bild noch bunter. Es wäre wünschenswert, wenn dieses vorläufige Verzeichnis dazu anregen würde, in Zukunft die reichen Bestände dieser Bibliothek ebenso für mathematikhistorische Arbeiten heranzuziehen wie die viel bekannteren Handschriften in London, Oxford, Cambridge, Paris, Wien, München, New York oder in den großen italienischen Bibliotheken.

ANMERKUNGEN

1. *Deutsche Staatsbibliothek 1661–1961.* I: *Geschichte und Gegenwart* (Leipzig: 1961), vor allem S.1–47, H. Kunze/W. Dube: "Zur Vorgeschichte der Deutschen Staatsbibliothek," und S.319–380, H. Lülfing: "Die Handschriftenabteilung." E. Paunel: *Die Staatsbibliothek zu Berlin. Ihre Geschichte und Organisation während der ersten zwei Jahrhunderte seit ihrer Eröffnung. 1661–1871* (Berlin: 1965). W. Schmidt: "Von der Kurfürstlichen Bibliothek zur Preußischen Staatsbibliothek. Geschichtlicher Überblick von 1661 bis 1945," in *Staatsbibliothek Preußischer Kulturbesitz, Festgabe zur Eröffnung des Neubaus in Berlin.* Hg.v. E. Vesper (Wiesbaden: 1978), S.1–94.
2. Rose 1; siehe Literaturverzeichnis.
3. Rose 2; siehe Literaturverzeichnis.
4. Degering 1–3; siehe Literaturverzeichnis.
5. S. Lemm: "Kurzes Verzeichnis der romanischen Handschriften," = *Mitteilungen aus der Königlichen Bibliothek*, Band *4*, (Berlin: 1918).
6. Beschreibende Verzeichnisse 1 und 5; siehe Literaturverzeichnis.
7. Boese; siehe Literaturverzeichnis.
8. Die Nummern beziehen sich auf mein Verzeichnis.

BESCHREIBUNGEN

1. **Frgm.47**, s.9, 5 Blätter (DSB)
   Provenienz/Vorbesitzer: Einband von Ms.lat.qu.931
   Inhalt:
   f.III–V sind Teile eines Blattes. Es enthält auf der Rückseite einen fragmentarischen Text über Gewichte.
   Lit.:
   K. Christ, in: *Deutsches Archiv für Geschichte des Mittelalters 1* (1937), S.293.

2. **Ms.germ.fol.59**, s.15 (1488), 238 Blätter (SBPK)
   Provenienz/Vorbesitzer: Daniel Sudermann
   Inhalt: Trojanerkrieg; Gesta Romanorum; Rezepte. Mathematisch sind:
   2.1 f.168r–187r: Deutscher Algorismus. Behandelt u.a. Grundrechenarten, Proben, Bruchrechnung, kaufmännische Mathematik. Beginnt mit der Aufzählung mathematischer Begriffe; endet unvollständig in der Progressio.
   Inc.: Numeracio—Erkantnis der linien und spacia . . .
   Prima per se, 2ª decies, 3ª centum, 4ªque mille.
   Die erst figur heißt numeracion . . .
   Expl.: . . . per modum multiplicacionis gar sch(on).
   2.2 f.190v–197r: Deutscher Bruchalgorismus.
   Inc.: Hie noch volget Ein hubscher Tractat von den siben species deß algrißmus ingebrochen . . .
   Expl.: . . . ist den vnglichen brůchen im crůcz zů globen.
   Explicit Tractatus Anno 1488.
   2.3 f.197r–200v: Aufgaben zur kaufmännischen Mathematik.
   Inc.: Einen divisor in $\frac{1}{2}$ in $\frac{1}{3}$ in $\frac{1}{4}$. Item Es sint dryg gesellen die hant zů teilen 12 guldin . . .
   Expl.: . . . und also hastu dise regel in dryerley wiß.
   Deo gracias.
   Lit.:
   Degering 1, S.6f.
   H. Hornung: *Daniel Sudermann als Handschriftensammler* (Dissertation: Tübingen, 1956), S.267–269.

3. **Ms.germ.fol.642**, s. 15, 119 Blätter (SBPK)
   Provenienz/Vorbesitzer: Franken; Jorg Ledarer (?); Dr. Schneider, Erfurt; Karl Hartwig Gregor von Meusebach
   Inhalt: Astrologisch-medizinische Sammelhandschrift. Mathematisch sind:
   3.1 f.98v–100v: Alexander de Villa Dei, Carmen de algorismo (ed. J. O. Halliwell: *Rara Mathematica* (London: 1841), S.73–83; R. Steele: *The earliest arithmetics in English* (London: 1922), S.72–80.—TK 597).
   Inc.: Hec algorismus ars presens dicitur in qua . . .
   Expl.: . . . Ad dextrum digitum servando prius documentum. Explicit Algorismus.
   3.2 f.101r: Drei zahlentheoretische Exempla.
   Inc.: Et nota quando aliquis numerus quadratus ducitur in se cubice . . .
   Expl.: . . . et minimi in maximum idem resultat exemplum.
   3.3 f.101rv: Über den Zahlenwert der Buchstaben (TK 1062).
   Inc.: Possidet A numero quingentos ordine recto . . .
   Expl.: . . . Ultima zeta canit finem bis mille tenebit.
   Es folgen zwei Zeilen mit Gewichtsnamen.

3.4 f.102r–108v: Auszüge aus der Geometria incerti auctoris (ed. Bubnov, S.317–365). Vorhanden sind wie in Vat.Reg.lat. 1661 die Kapitel IV prol., III 18.2.8.23.1.16.17, Zusatz 3, III 14.10.4.11.19, Zusatz 5a, III 12.13.5.15.3.26.

Inc.: Geometria. Geometrie talis (!) tractati diversitas premonstrandum est . . . (=S.336,3 Bubnov)

Expl.: . . . Rebus in obscuris oritur lux clara figuris. (=S.335,29 Bubnov)

3.5 f.109r–110v: Vermutlich neuer geometrisch-astronomischer Traktat, der sich wie 3.4 teilweise an die Geometria incerti auctoris anlehnt.

Inc.: Si volueris cum astrolabio lucente sole horas diei invenire . . .

Expl.: . . . pro mensura rei illius habeto. Et sic est finis. (=S.323,21 Bubnov)

Lit.:

Degering 1, S.69f.

Beschreibende Verzeichnisse 5, S.41–43

4. **Ms.germ.fol.1278**, s.14/15, 111 Blätter (SBPK)

Provenienz/Vorbesitzer: Österreich; Th. Phillipps (Nr. 11855, 11857, 11854, 11856)

Inhalt: Rechte und Freiheiten der Stadt Wien.

Teil 4, f.1r–8r: Sacrobosco, Algorismus (deutsch).

Inc.: Algarismus. Allew dinkck die von ersten wegunstnuzz der ding sind wekomen . . .

Expl.: . . . so heb an an der ersten figur zu wurchen etc. Hie hat der dawczscht Algarismus ein end got uns sein genad send etc. Amen.

Lit.:

Degering 1, S.171

Mitteilungen aus der Königlichen Bibliothek, Band 3: "Neue Erwerbungen der Handschriftenabteilung. II: Die Schenkung Sir Max Waechters 1912," (Berlin: 1917), S.63–67.

5. **Ms.germ.qu.20**, s.15, 184 Blätter (SBPK)

Inhalt: Astronomisch-astrologisch-medizinisches Handbuch.

5.1 f.40r–41r: Quadrant-Traktat (deutsch).

Inc.: Wieman Sunnur machen sol. Ain quadrantt ist geristen am leczisten plat dis büchlins . . .

Expl.: . . . Die stund sind leicht ze erkennen durch das obgemelten zeygers schaten.

5.2 f.41v–43r: Quadrant-Traktat (deutsch).

Inc.: Von der andern beraitten Sunnur. Es ist ain andre sunnur im ußgang diß püchlins . . .

Expl.: . . . und taüglichait nach meinen slechten teütsche und dainen vermügen.

f.43v–44r: Figuren zum Quadrant.

Lit.:

Degering 2, S.4

6. **Ms.germ.oct.375**, s. 15, 129 Blätter (SBPK)

Provenienz/Vorbesitzer: Prof. Crecelius, Elberfeld

Inhalt:

f.1r–127r: Rechenbuch (deutsch). Behandelt Rechnen mit ganzen Zahlen, mit Brüchen und kaufmännische Mathematik.

Inc.: Salomon spricht yn dem puch der weißheyt Sapientie am aylften. Got hat alle ding beschaffen zu tzal . . .

Expl.: . . . und kumen gerad 390 gulden und ist gemacht und richtig.

Auf f.127v–128v sind drei Rechenaufgaben in italienischer Sprache angehängt.

Lit.:

Degering 3, S.126

7. **Ms.germ.oct.381**, s.15, 125 Blätter (SBPK)
Provenienz/Vorbesitzer: Catharina Woreen (16.Jh.); Catharina Tide; Geh. Justizrat Barnheim, Insterburg
Inhalt: Stundenbuch (niederdeutsch).
f.124v–125r: Zahlentabelle (arabische Zahlen von 0 bis 90000 und einige weitere Zahlen; Füllsel)
Lit.:
Degering 3, S.132

8. **Ms.Hamilton 213**, s.15, 68 Blätter (DSB)
Provenienz/Vorbesitzer: Italien; Hamilton
Inhalt: Priscianus, Periegesis; Vita Dionysii.
f.41r–58v: Sacrobosco, Algorismus (ed. M. Curtze: *Petri Philomeni de Dacia in Algorismum vulgarem Johannis de Sacrobosco commentarius* (Kopenhagen: 1897), S.1–19.—TK 991).
Inc.: Omnia quae a primeva rerum origine processerunt . . .
Expl.: . . . hec de Radicum extractione sufficiant tam in numeris cubicis quam in quadratis. Explicit. Algorismus. Amen.
Lit.:
Boese, S.105f.

9. **Ms.Hamilton 692**, um 1500, 298 Blätter (DSB)
Provenienz/Vorbesitzer: Oberitalien (Pavia?); Hamilton
Inhalt:
9.1 f.1r–275v: Lehrbuch der Arithmetik, Algebra und Geometrie in mehreren Abschnitten (italienisch), darunter Übersetzung von Jordanus Nemorarius, De elementis arismetice artis.
Inc.: Nota che questo libro e stato facto per Zordano philosopho . . .
Expl.: . . . et diray che sono mogia 616 et he facta.
Auf f.276rv folgen Umrechnungen von Maßen.
9.2 Innerhalb des italienischen Textes gibt es zwei Isidor-Exzerpte:
f.192rv: Isidor, Etymol. XV 15, 1–7 und XV 16, 1–8 Lindsay (TK 870).
Inc.: De mensuris agrorum capitulum XV Isidori libro XVI. Mensura est quidquid pondere capacitate longitudine altitudine latitudine animoque finitur . . .
Expl.: . . . est enim locus transitu facilis: unde et appelamus aditum.
9.3 f.279r–291v: Alchwarizmi, Algebra, übersetzt von Gerhard von Cremona, mit Anhängen (ed. G. Libri: *Histoire des sciences mathématiques en Italie* . . . , Band 1, (Paris: 1838), S.253–297.—TK 624).
Inc.: In nomine dei eterni Incipit liber mauchumeti in Algebra et almuchabula qui est origo et fundamentum totius scientie Arismetice. Hic post laudem dei et ipsius exaltationem inquit . . .
Expl.: . . . et proveniunt 25, cuius radix est 5.
Lit.:
Boese, S.334f.
Thomson, S.114

10. **Ms.lat.fol.49**, s.15, 236 Blätter (SBPK)
Provenienz/Vorbesitzer: Leipzig—Heidelberg
Inhalt: Humanistisch-poetische Sammelhandschrift.
f.44r–45v: Über die Zahlwörter.

Inc.: Tracturi (!) de nominibus numeralibus scire debemus quod species numeralium nominum . . .
Expl.: . . . Et hec de numeralibus nominibus dicentibus (?) sufficiant. Finis.
Lit.:
Rose 2, S. 1267–1272 (Nr.991)
A. Sottili: "I codici del Petrarca nella Germania Occidentale. VII: Appendice," in: *Italia medioevale e umanistica, 18* (1975), S.1–72 (hier: S.17 = Nr.225)

11. **Ms.lat.fol.192**, s.15, 225 Blätter (SBPK)
Provenienz/Vorbesitzer: Trier, St. Maximin
Inhalt: Astronomische Sammelhandschrift.
11.1 f.1r–3r: Quadrant-Traktat (TK 1470).
Inc.: Si vis facere quadrantem fac semicirculum super dyametrum . . .
Expl.: . . . reduc in gradus equales (. . .: 2 Worte unklar)
11.2 f.9v–12r: Sacrobosco, Algorismus (siehe 8.)
Inc.: Omnia que a prima rerum origine processerunt . . .
Expl.: . . . tam in numeris quadratis quam cubicis.
11.3 f.12r–14v: Robertus Anglicus, Quadrans vetus (ed. Paul Tannery: "Le traité du quadrant de Maître Robert Anglès," in: *Notices et extraits . . . 35* (1897), S.561–640; Neudruck in: *Mémoires scientifiques 5* (Paris/Toulouse: 1922), S.118–197.—TK 585).
Inc.: Geometrie duo sunt partes theorica et practica . . .
Expl.: . . . et productum dabit capacitatem.
11.4 f.192rv: Quadrant-Traktat.
Inc.: Quadrans est instrumentum continens quartam partem circuli . . .
Expl.: . . . ad que tu vis valere quadrantem.
11.5 f.192v–197r: Robertus Anglicus, Quadrans vetus, z.T. bearbeitet (siehe 11.3).
Inc.: Geometria due habet species theorica et practica . . .
Expl.: . . . et productum dabit eius (?) capacitatem. Explicit textus de mensuracionibus etc.
Es folgt auf f.197v eine Tabula antiqui quadrantis.
11.6 f.198rv: Beda, De computo digitorum (=De temporum ratione, Kap.1; ed. C. W. Jones: *Bedae opera de temporibus* (Cambridge/Mass.: 1943), S.179,2–180,62.—TK 393).
Inc.: De temporum ratione domino iuvante diuturni (!) necessarium duximus . . .
Expl.: . . . insertis invicem digitis implicabis. Explicit quedam Ars numerandi ysidori secundum In libro ethimologiarum secundum signa manuum. etc.
Lit.:
Rose 2, S.1199–1209 (Nr.963)

12. **Ms.lat.fol.246**, s.15 (1443–1458), 263 Blätter (SBPK)
Provenienz/Vorbesitzer: Größtenteils geschrieben von Ludolphus Borchtorp de Brunswik in Erfurt, Padua und Braunschweig
Inhalt: Astronomisch-mathematische Sammelhandschrift.
12.1 f.53r–56r: Johannes de Lineriis, Algorismus de minuciis (ed. H. Busard: "Het rekenen met breuken in de middeleeuwen, in het bijzonder bij Johannes de Lineriis," in *Mededelingen van de Kon. Vl. Acad. voor Wetenschappen, Letteren en Schone Kunsten van België, Kl. Wetensch., 30* (Brüssel: 1968, Nr. 7), S.21–36.—TK 878).

Inc.: Algorismus de minuciis. Modum representacionis minuciarum volgarium et phisicarum preponere . . .
Expl.: . . . ad propositum nostrum sufficiant minuciarum volgarium et phisicarum. Explicit algorismus de minuciis pro quo laus et gloria sit christo.

12.2 f.56v–59r: Robertus Anglicus, Quadrans vetus (siehe 11.3).
Inc.: Geometrie due sunt partes theorica et practica . . .
Expl.: . . . et productum dabit capacitatem. et sic est finis huius.

12.3 f.70v–71v: Profatius Judaeus, Compositio novi quadrantis, mit Randglossen (TK 241).
Inc. (Text): Compositurus novum quadrantem accipe tabulam planam . . .
Inc. (Komm.): Si vis scire umbram rectam per solis altitudinem . . .
Expl.: . . . ad cognoscendum predicta de dorso. et sic finitur composicio quadrantis novi.

12.4 f.71v–73r: Johannes Eligerus de Gondersleuen, De utilitate quadrantis (TK 1627).
Inc.: Utilitates novi quadrantis breviter et lucide colligere visum est . . .
Expl.: . . . et sic esset finis si ma[a] esset correcta. corrigatur in posterum quia multum deficit et (. . .) scripta est propter maliciam exemplaris in fine (?) deficientis hec canones de mensuracionibus trium dimensionum per quadrantem hos igitur require supra in canonibus quadrantis antiqui.

12.5 f.79v: Kurzer Visiertraktat.
Inc.: Pro arte visandi mensura fundum in longitudinem et latitudinem . . .
Expl.: . . . tabulam tunc mensuretur cuius medietas primo postea a[i]ame[tas].
Es folgt eine Tabula proportionum.

12.6 f.79v–81r: Johannes de Muris, Canones de tabula proportionum (TK 1461).
Inc.: Canones tabule tabularum Io de muris. Siquis per hanc tabulam tabularum tabulam proporcionum alio nomine nuncupatam . . .
Expl.: . . . propter amorem sciencie sollempniter exaltare. Explicit canon tabule tabularum edite a magistro iohanne de muris anno domini 1321.

12.7 f.81r: Zwei Nachträge über den Wert der geometrischen Reihe 1, 2, 4, 8, . . . (datiert 1447)

12.8 f.82rv: Visiertraktat.
Inc.: Tractatus de virga visoria. In confectione virge visorie accipe circinum ad magnitudinem . . .
Expl.: . . . et tantum sit dictum de virga visoria. etc. est finis.

12.9 f.135r: Johannes de Lineriis, Sinustafel
Inc.: Incipiunt tabule magistri iohannis de lineriis. Tabula sinus.
Es folgen astronomische Tafeln

12.10 f.155r: Johannes de Lineriis, De mensurationibus (TK 1137).
Inc.: Excerptum Tractatus de mensuracionibus iohannis de lineriis. Profunditatem putei mensurare fac 2 signa in oppositis partibus diametri putei . . .
Expl.: . . . et aggrega mensuras omnium adinvicem et habebis propositum.

12.11 f.156r–157r: Quadrans novus (TK 1267).
Inc.: Incipit composicio quadrantis novi multum bona. Quoniam conceditur opus quadrantis prevalere . . .
Expl.: . . . hiis igitur omnibus impletis completum est instrumentum quadrantis et ad suas utilitates aptum deo gracias.

12.12 f.175r: Auszüge aus Johannes de Lineriis, Canones tabularum primi mobilis (ed. M. Curtze: "Urkunden zur Geschichte der Trigonometrie im christlichen

Mittelalter," in: *Bibliotheca Mathematica* ^{3}*1*(1900), S.321–416; hier S.391–403.—Vgl. TK 1509).

Inc.: Notandum pro sinibus quod dux est sinus scilicet rectus et versus. Sinus rectus est medietas corde porcionis arcus duplicate . . .

Expl.: . . . et quod provenerit minue a 90 et proveniet arcus etc. (?)

12.13 f.206r–215v: Fortolfus, Rithmimachia (ed. R. Peiper: "Fortolfi Rythmimachia," in: *Abhandlungen zur Geschichte der Mathematik 3* (1880), S.169–197.—TK 1296).

Inc.: Incipit tractatus Rithmimathie id est de pugna numerorum ars pulchra. Quoniam igitur huius artis sciencia ab ignorantibus contempnitur . . .

Expl.: . . . Et in hoc terminatur hic tractatus totus qui Rithmimathia intitulatur ad laudem dei.

Lit.:

Zinner, S.14–17

W. Wattenbach, in: *Neues Archiv der Gesellschaft für ältere deutsche Geschichtskunde 9* (1884), S.624–630.

13. **Ms.lat.fol.307**, s.12, 40 Blätter (SBPK)

Provenienz/Vorbesitzer: Frankreich; Havelberg

Inhalt: Mathematisch-astronomische Sammelhandschrift.

13.1 f.12r–17v: Auszüge aus der Geometria incerti auctoris (siehe 3.4): Kapitel III 5 Ende.6–23.25.24, IV prol., III 26, IV 1–26.

Inc.: /sunt ad DE . . . (S.321,18 Bubnov)

Expl.: . . . qui in verticis habeat circuitu. pedes CCC in ascensu. (S.350,11 Bubnov)

13.2 f.17v–19r: Gerbert, Geometrie (ed. Bubnov, S.48–97.—TK 714). Unvollständig; endet in Kap. III 11 (S.60, 9 Bubnov).

Inc.: Hic incipiunt ysagoge geometriae. In quatuor matheseos ordine disciplinarum tercium . . .

Expl.: . . . dicta pertica quasi portica./

Lit.:

Rose 2, S.1177–1185 (Nr.956), mit Nachträgen S.1347f.

Bubnov, S.XX–XXII

Archiv der Gesellschaft für ältere deutsche Geschichtskunde 8 (1843), S.832f.

14. **Ms.lat.fol.436**, s.15, 2 Blätter (SBPK)

Inhalt:

f.1r–2v: Beda, De computo digitorum (siehe 11.6).

Inc.: Beda pater morum numerum docet hic digitorum. Inter ceteras artes perutile est promptissimam flexus digitorum nosse solertiam . . .

Cum ergo dicis unum . . .

Expl.: . . . insertas invicem digitis implicabis. laus iesu christo. Iam requiē donem. numeri legis hic rationem.

Es folgen Abbildungen der Fingerhaltungen von 1 bis 1000000.

15. **Ms.lat.fol.438**, s.15 (1421), 163 Blätter (SBPK)

Provenienz/Vorbesitzer: Johannes Sellator, notarius und capellanus zu Burkaym (1448)

Inhalt: Theologische Sammelhandschrift mit einigen astronomisch-mathematischen Texten.

15.1 f.118r–121v: Sacrobosco, Algorismus, mit Randbemerkungen (siehe 8). Unvollständig; endet mit der Progressio (S.13,33 Curtze).

Inc.: Omnia que a primeva rerum origine mundi processerunt . . .
Expl.: . . . et erit novenarius summa tocius progressionis. Anno domini 1421 finitus est iste liber in die Eufemie virginis hora 5ta post vesperas.

15.2 f.121v: 6 Verse über den Stellenwert der Ziffern.
Inc.: Unum primo secundo decem dat tercia centum . . .
Expl.: . . . Sequensque millesies mille millena milia centum.

16. **Ms.lat.fol.456**, s.14, 272 Blätter (SBPK)
Inhalt: Philosophisch-theologische Sammelhandschrift.
f.174r: Kurzer Text über Proportionen (Füllsel).
Inc.: Differencia (?) inter proporcionem et proporcionalitatem scilicet ut habitudo duorum ad 4 vocatur . . .
Expl.: . . . quod proporcionalitas est in iusticia (?) distributiva. arismetica in iusticia (?) commutativa.

17. **Ms.lat.fol.600**, s.15 (ca.1460), 16 Blätter (SBPK)
Provenienz/Vorbesitzer: William Morris
Inhalt: Musiktraktate von Johannes de Muris.
f.12r–16v: Johannes de Lineriis, Algorismus de minuciis (siehe 12.1).
Inc.: Modum representacionis minutiarum vulgarium et phisicarum proponere . . .
Expl.: . . . ad propositum nostrum sufficiunt. Expliciunt Canones Minutiarum.

18. **Ms.lat.fol.601**, s.11, 140 Blätter (SBPK)
Inhalt: Vitruvius, De architectura (ab f.68).
f.1r–66v: Boethius, De institutione arithmetica (ed. G. Friedlein: *Anicii Manlii Torquati Severini Boetii De institutione arithmetica libri duo, de institutione musica libri quinque. Accedit geometria quae fertur Boetii* (Leipzig: 1867), S.1–173.—TK 669). Bricht im letzten Kapitel ab (II 54; S.171, 10 Friedlein).
Inc.: Domino Patri Simacho Boetius. In dandis accipiendisque muneribus ita recte officia . . .
Expl.: . . . medio termino duplus. In his ergo geo/

19. **Ms.lat.fol.610**, s.13 (1276/77), 250 Blätter (SBPK)
Provenienz/Vorbesitzer: Paris (?); Johannes de Harlebeke (?); Th. Phillipps (Ms.254)
Inhalt: Astronomische Sammelhandschrift.

19.1 f.15r–21r: Sacrobosco, Algorismus (siehe 8.), mit umfangreichen Randglossen. Über der ersten Glosse die Jahreszahl 1284.
Inc. (Text): Incipit algorismus. Omnia que a primeva rerum origine processerunt . . .
Inc. (Komm.): In hoc prohemio tangit auctor breviter . . .
Expl.: . . . Et hec de radicum extractione sufficiant tam in numeris quadratis quam cubicis. amen.

19.2 f.56v–62r: Robertus Anglicus, Quadrans vetus (siehe 11.3).
Inc.: Geometrie due sunt partes. theorica et practica . . .
Expl.: . . . et productum dabit eius capacitatem.

Lit.:
Zinner, S.12–14
Zimelien. Abendländische Handschriften des Mittelalters aus den Sammlungen der Stiftung Preußischer Kulturbesitz Berlin (Berlin: 1975), S.102f. (Nr.87)
Bibliotheca Phillippica (1899), S.10 (Nr. 78).

20. **Ms.lat.fol.629**, s.14, 28 Blätter (SBPK)
Provenienz/Vorbesitzer: England; Th. Phillipps (Ms. 3074)
Inhalt: Astronomische Sammelhandschrift.
f.17r–25r: Robertus Anglicus, Quadrans vetus (siehe 11.3). Der Text ist vor allem im Abschnitt De utilitatibus quadrantis gegenüber Tannerys Fassung erweitert; u.a. ist ein Abschnitt Quadrantis in astralabio constituti sunt duo latera . . . (TK 1156) hinzugefügt.
Inc.: Geometrie due sunt partes scilicet theorica et practica . . .
Expl.: . . . talis est comparacio stature tue ad planiciem etc.
Lit.:
Bibliotheca Phillippica (1903), S.8 (Nr.57).

21. **Ms.lat.fol.633**, s.15, 55 Blätter (SBPK)
Provenienz/Vorbesitzer: Mailand; Th. Phillipps (Ms.6568)
Inhalt:
f.1r–47v: Theodosius, De spheris (siehe J. L. Heiberg: "Theodosius Tripolites, Sphaerica," in *Abhandlungen der Gesellschaft der Wissenschaften zu Göttingen, philosophisch-historische Klasse, Neue Folge*, Band XIX, 3 (Berlin: 1927), S.VIII-XII.—TK 1523). Es handelt sich um die kürzere Version (22 + 22 + 14 Sätze). Die subscriptio entspricht dem Druck Venedig 1518.
Inc.: Spera est figura corporea una quidem superficie contenta intra quam unum punctorum existit . . .
Expl.: . . . maior proportione anguli COH ad angulum CQH. Expletus est tractatus tertius theodosii de speris.
Es folgt auf f.47v–55r Theodosius, De habitationibus (TK 660).
Lit.:
Bibliotheca Phillippica (1895), S.174 (Nr. 1099).
Phillipps Manuscripts, S.97

22. **Ms.lat.fol.651**, s.15, 109 Blätter (SBPK)
Inhalt: Humanistische Sammelhandschrift (Cicero, Hyginus, grammatische Schriften).
22.1 f.68r–71r: Remus Favinus, De ponderibus (ed. F. Hultsch: *Metrologicorum scriptorum reliquiae*, Bd. *2* (Leipzig: 1866), S.88–98 = Nr.120, und *Anthologia Latina*, Pars 1, fasc. 1 (Leipzig: 1870), S.27–37 = Nr.486.—TK 1059). Unvollständig; endet in Z.163.
Inc.: Remi Sabini Poetae de ponderibus et mensuris libellus. Pondera peoniis veterum memorata libellis . . .
Expl.: . . . Necnon et sine aquis/
22.2 f.109v: Umrechnungstabelle für Gewichte.
Inc.: Siliqua obolus siliquas duas idem quod scripulus . . .
Expl.: . . . fuisse siliquarum XXIIII id est scryp̄ IIII.

23. **Ms.lat.fol.852**, s.14, 225 Blätter (SBPK)
Provenienz/Vorbesitzer: Erfurt, Königl. Bibliothek (MS.f.51)
Inhalt: Sammelhandschrift mit zahlreichen Texten zur Logik und Physik; darunter
23.1 f.11v–13v: De proportionibus.
Inc.: Omnis proporcio vel est communiter dicta vel proprie dicta . . .
Expl.: . . . sed si talis sic esset nōeēt motus igitur etc. Expliciunt proporciones bone et utiles.
23.2 f.71r–75v: Bradwardine, Tractatus de proportionibus (ed. H. Lamar Crosby: *Thomas of Bradwardine: His Tractatus de Proportionibus* (Madison: 1961), S.64–141.—TK 984).

Inc.: Omne motum successivum alteri in velocitate proporcionari quapropter phylosophia naturalis . . .
Expl.: . . . Sic ergo (?) quod volumus luculenter demonstratur. Expliciunt proporciones beroardine.

23.3 f.76r–79r: Heytesbury, De maximo et minimo (TK 206).
Inc.: Circa finem seu terminum tam potentie active . . .
Expl.: . . . quod ipsa generaliter (?) sufficiant in hac parte. Explicit.

23.4 f.79v–81r: Jacobus de Sancto Martino, De latitudinibus formarum (TK 1218).
Inc.: Quia formarum latitudines multiplices variantur . . .
Expl.: . . . Explicit tractatus de latitudine formarum deo gracias.
Auf f.81v zeichnerische Darstellungen von Formlatituden und geometrische Figuren.

23.5 f.82r–84r: Termini naturales (TK 901).
Inc.: Natura est principium motus et quietis in quo est prius et per se . . .
Expl.: . . . Et sic est finis terminorum philosophalium compilati a reverendo magistro wilhelmo esbri anglico spoliati.

23.6 f.85r–92r: Euklid, Elemente I–II 7 nach Campanus (TK 1152).
Inc.: Punctus est cuius pars non est . . .
Expl.: . . . equalis est toti quadrato AD patet propositum.

24. **Ms.lat.qu.23**, s.15, 399 Blätter (Kriegsverlust)
Provenienz/Vorbesitzer: Ruppin
Inhalt: Astronomische Sammelhandschrift.

24.1 f.151v: Nota de progressione naturali in algorismo de integris

24.2 f.233r–241r: Sacrobosco, Algorismus (siehe 8.), mit Anfang einer Randglosse.
Inc.: Omnia que a primeva rerum origine processerunt . . .
Expl.: . . . Et hec sufficiant de radicum extractione tam in numeris quadratis quam in cubicis.

Lit.:
Rose 2, S.1187–1192 (Nr. 959)

25. **Ms.lat.qu.33**, s.14, 46 Blätter (SBPK)
Inhalt: Astronomische Sammelhandschrift.
f.42v–44v: Campanus, Quadrant (TK 1405).
Inc.: Scire debes quod circulus solis habet duas medietates . . .
Expl.: . . . profunditates quaslibet et planicies. hec quoque de practica quadrantis causa introductionis breviter sufficiant.

Lit.:
Rose 2, S.1185f. (Nr. 957)

26. **Ms.lat.qu.46**, s.15 (1449–1455), 302 Blätter (SBPK)
Provenienz/Vorbesitzer: Osnabrück
Inhalt: Astronomische Sammelhandschrift.

26.1 f.90r–97r: Algorismus-Traktat (TK 953).
Inc.: Circa artem algoris. Notandum quod tota ars numerandi novem speciebus perficitur . . .
Expl.: . . . Et hec dicta de octo speciebus algorismi sufficiant.

26.2 f.125r–128v: Über das Wurzelziehen, mit Randglossen.
Inc. (Text): Sequitur de radicum extractione Et primo in numeris quadratis . . .
Inc. (Glosse): Ista est nona species artis algoristice scilicet extractio radicum . . .
Expl. (Text): . . . et triplatum ponemus sub proxima figura 3[a] versus dextram.
Expl. (Glosse): . . . 98 Residuum.

26.3 f.141r–173v: Kommentar zum Algorismus, wohl auf Sacrobosco bezogen.
Inc.: Si quis sciencias mathematicas pretermiserit . . . Sequitur Omnia que. Iste liber cuius subiectum est numerus mathematicus . . .
Expl.: . . . sit benedictus in secula seculorum Amen amen Et sic est finis.

26.4 f.185r–212r: Sacrobosco, Algorismus (siehe 8.), mit Kommentar und Interlinearglossen.
Inc. (Komm.): Ve anime peccatrici que non habet virtutem redeundi in locum suum . . .
Inc. (Text): Omnia que a primeva rerum origine processerunt . . .
Expl.: . . . Et hec de radicum extractione dicta sufficiant et de aliis speciebus practicis. Et sic est finis amen.

Lit.:
Rose 2, S.1192–1196 (Nr. 960)

27. **Ms.lat.qu.175**, s.15 (1474), 216 Blätter (SBPK)
Provenienz/Vorbesitzer: Krakau (Schreiber: Johannes de Schydlow)
Inhalt: Astronomisch-musikalische Sammelhandschrift; Pecham, Perspectiva communis.
f.145r–155v: Johannes de Muris, Arithmetica speculativa (ed. H. L. L. Busard: "Die 'Arithmetica speculativa' des Johannes de Muris," in: *Scientiarum Historia 13* (1971), S.103–132.—TK 959). Der Text ist am Ende unvollständig (er endet auf S.124 oben Busard) und weist Rand- und Interlinearnoten auf.
Inc.: Numerus est duplex scilicet mathematicus qui dicitur numerus numerans . . .
Expl.: . . . ad species simplicis superparcientis ita quod/

28. **Ms.lat.qu.181**, s.15, 64 Blätter (SBPK)
Inhalt: Astronomisch-medizinische Sammelhandschrift.

28.1 f.1r–6v: Sacrobosco, Algorismus (siehe 8.), mit Interlinearglossen.
Inc.: Omnia que a primeva rerum origine processerunt . . .
Expl.: . . . et hec de radicum extraccione tam in numeris quadratis quam cubicis dicta sufficiant (?). Explicit liber artis algoristice etc. Explicit liber artis algoristice per manus etc.

28.2 f.7r: Kurzer Zusatz über Progressionen.
Inc.: De progressione interscisa similiter tales dantur regule . . .
Expl.: . . . Expliciunt regule de progressione etc.

28.3 f.9v: Zahlentafel: Zehnerpotenzen bis 10^9 und einige zusammengesetzte Zahlen. Darunter: . . . tabula limitum habens locum in secunda parte capituli de radicum extraccione.

29. **Ms.lat.qu.183**, s.15 (1490), 237 Blätter (SBPK)
Inhalt: Astronomisch-philosophische Sammelhandschrift; Tractatus de partibus orationis.
f.93v–110v: Sacrobosco, Algorismus (siehe 8.), mit Rand- und Interlinearglossen.
Inc.: Omnia que a primeva mundi rerum origine processerunt . . .
Expl.: . . . tam in numeris quadratis quam cubicis dicta sufficiant. Telos et finis huius (?) per me thomam skzyothla anno domini 1490.

30. **Ms.lat.qu.238**, s.15 (1457), 235 Blätter (SBPK)
Provenienz/Vorbesitzer: Huysburg (bei Halberstadt)
Inhalt: Theologie; Computus.
f.233r: Zahlübungen (24 Zeilen).
Inc.: Arismetrica. I II III IIII V VI . . .
Expl.: . . . $\overline{\text{CCC}}$ $\overline{\text{D}}$ $\overline{\text{DC}}$ $\overline{\text{DCCC}}$ III addendo.

31. **Ms.lat.qu.382**, s.15 (1478–80), 343 Blätter (SBPK)
Provenienz/Vorbesitzer: Schreiber: Johannes Gra(u)mug, Memmingen; Buxheim, Kartause
Inhalt: Humanistische Sammelhandschrift.
f.299r–310v: Sacrobosco, Algorismus (siehe 8.), mit Glossen, vor allem zu Beginn des Textes. Endet S.13,33 Curtze.
Inc.: Omnia que a primeva mundi rerum origine processerunt . . .
Expl.: . . . et erunt novem summa tocius progressionis.
Lit.:
A. Sottili: "I codici del Petrarca nella Germania Occidentale. VII: Appendice," in *Italia medioevale e umanistica 18* (1975), S.1–72 (hier: S.35–41 = Nr.232)

32. **Ms.lat.qu.485**, s.14 (1388), 12 Blätter (SBPK)
Provenienz/Vorbesitzer: Oxford (Schreiber: Willelmus Edlyngton); Th. Phillipps (Ms. 24600)
Inhalt: Computus ecclesiasticus; ferner
32.1 p.1–7: Alexander de Villa Dei, Carmen de algorismo (siehe 3.1). Endet S.81, 1 Halliwell = Z.258 Steele.
Inc.: Hic incipit Algorismus. Hec algorismus ars presens dicitur in qua . . .
Expl.: . . . Tali quesita radix patet arte reperta. Explicit algorismus versificatus.
32.2 p.22–24: Mathematische Aufgabensammlung (siehe M. Folkerts: "Mathematische Aufgabensammlungen aus dem ausgehenden Mittelalter," in *Sudhoffs Archiv 55* (1971), S.58–75).
Inc.: Hic incipiunt cautele algorismi. Patre suo natum si filia concipit virum . . .
Expl.: . . . tunc remanet una quarta.
Lit.:
Bibliotheca Phillippica (1895), S.2 (Nr.7).

33. **Ms.lat.qu.488**, s.15, 152 Blätter (SBPK)
Provenienz/Vorbesitzer: Agen, Jesuiten; Th. Phillipps (Ms.1007 = 2943)
Inhalt: Grammatische Sammelhandschrift.
f.89r–93v: Priscianus, De figuris numerorum (ed. H. Keil: *Grammatici Latini.* Band 3 (Leipzig: 1860), S.403–417).
Inc.: De figuris numerorum liber singularis prisciani grammatici caesariensis foeliciter incipit. Omni te symache nobilitatis splendore celebratum . . .
Expl.: . . . et similia bifurcus trifurcus quadrifurcus et similia. τέλοσ. De figuris numerorum liber singularis prisciani grammatici caesariensis finit.
Lit.:
H. Schenkl: "Bibliotheca patrum latinorum Britannica. IV," in *Sitzungsberichte der philosophisch-historischen Classe der Kaiserlichen Akademie der Wissemschaften, 126* (Wien: 1892), S.31f.
Bibliotheca Phillippica (1895), S.42 (Nr.251).
Phillipps Manuscripts, S.12.33.

34. **Ms.lat.qu.490**, s.15 (1432), 6 Blätter (SBPK)
Inhalt:
f.1v–4v: Johannes von Gmunden, Tractatus quadrantis (TK 1582).
Inc.: Tractatus quadrantis de horis diei equalibus et altitudinibus solis et stellarum . . .
Expl.: . . . erit profunditas putei cuius exemplum patet in illa figura. Explicit textus quadrantis sub anno domini 1432° feria sexta post festum francisci etc.

35. **Ms.lat.qu.510**, s.13, 211 Blätter (SBPK)
Provenienz/Vorbesitzer: England; Edward Langford (1611); G. Libri (Ms.665); Th. Phillipps (Ms. 16345)
Inhalt: Mathematisch-astronomische Sammelhandschrift.

35.1 f.1r–59v: Euklid, Elemente I–XV nach Adelhard II (TK 1152).
Inc.: Punctus est cuius pars non est . . .
Expl.: . . . Explicit liber euclidis philosophi de arte geometrica continens CCCCLXV proposita (?) et propositiones et XI porismata preter anxiomata singulis libris premissa. proposita quidem infinitivis proposiciones indicativis explicans. Deo gracias.

35.2 f.59v: Zusatz über die Linea rationalis.
Inc.: Diffinitio rationalis linee. Linea rationalis est que vel ab aliquo numero sive unitate . . .
Expl.: . . . equale superficiei mediali.
Es folgen auf f.60r–72v Euklid, De speculis und De visu (Version 1).

35.3 f.72v–77r: Jordanus Nemorarius, Demonstratio de algorismo (siehe Thomson, S.108f.; TK 558). Enthält 22 Definitionen und 34 Sätze.
Inc.: Incipit demonstratio magistri Jordani de algorismo. Figure numerorum sunt IX . . .
Expl.: . . . bis et in se semel quod erat proponi.

35.4 f.77r–81v: Jordanus Nemorarius, Demonstratio de minutiis (siehe Thomson, S.110f.; TK 1247).
Inc.: Incipit demonstratio de minutiis. Quidlibet intellectum resspectu partis aut parcium . . .
Expl.: . . . quesitam radicem efficient per XX et XXIIII. Explicit demonstratio de minuciis.
Es folgt auf f.81v–88v Gerhard von Brüssel, De motu.

35.5 f.89r–90r: Archimedes, De mensura circuli, übersetzt von Gerhard von Cremona (ed. M. Clagett, in *Archimedes in the Middle Ages*. Bd.I (Madison: 1964), S.40–54.—TK 996), mit Zusatz (siehe Clagett, S.54, Anm.157).
Inc.: Incipit de quadratura circuli. Omnis circulus ortogonio triangulo est equalis . . .
Expl.: . . . ad alterum laterum ita reliquum ad perpendicularem.

35.6 f.90r–94v: Johannes de Tinemue, De curvis superficiebus (ed. M. Clagett, in *Archimedes in the Middle Ages*. Bd.I. (Madison: 1964), S.450–506.—TK 277).
Inc.: Incipit liber magistri iohannis de tinemue de curvis superficiebus. Cuiuslibet rotunde piramidis curva superficies est equalis . . .
Expl.: . . . archimenides. Remigii. et Iohannes navigationis summo grates agunt creatori. Deo gracias.

35.7 f.94v: Quadratura circuli per lunulas, Version II (ed. M. Clagett, in *Archimedes in the Middle Ages*. Bd.I (Madison: 1964), S.618–624.—TK 1156).
Inc.: Quadratura per lunulas hoc modo est . . .
Expl.: . . . igitur et toti circulo. Hec igitur est quadratura per lunulas.

35.8 f.94v–112v: Theodosius, De spheris (siehe 21.), kürzere Version (22 + 22 + 11 Sätze).
Inc.: De speris liber I. Spera est figura corporea una quidem superficie contenta intra quam unum punctum ipsius existit . . .
Expl.: . . . ergo basis HO est equalis basi HF.
Es folgt auf f.114r–175v Geber, Almagestus minor.

35.9 f.175v–178v: Jordanus Nemorarius, De proportionibus (ed. H. L. L. Busard: "Die Traktate De Proportionibus von Jordanus Nemorarius und Campanus," in *Centaurus 15* (1971), S.193–227; hier S.205–213.—Thomson, S.124f.; TK 1139).
Inc.: Proporcio est rei ad rem determinata secundum quantitatem habitudo . . .
Expl.: . . . ut omnes pariter fiant triginta sex.

35.10 f.178v–190v: Euklid, Data, aus dem Griechischen übersetzt (TK 363).
Inc.: Data magnitudine dicuntur et spatia et linee et anguli . . .
Expl.: . . . et per eum ad circulum/ (bricht ab)
Es folgt auf f.194r–209r Alfraganus, Liber differentiarum.

Lit.:

M. Clagett: *Archimedes in the Middle Ages*. Bd.I (Madison: 1964), S.XXIV (unter der falschen Signatur Q.150)
Bibliotheca Phillippica (1896), S.153–155 (Nr.870).
Phillipps Manuscripts, S.316.

36. **Ms.lat.qu.514**, s.14, 85 Blätter (SBPK)
Provenienz/Vorbesitzer: Guillelmus Tabovotus; Th. Phillipps (Ms.820).
Inhalt: Pecham, Tractatus de perspectiva und Perspectiva communis.
f.1r–56v: Euklid, Elemente I–VI nach Campanus (siehe 23.6).
Inc.: Punctus est cuius pars non est . . .
Expl.: . . . idem intellige in eodem circulo. Expliciunt sex libri geometrie euclidis cum comento campani.

Lit.:

Bibliotheca Phillippica (1897), S.45 (Nr.321)

37. **Ms.lat.qu.515**, s.14, 146 Blätter (SBPK)
Provenienz/Vorbesitzer: Ottery St. Mary (Devon); John Maynard; Th. Phillipps (Ms.3121)
Inhalt: Sammelhandschrift (u.a. Johannes de Muris, De musica; Aesop; Geoffry Wynsafe).

37.1 f.35v–36r: Drei kurze Texte über Proportionen.
Inc.: Nota ad inveniendum proporcionem duplam ad aliam proporcionem . . .
Expl.: . . . operandum est ad proporcionem triplam inveniendum etc.

37.2 f.36v, 35r: Multiplikationsregeln (TK 498).
Inc.: Regule ad multiplicandum sive anterioracione (?) vel delecione figurarum. En brevis interius datur ars nova multiplicandi . . .
Expl.: . . . Quam si didiceris certum pro laude mereris. Es folgen zwei Multiplikationsfiguren.

Lit.:

Bibliotheca Phillippica (1897), S.68F (Nr.538).
Phillipps Manuscripts, S.36

38. **Ms.lat.qu.526**, s.14, 52 Blätter (SBPK)
Provenienz/Vorbesitzer: Baptistus de Poleziis (1480); Th. Phillipps (Ms.896)
Inhalt: Astronomische Sammelhandschrift.

38.1 f.47r–50v: Alexander de Villa Dei, Carmen de algorismo (siehe 3.1), mit einigen Interlinear- und Randglossen.
Inc.: Hec algorismus ars presens dicitur. in qua . . .
Expl.: . . . Quadratus fuerat de dupla quod duplicasti. 66584641600.258040

38.2 f.50v–51v: Johannes de Elsa, Rechentafel.
Inc.: Iohannes de elsa canonicus et magister scolarum Burdeḡ. invenit hoc opus. Inveni quendam modum apertius et facilius computandi . . .
Expl.: . . . que necessaria essent ad computaciones huiusmodi faciendas. Et hec est forma plumbi. (Es folgt eine Figur.)

38.3 f.52rv: Bruchalgorismus, unvollständig.
Inc.: Numeri parcium denominatos possunt fieri infinitis modis secundum infinitos numeros . . .
Expl.: . . . id est ad minuta multiplicavimus/

Lit.:
Bibliotheca Phillippica (1898), S.7 (Nr. 52)

39. **Ms.lat.qu.528**, s.12, 37 Blätter (SBPK)
Provenienz/Vorbesitzer: Baldassarre Boncompagni (Ms.170)
Inhalt:
f.2r–36r: Boethius, De institutione arithmetica (siehe 18.).
Inc.: Incipit prologus in arismeticam Boetij. In dandis accipiendisque muneribus ita recte officia . . .
Expl.: . . . Huius descripcionis subter exemplar subiecimus. Es folgen 4 Figuren.
Lit.:
Narducci, S.98f.

40. **Ms.lat.qu.529**, s.15, 105 Blätter (SBPK)
Provenienz/Vorbesitzer: Carlo Morbio; B. Boncompagni (Ms.179)
Inhalt: Mathematische Sammelhandschrift, z.T. in italienischer Sprache.

40.1 f.1r: Zwei arithmetische Aufgaben (ital.), s.16; Anfang durch Feuchtigkeit nicht lesbar

40.2 f.2r–16v: Alchwarizmi, Algebra, übersetzt von Gerhard von Cremona, mit Anhängen (siehe 9.3).
Inc.: Hic post laudem dei et ipsius exaltationem inquit . . .
Expl.: . . . et proveniunt 25 cuius radix est quinque.

40.3 f.16v–17v: Über Quadratwurzeln.
Inc.: Si vis invenire radicem quadratam 60 vel alterius numeri . . .
Expl.: . . . et residuum esset radix prout dictum est.

40.4 f.17v–19v: Aufgabensammlung (vgl. 32.2).
Inc.: Quidam habuit libras 100 de quibus et eorum proficuo lucratus est . . .
Expl.: . . . vel quadratum quo ad secundum casum significata est.

40.5 f.20r–24r: Trigonometrie (Über Bogen und Sehne).
Inc.: Quia pro multiplicibus demonstrationibus requiritur tabula cordarum et arcuum . . .
Expl.: . . . modicum utilitatis ei propterea supersedeo.

40.6 f.25rv: Campanus, De proportione et proportionalitate (ed. H. L. L. Busard: "Die Traktate De Proportionibus von Jordanus Nemorarius und Campanus," in *Centaurus 15* (1971), S.193–227; hier S.213–222.—TK 1139). Unvollständig; endet S.215 oben Busard.
Inc.: Incipit tractatus aureus alchindi de proportione et proportionalitate. Proportio est duarum quantitatum eiusdem generis ad se invicem habitudo . . .
Expl.: . . . in quarum prima A comparatur ad reliquos/

40.7 f.26r–34v: Algebra, am Anfang unvollständig (?).
Inc.: Numerus habens radicem est qui provenit ex ductu alicuius . . .
Expl.: . . . 135 Res et R_p qui equalis q^o.

40.8 f.34v–58v: Algebra (ital.)
Inc. (Tabellen): 1 Cossa engual a numero . . .
(Text): Fame de 10 do tal parte . . .
Expl.: . . . sera el valor del zenso.

40.9 f.60r-66v: Aufgabensammlung (ital.).
Inc.: Fame questa raxone 4 ove valleno 5 danari . . .
Expl.: . . . anoverati a uno modo.
40.10 f.66v–67r: Bewegungsaufgabe.
Inc.: A roma usque ad mediolanum sunt miliaria 330 . . .
Expl.: . . . miliaria 15 et proficit per 15 diebus.
40.11 f.67r–71r: Geometrie (ital.).
Inc.: Qui te voglio asumare a fare alchune raxon . . .
Expl.: . . . se fano le simile raxon.
40.12 f.71r–78r: Arithmetik; Vermessungstraktat (ital.).
Inc.: Uno sia comprato chastagne . . .
Expl.: . . . de zuchade chomo de sopra.
40.13 f.79v–104r: Teilweise Wiederholung der Texte auf f.60–78 (ital.).
Lit.:
Narducci, S.106f.

41. **Ms.lat.qu.530**, s.14, 89 Blätter (SBPK)
Provenienz/Vorbesitzer: B. Boncompagni (Ms.535)
Inhalt: Pecham, Perspectiva communis.
41.1 f.31r–64v: Bradwardine, Geometria speculativa (ed. A. G. Molland: *The Geometria speculativa of Thomas Bradwardine* (Dissertation: Cambridge, 1967.)—TK 584).
Inc.: Geometria assecutiva est arismetrice. Nam et posterioris ordinis . . .
Expl.: . . . per hanc coniunctionem evidenter. Et in hoc completa est 4ª et ultima pars istius tractatus . . . Explicit tractatus geometrie editus a petro de dacia et scriptus per fratrem marcholinum de cor^uo provincie sancti antonii peruss. 1365. die quo supra de vanis (?) malis ex^is.
41.2 f.65r–79r: Kommentar zu Sacrobosco, Algorismus.
Inc.: Rerum generat cognitionem que quidem cognicio est certissima . . .
Expl.: . . . alia omnia satis sunt plana.
41.3 f.81r–89v: Sacrobosco, Algorismus (siehe 8.). Endet S.17,29 Curtze; es fehlt ein Blatt.
Inc.: Omnia que a primeva rerum origine processerunt . . .
Expl.: . . . extrahere est maximi cubici/
Lit.:
Narducci, S.307

42. **Ms.lat.qu.561**, s.14, 50 Blätter (SBPK)
Inhalt: Astronomische Sammelhandschrift; Cicero, Ad Herennium.
42.1 f.1rv: Aufgabensammlung: Unterhaltungsmathematik (vgl. 32.2).
Inc.: Ponatur quod duo peregrini peregrinantes ad beatum iacobum . . .
Expl.: . . . et habebitur summa denariorum.
Es folgen die arabischen Zahlen von 1 bis 900 und ein kurzer Text über das Stellenwertsystem: Nota quod quelibet figura primo loco posita . . . et sic procedas semper multiplicando.
42.2 f.2r–12v: Alexander de Villa Dei, Carmen de algorismo (siehe 3.1), mit Rand- und Interlinearglossen.
Inc.: Hec algorismus ars presens dicitur in qua . . .
Expl.: . . . Si par per medium sibi multiplicato propinquum. Explicit algorismus id est ars numerandi.
42.3 f.13rv: Verse über die Multiplikation.
Inc.: Articulum si per reliquum vis multiplicare . . .
Expl.: . . . Multiplicandorum de normis sufficit hec.

42.4 f.14rv: Aufgabensammlung: Unterhaltungsmathematik (vgl.32.2).
Inc.: Si quis intret in ecclesiam veniens primo . . .
Expl.: . . . bene feceris nec ne multiplica etc.

42.5 f.15r–26r: Sacrobosco, Algorismus (siehe 8.). Der Abschnitt über die Progressio folgt nach dem Wurzelziehen.
Inc.: Omnia que a primeva rerum origine precesserunt . . .
Expl.: . . . summa tocius progressionis. Explicit.

43. **Ms.lat.qu.576**, s.15, 197 Blätter (SBPK)
Provenienz/Vorbesitzer: Oberitalien; Alessandro Padoani (um 1600); G. Libri (Ms.507); B. Boncompagni (Ms.121)
Inhalt: Astronomisch-geometrische Sammelhandschrift.

43.1 f.84r–165r: Trattato di geometria (ital.). Das erste Blatt fehlt.
Inc.: /Superficia piana e quelo che le suoy liney . . .
Expl.: . . . e da 70 a ℞ 4500 ed e facta.

43.2 f.172v–187r: Praktische Geometrie (ital.).
Inc.: A lauda delo onipotente dio . . .
Expl.: . . . e quanto $\frac{1}{4}$ e $\frac{1}{13}$ $\frac{1}{4}$ $\frac{1}{13}$ fa $\frac{1}{52}$.

Lit.:
Narducci, S.75–77

44. **Ms.lat.qu.577**, s.15, 141 Blätter (SBPK)
Provenienz/Vorbesitzer: Paderborn (Benediktinerabtei Abdinghof); B. Boncompagni (Ms.350)
Inhalt: Astronomisch-mathematische Sammelhandschrift.

44.1 f.2r–9r: Sacrobosco. Algorismus (siehe 8.), in niederdeutscher Übersetzung.
Inc.: Incipit algorismus. Alle dink de von den ersten beghintnisse der dinge sint gekomen . . .
Expl.: . . . so heve an der lesten figuren an to hevende. Explicit algorismus.

44.2 f.9v–23r: Sacrobosco, Algorismus (siehe 8.).
Inc.: Incipit algorismus. Omnia que a primeva rerum origine processerunt . . .
Expl.: . . . qui modus operandi idem est cum priori et predicto. Explicit algorismus.

44.3 f.23v: Zahlentabelle: Wert der Ziffern im Stellenwertsystem (von 10^0 bis 10^8).

44.4 f.24r: Multiplikationstabelle (1×1 bis 20×20).

44.5 f.95v: Über den Zahlenwert der Buchstaben (siehe 3.3).
Inc.: Possidet a numero quingentos ordine recto . . .
Expl.: . . . Z ultima canit finem bis mille tenebit.

44.6 f.95ar–96r: Linienalgorismus; unvollständig.
Inc.: Italice protractis lineis quinque vel septem ut supra secundum beneplacitum . . .
Expl.: . . . sic in divisione descenditur et prolongatur. Sequitur de progressione et scita ista practica sciatur et progressio./

Lit.:
Narducci, S.206f.

45. **Ms.lat.qu.578**, s.9–13, 14 Blätter (DSB)
Provenienz/Vorbesitzer: Frankreich; P. Pithou; Le Peletier; Duc de Berry; Bibl. de Rosny; Ashburnham (Sale 1901, Nr.313).
Inhalt: Besteht aus drei Teilen. Teil 1 enthält auf f.1v–4v Laterculi notarum (s.11).

45.1 f.1r (s.9): Bruchstück aus Boethius, De institutione arithmetica (siehe 18.): II 54 = S.170, 23–172, 7 Friedlein.
Inc.: /non dubium est VI enim nascuntur . . .
Expl.: . . . Huius autem descriptionis exemplar subter adiecimus.

45.2 f.1r (s.11): Gerbert, Scholium ad Boethii arithmeticam institutionem II 1 (ed. Bubnov, S.32–35.—TK 623).
Inc.: Expositio gerberti Hic locus quem quidam invictum estimant sic resolvitur . . .
Expl.: . . . Hec est igitur vera natura numerorum.
45.3 f.1r (s.11): Vermessungstraktat, Bruchstück.
Inc.: Si latitudinem cuiuslibet rei vis scire in plano loco . . .
Expl.: . . . donec cacumen ipsius rei videas et tot (?) erit altitudinis (?)/

Lit.:
P. Lehmann: Mitteilungen aus Handschriften. IV. In: *Sitzungsberichte der Bayerischen Akademie der Wissenschaften, philosophisch-historische Abteilung, 1933, Heft 9,* (München: 1933), S.17–27.

46. **Ms.lat.qu.587**, s.15, 241 Blätter (SBPK)
Provenienz/Vorbesitzer: Geschrieben Erfurt 1446 von Nikolaus Currificis de Friberga; G. Libri; B. Boncompagni (Ms.404)
Inhalt: Astronomisch-mathematische Sammelhandschrift.
46.1 f.2v: Über Längenmaße.
Inc.: Nota cubicus geometricus habet 6 cubicus . . .
Expl.: . . . 16 stadia faciunt leucam.
46.2 f.119r–122r: Aufgabensammlung: Unterhaltungsmathematik (vgl.32.2).
Inc.: Si cupis scire numerum denariorum in bursa . . .
Expl.: . . . Et patet quod fuit 20 annorum etc. Enigmata finita sub anno domini 1440 proxima dominica post festum martini in hora sexta ante diluculum etc. Deo gracias.
46.3 f.122v: Maße und Gewichte (deutsch).
Inc.: 100 lb ist 1 Cf, 32 lot ist 1 lb . . .
Expl.: . . . 400 lb ist 1 karch zcu venedige.
46.4 f.123r–159v: Sacrobosco, Algorismus (siehe 8.), mit Kommentar und Interlinearglossen. Unvollständig; endet mit der Progressio (S.13, 33 Curtze). Fortsetzung: 46.14.
Inc. (Komm.): Macrobius scribit in libro suo ci ciceronis hominis . . .
Inc. (Text): Omnia que a primeva rerum origine processerunt . . .
Expl. (Text): . . . erit novenarius summa tocius progressionis. Et sic est finis huius operis.
Expl. (Komm.): . . . summa tocius progressionis Et hoc de progressione etc. de quo deus gloriosus sit benedictus in seculorum secula Amen.
46.5 f.133v: Zahlentabelle: Wert. der Ziffern im Stellenwertsystem (von 10 bis 10^{10}), wohl zum Sacrobosco-Kommentar gehörig (siehe 44.3).
46.6 f.159v: Multiplikationstabelle (1×1 bis 10×10); darüber: Digitus 1 Unum, Articulus 2 Viginti, . . . , Millenarius cum mille milibus 10 Millesies mille.
46.7 f.160rv: Bemerkungen zur Progressio.
Inc.: Progressio naturalis autem (?) rimatur in numerum $\left\{\begin{matrix}\text{parem}\\ \text{imparem}\end{matrix}\right.$ . . .
Expl.: . . . 1296.7776. suam 9331.
46.8 f.161r–163v: Novus algorismus (TK 1295).
Inc.: Quoniam quibusdam iuvenibus sciencia calculandi videtur difficilis et penalis . . .
Expl.: . . . summa progressionis. Et sic gracia dei est finis huius algorismi prosayci novi de integris per me nicolaum currificis maguncie XI die mensis julii anno domini etc. lii.

46.9 f.164r–169r: Ricardus Anglicus, Tractatus minuciarum vulgarium (TK 875).
Inc.: Minuciarum vulgarium scribes superius numeratorem . . .
Expl.: . . . iterum ex integris dividentibus fac minutiam ut nunc dictum est.

46.10 f.169v–170v: Ewaldus, Kaufmännischer Computus (deutsch).
Inc.: Computus magistri ewaldi erffordiensis etc. Kompt dir abir fur, das vorne und mitten gancz were und zcu dem letcztin gebrochen were . . .
Expl.: . . . et videns quot cuilibet veniunt.
Es folgen von anderer Hand zwei weitere Aufgaben.

46.11 f.171rv: Bruchalgorismus.
Inc.: Modum reductionum numerorum dissimilium denominationum . . .
Expl.: . . . demum ad divisionem procedat.

46.12 f.171v: Aufgabensammlung.
Inc.: Sequuntur aliqua enigmata pro maiore declaratione regularum de minuciis prius scriptarum. Exemplum primum. Emi I ulnam . . .
Expl.: . . . quod veniunt pro 1113 gr. et $\frac{1}{3}$ unius gr.

46.13 f.175r–187v: Johannes de Lineriis, Algorismus de minuciis (siehe 12.1), mit wenigen Randglossen.
Inc. (Glosse): Causa efficiens fuit iste Johannes de sacrobusco . . .
Inc. (Text): Modum representacionis minuciarum volgarium et phisicarum contingit declarare . . .
Expl.: . . . ad propositum nostrum sufficiunt minuciarum volgarium et phisicarum. finitus per bacc. ysen 1466 in profesto marie mag^{de}.
Es folgen Rechenbeispiele.

46.14 f.188r–191v: Sacrobosco, Algorismus (siehe 8.), nur der Abschnitt über die Wurzeln (S.14–19 Curtze).
Inc.: Extractio Radicum. Sequitur de radicum extractione et primo in numeris quadratis . . .
Expl.: . . . Et hec de radicum extractione sufficiunt dicta tam in numeris quadratis quam cubicis. 1466 die vero 26 mensis junii.

46.15 f.192r–193v: Aufgabensammlung: Unterhaltungsmathematik (vgl.32.2).
Inc.: Enigmata. Si vis scire in quo digito annulus sit absconsus . . .
Expl.: . . . annorum numerum noscito per medium.

46.16 f.193v: Über den Stellenwert der Ziffern (TK 938).
Inc.: Nota quod omnis figura in primo loco posita . . .
Expl.: . . . Extrahe radicem dupla sub parte sinistra.

46.17 f.193v–194r: Über die Progressio, mit Kommentar (TK 1137).
Inc. (Text): Progressio est numerorum secundum equales excessus augmentorum aggregacio . . .
Expl. (Text): . . . et patebit summa progressionis et sic est finis.
Inc. (Komm.): Progressio naturalis sive continua est . . .
Expl. (Komm.): . . . dividatur ultimus numerus per 5 ut prius et sic deinceps.

46.18 f.195r–196r: Bruchalgorismus.
Inc.: Minucie duplices sunt scilicet phisice et volgares. Phisice sunt sexagesime . . .
Expl.: . . . et sic de aliis est faciendum suo modo.

46.19 f.196rv: Linienalgorismus; Progressio.
Inc.: Per denarios proiectiles facile est addere subtrahere duplare et dimidiare . . .
Expl.: . . . subtraxeris numerum primi loci scilicet I iam habes totam summam.

46.20 f.197r–201r: Kaufmännische Aufgabensammlung.
Inc.: Istis notatis ponende sunt aliquot regule per quas expedientur questiones facte per numerum . . .

Expl.: ... hoc est sue denominacionis proporcionalis tunc operare de tri.
Es folgen auf f.201rv weitere Aufgaben und Rechnungen, die wohl später nachgetragen sind.

46.21 f.201v–203r: Johannes Langheym, Visiertraktat.
Inc.: Johannes langheym profunditas. Notandum quod in arte visorandi due dimensiones sunt ...
Expl.: ... et sic procede per omnia spacia.

46.22 f.205r–210r: Robertus Anglicus, Quadrans vetus (siehe 11.3). Der Text weicht im 2.Teil von der Fassung bei Tannery ab.
Inc.: Ad faciendum quadrantem cum cursore. Geometrie due sunt species scilicet theorica practica ...
Expl.: ... et exibit profunditas putei et sic est finis de quo sit trinitas benedicta ... Anno domini 1446 in vigilia annunciationis beatissime Marie virginis.

46.23 f.210v–211r: Quadrant-Traktat.
Inc.: Compositio quadrantis simplicis. Ad faciendum quadrantem capiatur asser ex utraque parte ...
Expl.: ... finitum anno domini 1446.

46.24 f.227v–243r: Verschiedene Texte zur Visierkunst; vielleicht Teile eines größeren Traktates.
Inc.: (f.227v) Compositio virge visorie. Si vis componere virgam visoriam hoc modo precede ...
(f.230r) Stich irst dyn ruth yn den boden ...
(f.231r) De modo mensurandi dantur 3 regule. Prima est quod debemus notare profunditatem vasis ...
(f.232r) Si vis virgam visoriam conficere accipe circinum ...
(f.238r) Secuntur cambia. Alzo du hast gemacht dy tiffe und dy lenge der visierrutten ...
(f.242v) De modo mensurandi dantur 3 regule.
Prima est quod debemus notare profunditatem vasis ...

Lit.:
Narducci, S.246–249

47. **Ms.lat.qu.787**, s.15, 228 Blätter (SBPK)
Provenienz/Vorbesitzer: Buxheim, Kartause
Inhalt: Regimen sanitatis; Rhetorik; Grammatik.
f.97r–106r: Sacrobosco, Algorismus (siehe 8.), mit Kommentar. Unvollständig; endet nach der Subtraktion (S.5,29 Curtze).
Inc. (Komm.): Quisquis namque scientias mathematicas praetermiserit ...
Inc. (Text): Omnia que a primeva rerum origine processerunt ...
Expl.: ... est enim subtractio addicionis probacio et econtra. etc.

48. **Ms.lat.qu.846**, s.14 (1381), 56 Blätter (SBPK)
Provenienz/Vorbesitzer: Erfurt
Inhalt: Medizin; Theologie; Compendium naturae.
48.1 f.1r–4r: Novus algorismus (siehe 46.8).
Inc.: Incipit novum compendium de speciebus algoristicis. Quoniam quibusdam iuvenibus sciencia calculandi videtur difficilis et penalis ...
Expl.: ... laus et honor est cum gloria in seculorum secula. dm̄ explicit deo gracias. Explicit tractatus et compendium algorismi de integris.

48.2 f.9r–10v: Aufgabensammlung: Unterhaltungsmathematik (vgl. 32.2).
Inc.: Incipiunt enigmata de speciebus algoristicis. Ponatur quod sint 2 socii quorum unus dicat alteri . . .
Expl.: . . . Ponatur quod sint 3 socii vel/ (bricht plötzlich ab)

49. **Ms.lat.qu.886**, s.14, 237 Blätter (SBPK)
Provenienz/Vorbesitzer: England; Th. Phillipps (Ms.9350)
Inhalt: Aegidius Romanus, De regimine principum.
f.224r–237v: Mehrere kurze mathematisch-naturwissenschaftliche Traktate.
Inc.:
(f.224r) Omne quod expetitur ab homine aut expetitur propter corpus aut propter animam aut propter utrumque . . .
(f.227v) In principio autem cuiuslibet sciencie VII admodum solent inquiri . . .
(f.229v) In principio geometrie sicut in principio cuiuslibet sciencie inquirendum sit quid sit et cuius generis . . .
(f.232r) Intencio geometrie est declaratio quantitatum et figurarum . . .
(f.235r) Practica arsmetrice sive arbari sive algorismius primo dividitur in duas partes magnas . . .
Expl.: . . . est virtus et rithmos numerus quasi virtus numeralis.

50. **Ms.lat.qu.898**, s.14, 100 Blätter (SBPK)
Provenienz/Vorbesitzer: Trier, St. Maximin
Inhalt: Astronomische Sammelhandschrift.
50.1 f.4v: Glosse zu Sacrobosco, Algorismus (?).
Inc.: Liber iste cuius subiectum est numerus numerans dividitur (?) principaliter in 2 partes in prohemium et execucionem . . .
Expl.: . . . et hoc per rare sedem (?).
50.2 f.5r–13r: Sacrobosco, Algorismus (siehe 8.).
Inc.: Incipit ars numerandi algoristica. Omnia que a primeva rerum origine processerunt . . .
Expl.: . . . qui modus operandi est idem predicto. Et hec de radicum extractione sufficiant.
50.3 f.83r–88v: Campanus, Quadrant (siehe 25.).
Inc.: Scire debes quod circulus solis duas habet medietates . . .
Expl.: . . . Hec quoque de practica quadrantis dicta sufficiant. Explicit quadrans magistri Campani.
50.4 f.91v–97r: Sacrobosco, Quadrant (TK 1003).
Inc.: Omnis sciencia per instrumentum operatam instrumenti sui noticiam . . .
Expl.: . . . et hec mihi dicta de simplici et composito quadrante sufficiant. Explicit quadrans magistri Jo. de sacro bosco.
Lit.:
Zinner, S.11

51. **Ms.lat.qu.926**, s.15, 96 Blätter (SBPK)
Inhalt: Regulae grammaticales.
51.1 f.53r: Zwei kurze Bemerkungen über die Mathematik.
Inc.: Pro recommendacione Arithmetice et linee (?) ayt ysodorus (!) ethymoloyarum 3° Arithmeticam numerorum . . .
51.2 f.54r–75r: Sacrobosco, Algorismus (siehe 8.), mit Rand- und Interlinearglossen.
Inc.: Omnia quae a primeva rerum origine processerunt . . .

Expl.: ... Et hec de radicum extractione tam in numeris cubicis quam quadratis sufficiunt.

Auf f.75v folgt eine kurze Bemerkung über die mittlere Proportionale.

51.3 f.76r–82r, 85r: Kaufmännische Mathematik und mathematische Notizen.

Inc.: Sequuntur Regulae mercatorum. Prima regula est detri apud Italos ...

51.4 f.86r–88r: Algorismus.

Inc.: Incipiunt species algorithmi. Pro (?) brevi et facili cognicione specierum algorithmi integrorum sunt tria premittenda ...

Expl.: ... Sequitur de radicum Extractione./ (Der Rest fehlt)

52. **Ms.lat.oct.153**, s.13, 68 Blätter (SBPK)

Provenienz/Vorbesitzer: Th. Phillipps (Ms.6904)

Inhalt:

52.1 f.1r–67r: Jordanus Nemorarius, De elementis arismetice artis (Thomson, S.113–115; TK 1600), unvollständig und falsch gebunden (richtige Reihenfolge: f.3r–18v. 27r–34v.19r–26v.1rv.2rv.35r–67r).

Inc.: Unitas est esse rei per se discretio ...

Expl.: ... tres medios assignare sit possible. Explicit distinctio decima et cum ea finitur liber Iordani de elementis arismethice. CCCC et XVII proposiciones continens.

52.2 f.67rv: Algorismi propositiones.

Inc.: Figura dicitur quo numeros secundum libitum representamus ...

Expl.: ... a primis eorundem dicuntur equidistare.

52.3 f.68rv: Zusätze zu Jordanus' Arithmetik. Am Rande: Hec proposicio debet esse post 12. sexti. Due vero sequentes debent esse in medietate post 44. decimi.

Lit.:

Bibliotheca Phillippica (1897), S.4 (Nr.21).

53. **Ms.lat.oct.155**, s.12, 86 Blätter (SBPK)

Provenienz/Vorbesitzer: England; Th. Phillipps (Ms.189)

Inhalt: Komputistische Sammelhandschrift.

53.1 f.42r: Text über die Minutien.

Inc.: Libra vel as sive assis duodecim unciae ...

Expl.: ... Scripulus sex siliquae.

Es folgen die Symbole vom as bis zum lentis.

53.2 f.42v: Text über die Minutien.

Inc.: Assis et unciarum divisionem nosse quae non minus temporibus rebusve aliis quam nummis est apta computandis non ignobilis intentio est ...

Expl.: ... et figuras earum paucis assignare curavimus.

Es folgen die Bruchzeichen vom as bis dodrans.

Lit.:

Bibliotheca Phillippica (1897), S.65 (Nr.500).

54. **Ms.lat.oct.157**, s.14, 52 Blätter (SBPK)

Provenienz/Vorbesitzer: Th. Phillipps (Ms.2229)

Inhalt: Komputistische Sammelhandschrift.

54.1 f.33r: Zahlentabelle: Wert der Ziffern im Stellenwertsystem (von 10^0 bis 10^7). Siehe 44.3.

54.2 f.46v–52v: Sacrobosco, Algorismus (siehe 8.). Endet nach der Progressio (S.13,34 Curtze).

Inc.: Incipit algorismus. Omnia que a primeva rerum origine processerunt ...

Expl.: ... imparibus media pars maior multiplicetur per medietatum posicionem multiplicentur extrema coniuncta.

Lit.:
Bibliotheca Phillippica (1897), S.80 (Nr.632).

55. **Ms.lat.oct.162**, s.12, 104 Blätter (SBPK)
Provenienz/Vorbesitzer: Trier, St. Eucharius-Matthias; August Naumann, Freiberg (Nr.18); B. Boncompagni (Ms.168)
Inhalt: Mathematische Sammelhandschrift.

55.1 f.1r–47v: Bernelinus, Liber abaci (ed. A. Olleris: *Oeuvres de Gerbert* (Clermont-Ferrand, Paris: 1867), S.357–400.—TK 876).
Inc.: Incipit prologus in abacum. Mirari pater sancte non desino exactionis tue instanciam ...
Expl.: ... Quod si an recte feceris dubitas. superiori argumento comproba.

55.2 f.47v–48v: Über die Multiplikation von Brüchen.
Inc.: Totus prior numerus et eius quarta pars in secundo invenitur tramite ...
Expl.: ... X in se C.

55.3 f.48v–60v: De minutiis (ed. Bubnov, S.228, 14–241,7.—TK 329).
Inc.: Cum passione contraria id est augmentatione vel diminutione ...
Expl.: ... minutie cum suis differentiis possint redintegrare.

55.4 f.61r–68r: Herigerus, Regulae in abacum (ed. Bubnov, S.210,3–221,5.—TK 442).
Inc.: Dividitur utique maior maior per minorem ...
Expl.: ... Haec sub exemplo posita mittunt ad reliqua.

55.5 f.69r–70r.71v–76r: "Boethius" Geometrie II (ed. M. Folkerts: "*Boethius*" *Geometrie II, ein mathematisches Lehrbuch des Mittelalters* (Wiesbaden: 1970), Auszüge (Z.927–946.530–550.429–518 Folkerts).
Inc.: Veteres igitur geometricae artis indagatores ...
Expl.: ... quod residuum sit ex dividendis.

55.6 f.70v–71r: Auszüge aus der Geometria incerti auctoris (IV 46; III 9; IV 48f. Siehe 3.4).
Inc.: Ad columnam faciendam. longitudinis septimam inferiori ...
Expl.: ... longa et lata fuerit, et alta amphora.

55.7 f.71v: Epaphroditus und Vitruvius Rufus, Kapitel 43 (ed. Bubnov, S.550,23–551,4).
Inc.: Arborem sive turrim vel quodcumque fuerit excelsum ...
Expl.: ... tot pedes sunt altitudinis eius.

55.8 f.77v–79v: Über die Maße.
Inc.: Haec de unciis et minutiis ceteris perscripsimus. et quid una quaeque significet prout competit monstravimus. Denique in omnibus disciplinis ...
Expl.: ... Nam et sextarius olei XX unciis et mellis XXX appenditur.

55.9 f.79v–81r: De mensuris in liquidis (ed. K. Lachmann in *Die Schriften der römischen Feldmesser 1*, (Berlin: 1848), S.374,23–376,13.—TK 870).
Inc.: Mensurarum in liquidis coclear est minima pars ...
Expl.: ... Duo chori culleum quod sunt modia LX.

55.10 f.81r–86v: Auszüge aus Isidor: Etymologiae XVI, 25–27 (TK 1059).
Inc.: Ponderum ac mensurarum iuvat cognoscere modum ...
Expl.: ... superiori littera coniuncta kenix est.

55.11 f.91r–104v: Gerlandus, De abaco (ed. P. Treutlein in *Bullettino Boncompagni 10* (1877), S.595–607.—TK 923).
Inc.: Opus magistri Gerlandi De abaco. Nonnullis arbitrantibus multiplicandi dividendique scientiam ...

Expl.: ... per celentim restituitur prima figura.

Lit.:

Narducci, S.95–98

M. Folkerts: "*Boethius*" *Geometrie II, ein mathematisches Lehrbuch des Mittelalters* (Wiesbaden: 1970), S.24f.

56. **Ms.lat.oct.193**, s.15, 226 Blätter (SBPK)
Provenienz/Vorbesitzer: Heiligenstadt, Jesuitenkolleg
Inhalt: Traktate zur Philosophie, Astronomie, Optik.
56.1 f.155r–183r: Geometrie.
Inc.: Geometria isagoge ad elementa Euclidis ortus obiectum nomen geometriae. Geometria quae ex puris Mathematicis ...
Expl.: ... esse aequale datae rectae lineae C. Finis coronat opus.
56.2 f.217r–225v: Arithmetik.
Inc.: Arithmetica. Quid nomen, quod obiectum arithmeticae? ...
Expl.: ... quae per magnos numeros exprimitur.

57. **Ms.lat.oct.259**, s.15, 142 Blätter (SBPK)
Provenienz/Vorbesitzer: Erfurt
Inhalt: Poetische und rhetorische Texte.
f.83v–85r: Aufgabensammlung: Unterhaltungsmathematik (vgl.32.2).
Inc.: Si duorum hominum pellencium greges ad forum unus alteri dicat ...
Expl.: ... non computato quod est ultra octonarios.

58. **Ms.lat.oct.267**, s.15, 150 Blätter (SBPK)
Inhalt: Sammelhandschrift (Grammatik, Musik, Theologie, Astronomie).
58.1 f.63r: Glosse, wohl zu Sacrobosco, Algorismus.
Inc.: Numerus a nummo imperatore romanorum dictus est ...
Expl.: ... et sine difficultate numero et distinctione computare potest.
58.2 f.63v–66r: Sacrobosco, Algorismus (siehe 8.), mit Kommentar und Glossen, unvollständig (endet S.2,33 Curtze).
Inc. (Komm.): Iste est liber de arte numerandi arismetrice sciencie subordinatus ...
Inc. (Text): Omnia que a primeva rerum origine processerunt ...
Expl.: ... Notandum eciam quod quelibet figura primo loco posita significat suum digi/

59. **Ms.lat.oct.416**, s.14 (1357–59), 118 Blätter (SBPK)
Inhalt: Astronomische Sammelhandschrift.
59.1 f.60v: Johannes Eligerus de Gondersleven, De utilitate quadrantis (siehe 12.4).
Inc.: Utilitas novi quadrantis breviter et lucide colligere ...
Expl.: ... extat latitudo tui orizontis.
59.2 f.85rv: Zwei Aufgaben der Unterhaltungsmathematik (vgl.32.2).
Inc.: Nota si vis narrare quod hallñ habeat socius tuus ...
Expl.: ... si 2 in indice, si 3 in medio etc.

60. **Ms.Magdeb.76**, s.15, 332 Blätter (DSB)
Provenienz/Vorbesitzer: Magdeburg, Domgymnasium
Inhalt: Theologische Sammelhandschrift.
f.318r–332bv: Isidor (?), De numeris (vgl. Migne, PL 81, 409).
Inc.: Incipit ysidorus hispalensis. De numeris ab unitate usque ad octo. Altissimo domino deo nostro adiuvante de numero et de eius mistico misterio ...
Expl.: ... in lybano id est in candidacione virtutum.

Lit.:
H. Dittmar: *Die Handschriften und alten Drucke des Dom-Gymnasiums* (Programm Magdeburg, 1878), S.43f.

61. **Ms.Magdeb.287**, s.15, 186 Blätter (DSB)
Inhalt: Theologisch-juristische Sammelhandschrift.
f.56v: Griechische Zahlen in lateinischer Umschrift (4 Zeilen): Enna dya tria . . . tryennes.

62. **Ms.Phillipps 1729**, s.12/13, 155 Blätter (DSB)
Provenienz/Vorbesitzer: Rouen; Paris, Jesuiten; G. Meerman; Th. Phillipps
Inhalt: Hrabanus Maurus, Allegoriae biblicae.
f.150v: Beda, De computo digitorum (siehe 11.6), nur S.179,11–180,47 Jones.
Inc.: Centesimus et LX et XXX fructus quamquam de una semente nascantur . . .
Expl.: . . . Mille autem facies in dextera quomodo unum in leva.
Lit.:
Rose 1, S.98f. (Nr.52)

63. **Ms.Phillipps 1805**, s.14, 148 Blätter (DSB)
Provenienz/Vorbesitzer: England (Chester); G. Meerman; Th. Phillipps
Inhalt: Grammatisch-astronomische Sammelhandschrift.
63.1 f.140v: 6 arithmetische Verse (siehe 15.2).
Inc.: Una prima secunda decem dat tertia centum . . .
Expl.: . . . Millesies decima dat millesies quoque mille.
63.2 f.140v: 7 Verse über den Algorismus.
Inc.: Septem sunt partes non plures istius artis . . .
Expl.: . . . Extrahe radicem semper sub parte sinistra.
Lit.:
Rose 1, S.433–435 (Nr.194)

64. **Ms.Phillipps 1810**, s.11, 39 Blätter (DSB)
Provenienz/Vorbesitzer: Paris, Jesuiten; G. Meerman; Th. Phillipps
Inhalt: Cicero, De inventione.
f.39v: Gerbert, Regulae de numerorum abaci rationibus (ed. Bubnov, S.6–22.—TK 1454), nur der Abschnitt über die Multiplikation (S.9,3–11,24 Bubnov).
Inc.: Si multiplicaveris singularem numerum per decenum, dabis unicuique digito X^{cem} et omni articulo centum . . .
Expl.: . . . Si decenum per decenum millenum dabis digito $\bar{C}$ et articulis mille milia.
Lit.:
Rose 1, S.437f. (Nr.198)

65. **Ms.Phillipps 1832**, s.9/10, 92 Blätter (DSB)
Provenienz/Vorbesitzer: Metz; Paris, Jesuiten; G. Meerman; Th. Phillipps
Inhalt: Beda, Komputistische und chronologische Schriften.
f.55v: Ps. Beda, De arithmeticis propositionibus (ed. M. Folkerts: "Pseudo-Beda: De arithmeticis propositionibus, eine mathematische Schrift aus der Karolingerzeit," in *Sudhoffs Archiv 56* (1972), S.22–43.—TK 1258), unvollständig (endet Z.73 Folkerts).
Inc.: Incipiunt numeri per quos potest qui voluerit alterius cogitationes de numero quolibet quem animo conceperit explorare. Quomodo numerus a quolibet animo conceptus . . .
Expl.: . . . tot quaternos divinator sumere debet.
Lit.:
Rose 1, S.289–293 (Nr.130)
Beschreibende Verzeichnisse 1, S.30–33

66. **Ms.Phillipps 1833**, s.10, 62 Blätter (DSB)
Provenienz/Vorbesitzer: Fleury; Paris, Jesuiten; G. Meerman; Th. Phillipps
Inhalt: Abbo von Fleury, Arithmetisch-chronologische Schriften.
66.1 f.1v–3r: Victorius, Calculus (ed. G. Friedlein: "Victorii calculus ex codice Vaticano editus," in *Bullettino Boncompagni 4* (1871), S.443–463.—TK 1601).
Inc.: Incipit prefatio de ratione calculi. Unitas illa unde omnis multitudo numerorum procedit . . .
Expl.: . . . usque ad $\overline{\text{IIII}}$ et sic usque ad finem.
Explicit praefatio. Incipit liber calculi quem Victorius conposuit.
Es folgen die Zahlentafeln, die aber nach der 15. Säule abbrechen.
66.2 f.7v–21v: Abbo, Kommentar zum Calculus des Victorius (teilweise ediert bei Bubnov, S.199–203.—TK 183).
Inc.: Incipit explanatio in calculo victorii. quam ysagogen arithmeticae placeat dicere. Calculum victorii dum quondam fratribus qui manu sancti desiderii pulsabant . . .
Expl.: . . . et decies XX sunt ducenti. et decies quini sunt L.
Lit.:
Rose 1, S.308–315 (Nr.138)
Beschreibende Verzeichnisse 1, S.21f.
Bubnov, S.XX

67. **Ms.Phillipps 1869**, s.9, 142 Blätter (DSB)
Provenienz/Vorbesitzer: Moselgebiet; Trier, St. Maximin; Paris, Jesuiten; G. Meerman; Th. Phillipps
Inhalt: Beda, De temporibus.
f.13r: Zahlenwert der griechischen Buchstaben (Füllsel).
A mia I alpha . . . $\bar{\Lambda}$ mrđ DCCCC.
Lit.:
Rose 1, S.293–295 (Nr.131)

68. **Ms.Phillipps 1870**, s.11, 148 Blätter (DSB)
Provenienz/Vorbesitzer: Verdun; Paris, Jesuiten; G. Meerman; Th. Phillipps
Inhalt: Hugo von Flavigny, Chronik.
f.55v–56r (am Rand): De ponderibus, z.T. Auszüge aus Isidor (Etymol. XVI,27; vgl. 55.10.—TK 1060, 183).
Inc.: Ponderum signa plerisque ignota sunt quapropter formas eorum et caracteres subiciamus . . .
Expl.: . . . in dextro brachio superiori O coniuncta kenix est.
Inc.: Calcus minima pars est ponderis scilicet III pars oboli . . .
Expl.: . . . Chorus XXX modios.
Lit.:
Rose 1, S.321–325 (Nr.142)

69. **Ms.theol.lat.fol.119**, s.12, 243 Blätter (SBPK)
Provenienz/Vorbesitzer: Lippstadt (?)
Inhalt: Hieronymus, Epistolae. Auf f.240r–242r (neue Zählung = 246r–247v,249r alte Zählung) stehen ohne Unterbrechung mehrere komputistische Exzerpte, darunter:
69.1 f.240r: Übersicht über die mit Fingern darstellbaren Zahlen (von 1 bis 1000000) in Form von Säulenfiguren. Zwischen den Säulen steht:
Differentia digitalis numeri. Tres digiti in sinistra manu . . . decies centena milia.
69.2 f.240v: Beda, De computo digitorum (siehe 11.6), S.179,22–180,62 Jones (gegen Ende abweichender Text).

Inc.: Cum ergo dicis unum. minimum in leva digitum inflectens . . .
Expl.: . . . Cum autem dicis decies centena milia. duae palmae apertae iuxta faciem eriges interpositae. sed ante aures retro respicientes.

69.3 f.240v–241v: Beda, De computo digitorum (siehe 11.6).
Inc.: De temporum ratione domino iubente dicturi necessarium duximus . . .
Expl.: . . . dominus adiuvare dignabitur exponenda veniamus.

69.4 f.241v: De numeris (TK 192); 7 Zeilen.
Inc.: De ratione numerorum. Cardinales sunt numeri I II III IIII . . .
Expl.: . . . Sunt item adverbialia ex his nascentia. ut simpliciter. dupliciter. tripliciter.

Lit.:
Rose 2, S.72–74 (Nr.294)

70. **Ms.theol.lat.fol.337**, s.12, 189 Blätter (SBPK)
Provenienz/Vorbesitzer: Liesborn
Inhalt: Augustinus, De civitate dei.
f.240r–243v: Beda, De computo digitorum (siehe 11.6).
Inc.: Cum dicis unum, minimum in leva digitum inflectens . . . (=S.179,22 Jones)
Expl.: . . . Decies autem centena milia cum dicis ambas sibi manus insertis invicem digitis implicabis. (=S.180,62 Jones)
Es folgen Abbildungen der Fingerstellungen bis 1000000.
Lit.:
Rose 2, S.83f. (Nr.302)

71. **Ms.theol.lat.oct.16**, s.15, 215 Blätter (SBPK)
Inhalt: Parabolae Salomonis.
f.215r: Geometrische Definitionen.
Inc.: Nota quod quedam sunt regule mathematicales breves et utiles. Quarum prima talis est. De numero linearum quedam sunt iacentes quedam cadentes . . .
Expl.: . . . Ad cognoscendum quomodo triangulus habeat tres angulos etc.
Lit.:
Rose 2, S.27–29 (Nr.255)

72. **Ms.theol.lat.oct.111**, s.15, 48 Blätter (SBPK)
Provenienz/Vorbesitzer: Staatsarchiv Berlin
Inhalt: Komputistische Texte.
f.35r–44r: Sacrobosco, Algorismus (siehe 8.)
Inc.: Algorismus. Omnia que a primeva rerum origine processerunt . . .
Expl.: . . . tam in numeris quadratis quam in numeris cubicis. Explicit algrissius (!) deo gracias etc. per manus iohannis de bacis. . . .
Lit.:
Rose 2, S.1186f. (Nr.958)

LITERATURVERZEICHNIS

Beschreibende Verzeichnisse 1, 5 = Beschreibende Verzeichnisse der Miniaturen-Handschriften der Preußischen Staatsbibliothek zu Berlin. 1.Band: Kirchner, J. 1926. *Die Phillipps-Handschriften* (Leipzig). 5.Band: Wegener, H. 1928. *Beschreibendes Verzeichnis der Miniaturen und des Initialschmuckes in den deutschen Handschriften bis 1500* (Leipzig).

Bibliotheca Phillippica = *Bibliotheca Phillippica. Catalogue of a portion of the famous collection of classical, historical, topographical, genealogical and other manuscripts and autograph letters &c. of the late Sir Thomas Phillipps* (London: Auktionskataloge Sotheby, Wilkinson & Hodge, 1895ff.).

Boese = Boese, H. 1966. *Die lateinischen Handschriften der Sammlung Hamilton zu Berlin* (Wiesbaden).

Bubnov = Bubnov, N. 1899. *Gerberti postea Silvestri II papae Opera Mathematica* (Berlin = Hildesheim, 1963).

Degering 1, 2, 3 = Degering, H. 1925–1932. *Kurzes Verzeichnis der germanischen Handschriften der Preußischen Staatsbibliothek, 1–3* (Leipzig; Mitteilungen aus der Preußischen Staatsbibliothek *7–9*).

Narducci = Narducci, E. 1892. *Catalogo di manoscritti ora posseduti da D. Baldassarre Boncompagni* (2.Auflage, Rom).

Phillipps Manuscripts = *The Phillipps Manuscripts. Catalogus librorum manuscriptorum in Bibliotheca D. Thomae Phillipps, Bt. Middlehill 1837–1871. With an introduction by A. N. L. Munby, Litt. D.* (London, 1968).

Rose 1, 2 = Rose, V. 1892–1905. *Verzeichniss der lateinischen Handschriften der Königlichen Bibliothek zu Berlin.* 1.Band: *Die lateinischen Meerman-Handschriften des Sir Thomas Phillipps in der Königlichen Bibliothek zu Berlin* (Berlin, 1892); 2.Band: *Die Handschriften der Kurfürstlichen Bibliothek und der kurfürstlichen Lande*, 1.–3.Abteilung (Berlin, 1901–1905). (Die Handschriften-Verzeichnisse der Königlichen Bibliothek zu Berlin, *12.13*).

Thomson = Thomson, R. B. 1976. "Jordanus de Nemore: Opera," *Mediaeval Studies, 38*, 97–144.

TK = Thorndike, L. und P. Kibre. 1963. *A catalogue of incipits of mediaeval scientific writings in Latin* (2.Auflage, London).

Zinner = Zinner, E. 1962. "Aus alten Handschriften," *Naturforschende Gesellschaft Bamberg, 38.Bericht*, 8–57.

NAMEN-UND SACHREGISTER

Die Zahlen beziehen sich auf die Texte in den Handschriftenbeschreibungen.

VERZEICHNIS DER INITIEN

Die Zahlen beziehen sich auf die Texte in den Handschriftenbeschreibungen.

Universität Oldenburg
Oldenburg, Bundesrepublik Deutschland

Mathematical Physics in France, 1800–1840: Knowledge, Activity, and Historiography*

I. GRATTAN-GUINNESS

The development of mathematical physics in France in the period 1800–1840 was one of the most important episodes in the history of the subject. Eighteenth-century Newtonian mechanics was transferred into the larger realm of mathematical physics, in which (apparently) non-Newtonian phenomena were mathematicized. Eighteenth-century algebraic calculus was similarly subsumed under mathematical analysis. In this paper I describe these developments, outline a historiography for mathematical physics, and describe the French community. I conclude with some remarks on the development of mathematical physics in some other countries, and a survey of problems in the primary and secondary literature.

Contents

* For K.-R. Biermann on his 60th birthday.

ISBN 0-12-204050-3

In art. 52 [of W. Whewell, *Of a liberal education in general* (1845)] we read, 'The general belief, for undoubtedly it is a general belief, that Mathematics is a valuable element in education, has arisen through the use of Geometrical Mathematics. If Mathematics had only been presented to men in an analytical form, such a belief would not have arisen. If, in any place of education, Mathematics is studied only in an analytical form, such a belief must soon fade away'. The last sentence in the preceding passage seems contradicted by experience; take for instance the mathematical education in the celebrated *Ecole Polytechnique* of France.

I. Todhunter 1876, Vol. 1, 159.

1. INTRODUCTION

Among historians of mathematics of our time, none have surpassed K.-R. Biermann in their concern with social and institutional factors as companion to the technical and conceptual development of mathematics. His work has dealt with the history of mathematicians and their social artifacts as much as with the history of mathematical knowledge. From his writings we have learnt much of the progress of German mathematics from the time of Leibniz until the mid-19th century, and under his stimulus numerous documents of past times have first seen the light of public day. In offering this paper as a token of both gratitude and expectation, I have attempted to wed the social aspects of a particular development in mathematics to its technical content in a way which, I hope, would gain the approval of my dedicatee. My period is 1800–1840, and my subject is mathematical physics, which are both encompassed by Biermann's interests; but my country is France, to which all other nations turned at that time for their scientific inspiration.

The paper is a report on work currently in progress. Thus it conveys impressions rather than draws conclusions, and is presented largely to encourage others to share the labor (and also the entertainment) of examining a development in both mathematics and physics which was of great significance for both disciplines and yet has been little studied historically. I am principally concerned with the mathematicization of a range of principal phenomena, both the mathematics used and the increasing range of branches of physics that were so treated. I shall not deal with the theoretical or experimental aspects of physics, and will confine myself largely to progress in France, which was then by far the most important centre for both teaching and research in this field.

Section 2 provides a "factual" survey of the developments, and introduces some of the most significant historical figures. Sections 3 and 4 perform a similar service in outlining, respectively, the institutional atmosphere (centred on the *Académie des Sciences*) and the educational aspects of the subject (especially the *Ecole Polytechnique*). Section 5 describes the methodology that was then usually followed in mathematical physics, and Section 6 outlines a spectrum of relationships between mathematics and physics that is essential for an historiography of the subject. Section 7 is concerned with a collection of companion philosophical questions on which a mathematical physicsist may have had views. Then, in terms of these two sections, the community in French mathematical physics is described in Section 8, and the nature of its decline during the latter part of the period 1800–1840 (and later) is discussed in Section 10. In between, Section 9 contains some general comments on the social history of science, with especial concern to the development discussed here. Section 11 supplies some notes on developments in other countries. Finally, Section 12 reviews the bibliographical situation for mathematical physics, concluding with references to some of the most important secondary literature. Many of these references are cited at suitable points in the text.

2. AN OUTLINE OF DEVELOPMENTS FROM 1800 TO 1840

In this section I shall provide a factual overview of the development of mathematical physics in this period. It is not specific to any particular country, although France is by far the most important one, and I shall name the principal French contributors.

Around 1800, mathematical physics was normally regarded as Newtonian celestial and terrestrial mechanics; in that the physical theories that could be treated mathematically were based on Newtonian principles of inter-particlate attraction. The principal concerns in celestial mechanics were the shapes of the planets, the motions of the heavenly bodies (including satellites and comets), and the stability of the planetary system. In terrestrial mechanics, the chief interests, in addition to the standard statics and dynamics, were fluid mechanics, elasticity, sound, tides, optics (for which there was a fairly widely held view that light was composed of fast moving projectiles which were susceptible to action by short-range forces) and various aspects of technology and engineering (arches, bridges, steam engines, hydraulics, the general theory of machines, and so on). Mechanics was founded on 'the principle of virtual velocities', and there was a vigorous "foundational" discussion of the status of the principle: its relationship with the lever and parallelogram laws, and the possibility of proving it or having to take it as a basic assumption (see Lindt 1904).

The mathematics involved was the calculus, cast in an algebraic form where the fundamental idea was the Taylorian expansion of a function (which a function allegedly always possessed, and which was always convergent except for cases where the function took infinite values): the 'derived functions' of the function were *defined* as the appropriate coefficients in the expansion, the integral function was defined as the converse of the derived function, and so on. The structure of the mathematical theory was claimed to be entirely algebraic; limits and infinitesimals were held to be unnecessary. Variational principles, again algebraic in form, were also very prominent. The principal use of the calculus was to form and solve differential equations (where power-series and functional solution forms were most greatly favoured); the theory of general and particular solutions to differential equations was also studied (see Rothenburg 1910, Sologub 1975). Difference and differentio-difference equations were sometimes used, and variational methods were quite prominent. The techniques of the calculus also applied to related problems, such as the summation of series and the evaluation of integrals.

The major figure was J. L. Lagrange, who was largely responsible for the algebraic cast of both mechanics and calculus. The other principal figures were L. Carnot, P. S. Laplace, A. M. Legendre, G. Monge, and R. de Prony. C. A. Coulomb exercised considerable influence on aspects of mechanics, engineering, and electricity, though his active career had more or less ended by 1800.

By 1840 a substantial amount of mathematicization had been introduced into several areas of physics (compare Kuhn 1976); heat (including some thermodynamics and gas dynamics), electricity, and magnetism. In optics new phenomena had been discovered and studied (including mathematical treatments), and the waval interpretation of light had replaced the particlate interpretation (see Chappert 1977, Silliman 1967). Newtonian mechanics itself advanced in several areas, especially sound, elasticity, hydrodynamics, pendulum theory, capillary action, attractions and potential theory, and the perturbations of orbits. Engineering was also given much mathematical attention, not only in the traditional areas mentioned but also in new ones such as railways (although not electric telegraphy, which got under way only in the 1840s).

In mathematics some new topics were introduced, especially Fourier analysis, complex variables, and elliptic functions. The traditional areas also advanced, but were radically affected by the new developments. In particular, the algebraic calculus was superceded by what we now call 'mathematical analysis', unified under the theory of limits: the derivative was now the limiting value of the difference quotient, the integral the limiting value of a sequence of partition sums, and so on. Careful attention was now paid to the convergence of infinite series. In addition, there was much progress in the theory of equations (where results on the resolvability of polynomials, and theorems on Sturm sequences, were particularly important), on some transcendental functions (especially Bessel functions and Legendre polynomials), in the evaluation of integrals, and in the calculus of variations. The principal absentees are vector and matrix algebra, which made only slight progress—although *we* can see that there was a great need for them at the time!

Throughout the period there was frequent discussion of the possible existence of imponderable fluids, such as aether and caloric, as carriers of (some of) the phenomena under study, and of the properties that these fluids might possess (see Schaffner 1972, Whittaker 1951). A similar discussion took place in the mathematics: namely, the possible existence of infinitesimals, and of the definitions in terms of which they could be introduced (see Cohen 1883).

All of the new advances in both mathematics and physics were well represented in print by the mid-1820s: the crucial period for publication was (roughly) 1815–1825. The principal French figures involved in the innovations were A.-M. Ampère, D. F. J. Arago, J. B. Biot, S. Carnot, A.-L. Cauchy, J. B. J. Fourier, A. J. Fresnel, E. L. Malus, and S.-D. Poisson. The years to 1840 (and beyond, in fact) were ones of consolidation: the main French contributors to 1840 were G. G. Coriolis, J. M. C. Duhamel, G. Lamé, J. Liouville, C. Navier, L. Poinsot, J. V. Poncelet, and C. Sturm.

The Newtonian "regime" is now often called 'Laplacian physics', because of Laplace's advocacy of it; and the change to mathematical physics is very much the story of the decline of Laplacian physics (see Fox 1974), to which Laplace himself contributed from the late 1810s. We may similarly speak of the change to mathematical analysis as being the story of the decline of Lagrangian mathematics (see Dickstein 1899, and Grattan-Guinness 1970).

The structural similarity of the parallel developments in both mathematics and physics is worth emphasizing. In each case a fairly unified body of knowledge around 1800 had become part of a larger whole by the mid-1820s: Newtonian mechanics had become part of mathematical physics, and the algebraic calculus part of mathematical analysis. The structural similarity is indicated in Tables 1 and 2, which present in schematic form the state of affairs in physics and mathematics, respectively, around 1840. In Table 1 Newtonian mechanics and part of optics roughly encompasses the range of knowledge in 1800; in Table 2 calculus was the principal topic, with some

TABLE 1

Relationships between Principal Topics in Physics, 1840

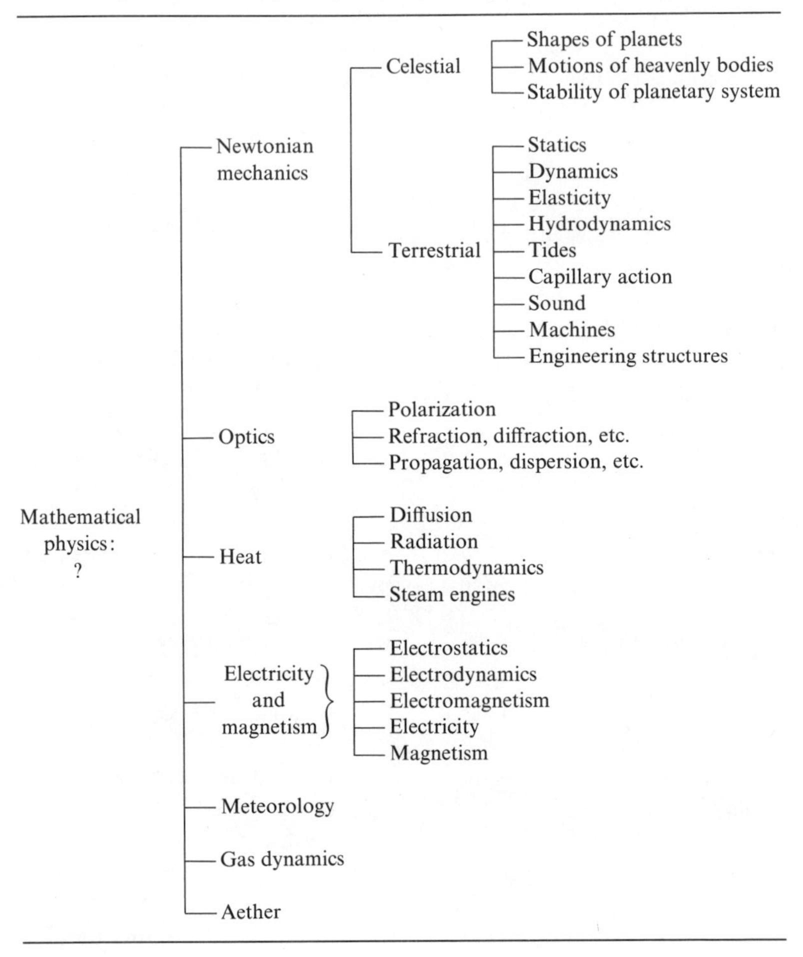

knowledge also in series, functions, complex numbers and the theory of equations.

In one important respect the similarity does not quite hold. The theory of limits held mathematical analysis together (more or less); but it was much less clear whether mathematical physics had any similarly unifying idea, or even whether it should have one. The advocacy of energy and its convertants to fulfil this role came almost entirely after 1840 (though see Subsection 10.2 below).

TABLE 2

Relationships between Principal Topics in Mathematics, 1840

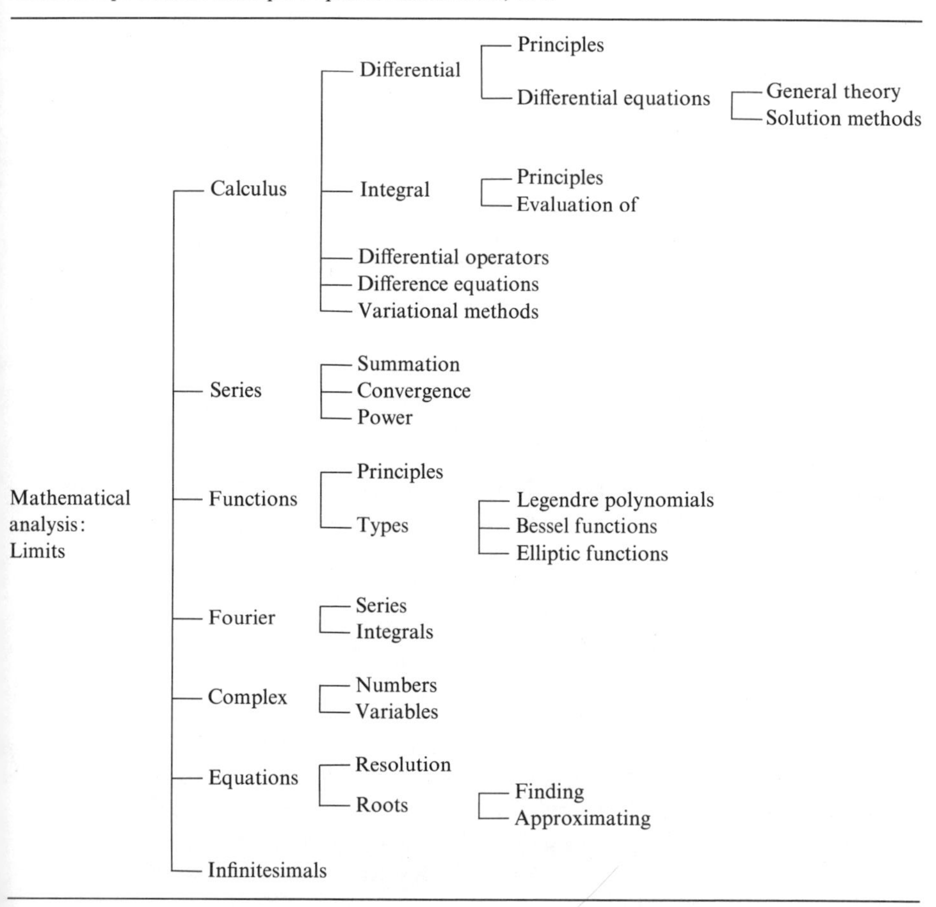

- Mathematical analysis: Limits
  - Calculus
    - Differential
      - Principles
      - Differential equations
        - General theory
        - Solution methods
    - Integral
      - Principles
      - Evaluation of
    - Differential operators
    - Difference equations
    - Variational methods
  - Series
    - Summation
    - Convergence
    - Power
  - Functions
    - Principles
    - Types
      - Legendre polynomials
      - Bessel functions
      - Elliptic functions
  - Fourier
    - Series
    - Integrals
  - Complex
    - Numbers
    - Variables
  - Equations
    - Resolution
    - Roots
      - Finding
      - Approximating
  - Infinitesimals

3. INSTITUTIONAL ASPECTS: THE *ACADEMIE DES SCIENCES*, AND OUTSIDE

The presentation and dissemination in France of the mathematical physics outlined in the last section was centered on the *Académie des Sciences*, which was known as the mathematical and physical class of the *Institut de France* during the revolutionary period of 1795–1815. (A useful source of information on the members of the *Académie* and the *Institut* is *Académie des*

Sciences 1979.) A notable aspect of the social history of science at this period is the emergence of a professional, state/education-supported scientific community. But the similarity with our modern situation should not be overstated. For example, the *Université de Paris*, founded in 1808, carried nothing like the importance that we expect today in universities (see also Subsection 10.4 below); patronage was strongly centred in the *Académie* and the status of *académicien*, and research pursued most closely in connection with the *Académie* and the *écoles*. Again, there is a notable lack of societies concerned with particular sciences; in other words, they were not becoming professionalized in the way that occurred for science as a whole. Nevertheless, France was beginning to set a style at this time which other countries followed in their own way: the chief differences lay in the balance between teaching and research in universities, colleges, and academies.

It is well known that during this period French science was frequently pockmarked by polemics, violent disagreements, and priority disputes. The *Académie* contained many ambitious, prolific geniuses: the 'Procès-verbaux' of their meetings, which for the period 1795–1835 (when the *Comptes rendus* started) were published as *Académie des Sciences* 1910–1922, reveal very clearly both the quality of the work and the manner in which it was often received.

Important manifestations of the atmosphere are the prize problems that were proposed from time to time by the *Académie*: the topics chosen, the timing, the phraseology of the problem, the membership of the examining board, the winners, the fate of their winning papers. Here are a few examples of the insights that study of these prizes can afford.

The heat problem for 1812 seems to have been a compromise between Fourier's supporters and his opponents; but his winning paper did not appear in the *Mémoires* for over a decade. An elasticity prize was set for the same day in 1812: eventually Sophie Germain won it in 1816, and in the interim there had not only been (justified) criticism of her mathematics but also a dispute about Poisson, one of the examiners, reading his own paper on the subject in the *Académie*. In 1815 Cauchy won a prize problem on water waves with a paper containing embryonic Fourier integral methods: this time Poisson had deposited papers on the topic to be read after the prize was awarded, but Cauchy waited 12 years for printing, by which time he, Poisson and Fourier had all extended these methods considerably. Finally, in 1819 Fresnel won a prize on diffraction with a paper on his *hypothesis* of light as a wave phenomenon, in response to a problem proposed by the Laplacian Biot which clearly expected an inductively grounded *secure* solution, and based on the particulate interpretation of light. Indeed, Fresnel's "victory" with his waval interpretation, of which the prize was an important part, was stage managed by Arago, who was deeply antipathetic to Biot because of earlier disputes.*

* Since doubts have been expressed to me about the extent or importance of these polemics, I shall describe the Arago–Biot relationship in more detail as an example. They both belonged to the Arcueil group of scientists led by C.-L. Berthollet and Laplace. Their disputes began in 1806, when Biot presented some joint research on refraction in gases under his own name. The second occurred in 1812, when they were both working on polarization and Arago was very sensitive about priority. On 1 June of that year Biot read his paper 1812 on reflection and polarization of light in crystalline media, in which he mentioned in a footnote that he had been absent from Paris on 11 August 1811 when Arago had read a paper 1811 on similar topics, and that meanwhile Arago had apparently neglected to deposit the dating of his discoveries with the *secrétaire perpétuel*. When the two papers were printed consecutively in the *Mémoires* of the *Institut* Arago objected to the footnote, and on 27 June 1813 they sent in a joint note requesting the recall of the copies of the journal and the replacement of the first sheet of Biot's paper, with the offending footnote replaced (see *Académie des Sciences* 1910–1922, Vol. 5, 224–225). This was done; but the copy of the *Mémoires* now in Aarhus University Library contains the *original* sheet. The offending footnote reads:

> Le manuscrit original de ce Memoire a été paraphé sur toutes les pages par M. Delambre, secrétaire perpétuel; il est déposé au secrétariat de l'Institut. A l'époque où il fut présenté à la Classe, M. Arago n'avait publié de ses recherches sur la lumière, que les extraits imprimés dans les nos 49, 50, 51, du Bulletin des Sciences [par la Société Philomathique de Paris], et dans le Moniteur du 31 août 1811. Me trouvant absent de Paris à l'époque où il avait présenté son Mémoire à la Classe, je n'en connaissais que ce qui avait paru dans les journaux que je viens de citer, et je n'en ai eu depuis aucune autre communication: c'est pourquoi, lorsque je voulus m'occuper de ce genre de phénomènes, j'invitai publiquement M. Arago, en présence du bureau des Longitudes, à faire parapher ses Mémoires par MM. les secrétaires de l'Institut, afin de constater invariablement les faits ou les théories qu'il pourrait avoir dès-lors découverts. Cette demande parut de toute justice, et M. Arago lui-même sembla y accéder; mais comme il a négligé depuis de remplir cette formalité, je crois devoir rappeler les extraits cités ci-dessus, comme renfermant tous les résultats qui étaient publiquement connus pour appartenir à M. Arago, à l'époque où j'ai lu mon Mémoire, le 15 février 1813. (B.)

Biot replaced this footnote by a new opening paragraph and a brief footnote citing Arago's immediately preceding paper. Note, by the way, the chronological difficulties of the period: Biot now says that his paper 'read 1 June 1812' was read on the following 15 February (actually 8 February, according to the record in *ibid.*, 166).

Although there were attempts at reconciliation later, the split between the two was basically irreparable (see Crosland 1967, esp. pp. 332–335). It manifested itself later not only in Arago's management of Fresnel but also in the election of the *secrétaire perpétuel* in 1822 to succeed Delambre, when Arago's intervention secured Fourier's election rather than Biot's. Although Arago's scientific achievements were minor, his entrepreneurial role was often significant (for example, he in turn succeeded Fourier as *secrétaire perpétuel* in 1830), and is a good example of the presence of sociopolitical aspects in science.

Another cause of the polemics was that the *Académie* could not afford to publish speedily in the *Mémoires* (or *Savan(t)s étrangers*) the papers that were being presented: even a major work might wait for years to appear. Thus, in order to stake his intellectual claim more promptly, an author would usually publish at least one summary paper in the satellite journals; if he was not a member of the *Académie* his work would be reported by members, and the report might well be published somewhere at the time (it is usually now available to us in the 'Procès-verbaux'); there may be a notice in the yearly report of the *secrétaire perpétuel* (I suspect that some of the mathematical ones were written by the authors of the articles themselves, thus producing another summary paper); and there could also be a review of the paper and/or its summaries, especially in the *Bulletin universel* which appeared between 1824 and 1831 under the general editorship of the Baron Ferrusac (see Taton 1947). Thus a major paper can produce several "offspring." These are often of considerable historical interest, for they reveal what an author regarded as the main points of his paper, and provide early responses to it by others.

Table 3 lists these principal satellite journals. Several of them would on occasion publish the main papers (in mathematics the *Journal de l'Ecole Polytechnique* regularly did so); and sometimes the *Académie* would then print them again in its own journals in due course. Every now and then the *Mémoires* of a provincial French *Académie* would publish a significant paper, when it would usually be reprinted in one of the satellite journals. Quite a number of papers (usually the summary ones) would also be published fairly soon afterwards in translation in British, German, and/or Italian journals.

Thus the literature of this period shows great complications and repetitions, which need unravelling in order to sort out peoples' motivations and progress. Chronology is a major problem, and the often nominal publication dates in the journals make the published literature unreliable. Sources such as the 'Procès-verbaux' are of great value here, because they record the dates of presentation of issues of journals in their meetings reports. Crosland 1967 is an irreplaceable source of information on institutional aspects of French science during the Revolutionary period.

Most of the work was published as papers, but there was a substantial amount of publishing of books; research treatises, textbooks (which are discussed in more detail in the next section), and pamphlets (a common form of publishing in those days). The enormous mass of this literature indicates that *the history of French scientific publishing* is a significant part of the social aspects of this history. Along with the emergence of a professional scientific community is seen the growth in prominence of *commercial scientific publishers*, who published most of the material in mathematical physics. An interesting example of contact between scientist and publisher was the

TABLE 3

Principal Satellite Journals for Mathematical Physics, 1800–1840[a]

Short title	Dates	Comments
{*Annales de chimie*	–1815	
{*Annales de chimie et de physique*	1816–	
Annales de mathématiques pures et appliquées	1810–1832	Published in Nismes; strong educational interests
Annales des sciences d'observation	1829–1834	
Annales des ponts et chaussées	1831–	
Annuaire du Bureau des Longitudes		Mostly astronomical papers
{*Bibliothèque britannique*	–1815	Significant Swiss journal; F. Maurice most relevant editor
{*Bibliothèque universelle*	1816–	
Bulletin de la Société d'Encouragement pour l'Industrie Nationale	1802–	Of relevance to engineering aspects
Bulletin, Société Philomathique of Paris		Sometimes '*Nouveau bulletin*'. '*Procès-verbaux*' from 1836
Bulletin universel des sciences et de l'industrie	1824–1831	Review journal
Comptes rendus de l'Académie des Sciences	1835–	
Connaissance des tem(p)s		Published papers related (if vaguely) to astronomy and navigation
Correspondance sur l'Ecole Impériale Polytechnique	1804–1816	Mainly for educational purposes
{*Gazette nationale*	–1810	Official newspaper
{*Le moniteur universel*	1811–	
L'institut	1833–	Short notices of work in academies and societies
Journal de l'Ecole Polytechnique		
Journal de mathématiques pures et appliquées	1836–	
Journal de physique		
{*Journal des mines*	–1815	
{*Annales des mines*	1817–	
Journal des savans	1816–	
Journal du génie civil	1828–	
{*Magasin encyclopédique*		
{*Annales encyclopédiques*	1817–1818	
{*Revue encyclopédique*		
Mémorial de l'artillerie	1825–	} Military engineering journals: not covered by Royal Society catalogue
Mémorial de l'officier du génie	1803–	}
Mémorial du dépôt générale de la guerre	1802–	}
Mercure de France		

[a] The curled brackets link together pairs of journals when the second *formally* succeeded the first. Dates are given only when the journal started after 1800 and/or finished before 1840.

founding by Liouville in 1836 of his important *Journal de mathématiques pures et appliquées*. The publisher was Bachelier, which was then headed by the member of the family who had been a fellow student of Liouville at the *Ecole Polytechnique* from 1825 to 1827.

One mathematical physicist whose activity embraces all matters discussed in this section is Poisson, who is a figure of quite peculiar importance (see Arnold 1977). One reason why I have chosen the period 1800–1840 is that it coincides exactly with his professional career. Before the end of the 1800s he was recognized as the mathematical heir to the aging group who had set up mathematical physics in its 1800 form, for he not only learnt up that form at the *Ecole Polytechnique* but also tried to maintain it throughout his career while all around him was changing. His considerable mathematical ability brought him a variety of significant contributions, especially to electrostatics and magnetism; but he was left behind by Fourier and Cauchy when their work began to be presented to the *Académie* in the late 1800s and the early 1810s, respectively, and lost out to them especially in the late 1810s over methods of solving linear partial differential equations. During his period of prominence he had used his influence over publications, and not always with perfect objectivity; for example, *he* wrote the summary papers of some of Fourier's and Cauchy's early important works, and obviously deliberately failed to convey their contents to the readers. In the last 20 years of his career, he fell away markedly in prestige and in relevance to current progress. In personality, and as an early example of the "new" professional career scientist, he strongly resembles his colleague Biot (see Frankel 1972, 1976); and as Laplacian physicists, they declined together in the late 1810s as mathematical physics and mathematical analysis took shape and replaced many of the doctrines which they held dear.

4. EDUCATIONAL ASPECTS: THE *ECOLE POLYTECHNIQUE*, AND OUTSIDE

Several branches of mathematical physics were taught extensively at this time, and many members of the community were active teachers. In fact, some of the innovations occurred in response to educational needs: Cauchy's advocacy of mathematical analysis based on the theory of limits is an important example, for it was largely developed in his teaching at the *Ecole Polytechnique* from the late 1810s onward.

Sometimes the truth is obvious. The importance of the *Ecole* is often stressed and its history has been (and is) studied extensively. (Valuable early histories include Fourcy 1828 and Hachette 1828; a significant successor is Pinet 1887, and fairly useful guides to its 19th-century members and achievements are provided in Marielle 1855 and *Ecole Polytechnique* 1895–1897.) I am sure that the claims of importance are entirely justified. It set both a style and a standard for all French higher education in science at this time, and its influence soon began to be noticed abroad. Most of the figures named in the last section were *polytechniciens*; the older ones were teachers and examiners, the younger ones were students, and several of these students became teachers and examiners there later on.

The *Ecole* started off auspiciously, in that its initial enrolment of 396 students in 1794 included Biot, B. Brisson, L. Francoeur, Malus, and Poinsot. Later the annual entry was usually between 75 and 150, with a total enrolment at any time normally between 200 and 350. A principal reason for my choice of 1800 as the starting date is that the state of mathematical physics can be well understood for that time, because much material for use at, or preparation for, the *Ecole* was then being produced.

In this connection I may mention S. F. Lacroix, who was not named in Section 2. He was the chief textbook writer in mathematics of the time; he exercised great influence at the educational level, and he was a respected member of the *Académie* although he conducted virtually no research. Fortunately for his students he seems normally to have been content to "go along" with developments as they happened rather than adopt a dogmatic stance for or against any one of them. For example, his three-volume *Traité de calcul différentiel et de calcul intégral* shows reasonable awareness of the innovations that were brought into mathematics between the appearance of the first edition (1797–1800) and the second (1810–1819). His writings also reflect his interest in the history of mathematics.

There were three levels of writing in educational material: rather elementary works, for those preparing to enter schools such as the *Ecole Polytechnique* or in their early years there; an intermediate level, seemingly for students in their later years; and advanced comprehensive *Traités*, designed for the education of anyone who could understand them and assessible in modern terms as 'postgraduate' texts. Lacroix's *Traité* is an example of the advanced level; most of his other books are at the intermediate level. Lagrange wrote at all three levels: a course of *Lecons élémentaires* delivered when the *Ecole Normale* ran briefly in 1795 as a preparatory school for the *Ecole Polytechnique*; an intermediate (though high-powered) series of *Leçons sur le calcul des fonctions* for the *Ecole Polytechnique*; and *Traités* such as his *Méchanique analitique* and the *Théorie des fonctions analytiques*. A feature of textbooks at all levels that strikes the modern eye is the absence of *examples* for the learning reader to try for himself.

In addition to textbooks there were also journals with educational interests, but their function was rather haphazardly executed. The *Journal de l'Ecole Polytechnique* began with educational material as one of its chief types of publication (both of 'Leçons' of various sorts, and of research papers which could be used for teaching purposes), but by the mid-1800s it had largely converted to a journal of research papers. In order to fill the gap thus left, the *Correspondance sur l'Ecole Impériale Polytechnique* began to appear under the editorship of J. N. P. Hachette, and published three useful small volumes between 1804 and 1816. Again, when J.-D. Gergonne began to publish his *Annales de mathématiques pures et appliquées* from Nismes in 1810, he emphasised in his preface to the first volume the need for journals in 'the exact sciences' related to educational needs, and he managed to maintain a reasonable proportion of educationally oriented mathematical articles. (There were many other features to its contents—one is mentioned at the end of Section 8—and the journal deserves an extended study.) Some of the journals attached to the other *Ecoles* published articles of educational interest. But the bulk of the educational literature appeared as textbooks, and there was little discussion in print of educational method. Similarly, the interest in producing encyclopaedias of science was slight: I discuss this matter in more detail in Subsection 11.2 below.

The *Ecole Polytechnique* deserves its prominence; but other schools are of significance in the period 1800–1840, especially for engineering education, and they deserve much more attention than they have received so far. In particular, I would emphasise the *Ecole Nationale des Ponts et Chaussées*, which had some important teachers (and students) in its time, especially de Prony and Navier. The *Ecole Polytechnique* was originally conceived as a successor to that *Ecole*, which is why it was known at first as the '*Ecole Centrale des Travaux Publics*'. In the late 1830s the *Ecole Normale* began to play a notable role; I describe this development briefly in Subsection 10.4 below.

In these last three sections I have surveyed the structure of mathematical physics, and French activity in it, in rather cinematographic style. The next three sections are addressed to the complementary task of describing the scientific and historiographical framework of the subject.

5. MATHEMATICAL PROCEDURE

The procedure in mathematics normally followed the 18th-century style of working in four distinguishable and mutually interacting stages:

5.1. choose the physical problem, and mould it into a mathematicizable form;

5.2. construct the mathematical expression (usually a differential equation) appropriate to represent that physical problem;

5.3. solve the differential equation (or some analogous procedure in the other cases); and

5.4. interpret the solution back into the physical problem.

Although this method did not change much during the period, there were substantial changes in the uses to which it was put. The range of physical phenomena to mathematicize greatly increased, of course. With regard to forming differential equations, there was a marked return to older methods using differential models in favour of variational methods, probably because the older methods are a much more useful means of obtaining *hitherto unknown* equations (which by definition was the situation in the new fields); variational methods tend to show their mettle as a rapid and general technique for obtaining equations which are already known. Fourier analysis and complex variables substantially replaced power series and functional solutions to differential equations. The new methods also led to a strong preference for physical models which could be expressed by *linear* differential equations.

There was also a noticeable concern with numerical methods, using difference rather than differential equations, for example, or obtaining approximate rather than exact solutions to differential equations. (Some of this work is surveyed in Goldstine 1977.) A main purpose was to optimize the comparison between theoretically and experimentally obtained results, and thus avoid the situation often found in 18th-century mechanics where the mathematical and the experimental sides were rather disparate activities. de Prony, a particularly neglected figure, advocated the new approach in mechanics and engineering with especial fervor, but it did not become habitual among many of his colleagues. For example, in theoretical astronomy, where the same sort of problem arises, the French seemed to prefer the tradition maintained by Laplace of obtaining "exact" solutions (compare Subsection 11.1 below on this matter).

The last point involves the use of mathematics in physics, to which I devote the next section.

6. THE USE OF MATHEMATICS IN PHYSICS: A SPECTRUM OF MODES

In order to clarify the nature of mathematical physics, I present below a spectrum of seven modes of research that can be carried out in both mathematics and physics. The spectrum passes from abstract mathematics through mathematical physics (both celestial and terrestrial) to experimental physics and engineering. I think that it applies to communities other than France and to periods other than 1800–1840, though I shall not discuss other uses here (compare Duhem 1954). The modes are as follows:

6.1. Mathematics pursued with no apparent interpretations in physics in mind (and none sought), although it may play a role in interpretable mathematics in some way.

6.2. Mathematics pursued with no interpretation in physics involved, but with an awareness that such interpretations exist. Indeed, they may have motivated the mathematics in the first place.

6.3. Mathematics pursued with an interpretation in physics initially in mind, but where the guidance of the physical problem has been lost. The result is, for example, a mathematical expression for a physical effect which cannot possibly be currently detected, or be computed from that expression with an error less than the effect itself; or the use of mathematical relationships between physical constants which make no physical sense; or the use of a hopelessly oversimplified physical model in the first place; or methods of reasoning in which physical concepts play little or no role.

6.4. Mathematics pursued with an interpretation in physics in mind, and where the guidance of the physical problem provides adequate control on the mathematics produced. The rigor of the mathematics may be suspect.

6.5. Theoretical analysis of (certain) physical phenomena, where the mathematical component is small or even nonexistent. Such work is often foundational in character, or concerned with physical constants.

6.6. Experimental work in physics, designed to test the (mathematical or nonmathematical) theory—including the theory of any instruments involved—that has been worked out already.

6.7. Engineering constructs, machines, and instruments: that is, the design of equipment and structures to fit particular situations or types of situations.

These modes are categories of knowledge rather than people; they relate to mathematical physics, not to mathematical physicists. But they are not *purely* epistemological, for they relate to particular stages of development of mathematics and physics. For example, a piece of mathematics that is uninterpretable at a given time (mode 6.1) may become interpretable later (6.2); or an effect which cannot be detected at the time of its calculation (6.3) may become so when the experimental situation improves (6.4, 6.6). For this reason the boundaries between modes are fuzzy rather than firm, although the distinctions themselves remain important. On the mathematical side, I have deliberately avoided using the customary terms 'pure' and 'applied', for I find them useless on philosophical grounds: *any* piece of mathematics is applic*able*, and the applications may be within 'pure' mathematics as much as outside it (calculus applied to number theory as opposed to calculus applied to hydrodynamics, say).

7. SOME COMPANION PHILOSOPHICAL ASPECTS

A mathematical physicist will be working within one or some of the modes described in the last section. In addition, he may hold certain views about the aim of physical theory, the role of mathematics within it, and the relationship between the two fields. In this section I present six aspects relating to the physics (not necessarily *mathematical* physics) followed by six concerned with the mathematics, adding some explanatory comments where necessary:

7.1. A physical theory is grounded in secure foundations, which are obtained by induction from basic facts. The development of the theory thereafter is deductive articulation, to be confirmed by experiment.

7.2. A physical theory is a hypothesis, based on foundations which are not necessarily secure: they have been chosen for other reasons (suitable for mathematical treatment, for example, or to fit in with some metaphysical world view). The development of the theory thereafter leads by deductive means to predictions, which can be tested by experiment.

7.3. (Some or all) branches of (mathematical) physics are unifiable under a basic idea or principle. Analogies of structure between theories in different branches are to be exploited.

7.4. Some branches of mathematical physics are autonomous from certain others (although this does not forbid the use of analogies between sciences).

7.5. The aim of physical theory is to explain the phenomena that we experience. Theories are to be understood as referring to this experientiable level, and are best formulated in terms of experientiable concepts.

7.6. The aim of physical theory is to reveal the ontology and teleology from which the world is constructed. (In the process, perhaps, the workings and revelations of God will be revealed.) Theories are *not* necessarily to be formulated in terms of experientiable concepts.

7.7. The mathematics of mathematical physics is to be conceived of in algebraic/geometric/analytical terms. For example, to an algebraist the integral is an expression produced via the antidifferential operator; to a geometer, it is an area; to an analyst, it is the limit (if it exists) of a sequence of partition sums.

7.8. Using mathematics increases the deductive character of physical theory.

7.9. The use of mathematics has no bearing on the deductive character of physical theory.

7.10. The structure of a mathematical expression reflects the structure of the physical theory in which it is being used. (For example, if $\sum u_r(x)$ says something about elasticity, then so does each $u_r(x)$ and '$\sum$' describes something about elasticity as well as indicating a mathematical summation.) Similarly, a unity of mathematical theories implies a unity of the physical theories in which they are used.

7.11. The structure of a mathematical expression does not necessarily reflect the structure of the physical theory in which it is being used. Similarly, mathematical unity does not imply physical unity.

7.12. The use of numerical and/or statistical methods in mathematics is caused by our inability to produce viable methods to describe *exactly* the phenomena that occur in this complicated world.

As in Section 6, these aspects are not *purely* epistemological, in that a mathematical physicist may change his mind over time. There is also the possible difference between what he says in public print and what he commits to private manuscripts.

8. THE FRENCH COMMUNITY IN MATHEMATICAL PHYSICS, 1800–1840

After the general information of Sections 2–4, and the more philosophical considerations of Sections 5–7, I shall now try to characterize the French community. I shall use the spectrum of seven modes outlined in Section 6 as the basic historiographical tool, and comment further on the community in terms of the aspects pointed out in Section 7.

The *Ecole Polytechnique* was planned to be an engineering school, but by the mid 1810s, if not earlier, it had moved significantly toward the more theoretical aspects of mathematical physics. This change is reflected in the community of French mathematical physicists, for *in terms of its intellectual interests it splits into two groups.*

I call the first the 'mechanics/engineering/calculus' group. They work chiefly in modes 6.2, 6.4, 6.6 and 6.7—that is, interpretable mathematics, mathematics with physical guidance, experimental physics, and engineering. They tended to concentrate on the traditional areas of mathematical physics (mechanics and the calculus, including the foundational aspects of each), even when the newer ones were emerging. (Gillispie 1971 provides a fine overview of the interests of L. Carnot, who is a fairly typical member; see also Taton 1951 on Monge, and Gillmor 1971 on C. A. Coulomb, who was something of a father figure to the group.) Several also worked in geometry, especially descriptive geometry and elements of projective geometry, which was useful for their fields of interest. They were often active teachers (especially at the *Ecole Polytechnique* and the *Ecole Nationale des Ponts et Chaussées*) and textbook writers (usually at the elementary and intermediate levels indicated in Section 4). Most were involved in engineering, and to a notable extent also in experimental physics and observational astronomy. By and large their creative mathematical ability was limited, although some of them tried to develop the numerical mathematics which I described at the end of Section 5. In this last respect they were new-fangled, although otherwise they were somewhat old-fashioned.

The other group is characterized by 'mathematical physics/mathematical analysis'. They were associated especially with modes 6.2–6.4 (interpretable mathematics and mathematics in physical problems, including the "notional" mode of 6.3), and to some extent in 6.1 (uninterpreted mathematics), 6.6, and 6.7 (theoretical and experimental physics). They worked "across the board" in mathematical physics; indeed, for some of them their *main* contributions were to the newer areas of mathematics and/or physics. Their creative mathematical ability was usually much greater than that found in the other group, although they seem to have been less interested in numerical mathematics. Their experimental and practical skills were rather limited, and in some cases their interests in those activities appear to have been nonexistent. Although also professional teachers, they were less productive of textbooks than were members of the other group, and the books that they did write tended to be rather high powered. In slipping into the "notional" mode of mathematical physics (mode 6.3) they were rather old-fashioned, although otherwise they were new-fangled.

I list the principal members of the two groups in Tables 4 and 5. I have omitted figures such as Arago and S. Carnot, for their work was chiefly in experimental and theoretical physics (modes 6.6 and 6.5, respectively), which I am not considering in detail in this paper. I have also not indicated whether members of the *Académie* were elected to the mathematics or the mechanics sections, or the *subjects* in which they held chairs at the *Ecole Polytechnique*; for these factors are not as useful a guide to intellectual interests as might be hoped, since appointments were often made to the next "suitable" (though not necessarily appropriate) vacancy.

Of course, the division into two groups must not mask a number of features that members of both usually have in common: status as professional scientists, for example. But even in their common areas they tend to differ: both groups deal with mechanics and calculus, but the first are more traditional in their treatment than the second, or more concerned with practical uses (hydraulics as opposed to hydrodynamics, say). In astronomy the first group are much more likely to look through telescopes, for example. That group shows rather less involvement with the *Ecole Polytechnique* and less reward from the *Académie*: an indication, perhaps, of the ways in which those institutions were developing in this period. It is also the smaller group numerically.

TABLE 4

Principal Members of the Mechanics/Engineering/Calculus Group[a]

Name	*Ecole Polytechnique* career	Election to *Académie* (as full member)
Bobillier, E. (1798–1840)	1817–1818	
Bossut, C. S. J. (1730–1814)	E	1779
Carnot, L. N. M. (1753–1823)		1796–1797, 1800–1816
Clapeyron, B.-P.-E. (1797–1864)	1816–1818	1858
Combes, C.-P.-M. (1801–1872)	1818–1820	1847
Coriolis, G. G. de (1792–1843)	1808–1810, T, D	1836
Delambre, J.-B.-J. (1749–1822)		1795
Dupin, F. P. C. (1784–1873)	1801–1803	1818
Francoeur, L.-B. (1773–1849)	1794–1897, T, E	1842
Girard, P.-S. (1765–1836)		1815
Hachette, J. N. P. (1769–1834)	T, E	1831
Monge, G. (1746–1818)	T, Director	1780
Morin, A.-J. (1795–1880)	1813–1817	1847
Navier, C.-L.-M.-H. (1785–1836)	1802–1804, T	1824
Olivier, T. (1793–1853)	1810–1815, T	
Poncelet, J. V. (1788–1867)	1807–1810, Commandant	1834
de Prony, G. C. F. M. Riche (1755–1839)	T, E	1795
Puissant, L. (1769–1843)		1828

[a] The dates in the second column are the years of studentship at the *Ecole Polytechnique*; the letters 'T' and 'E' indicate if teaching posts and/or examinerships at some level were held there later, and 'D' that the post of director of studies was obtained.

In addition, the membership of some figures to their groups needs careful weighing of factors. For example, Navier's creative mathematical ability was greater than that of the other members of his group, and more closely related to the innovations; but his major work on machines, canals, bridges and railways decisively determine his group membership. Conversely, the bulk of Lamé's life's work was devoted to certain parts of mathematical physics and mathematical analysis; but he also contributed significantly to some topics in engineering, usually in collaboration with Clapeyron, who belongs to the other group. (The close friendship between Lamé and Clapeyron is an excellent example of my division into groups being by intellectual concern, not personal connections: see further in Subsection 11.6.) As for Delambre, his prime concern with astronomy is valid for membership to either group; but his special interests in weights and measures, numerical mathematics, and geodesy mark him as a fellow member of the group containing de Prony (with whom he worked closely around 1800) and Puissant.

A social factor underlying the tendency of the theoretical branches of mathematical physics to supplant the practical ones may have been the

TABLE 5

Principal Members of the Mathematical Physics/Mathematical Analysis Group[a]

Name	*Ecole Polytechnique* career	Election to *Académie* (as full member)
Ampère, A.-M. (1775–1836)	T	1814
Binet, J. P. M. (1786–1856)	1804–1806, T, E, D	1843
Biot, J. B. (1774–1862)	1794–1895, E	1803
Cauchy, A.-L. (1789–1857)	1805–1807, T	1816
Chasles, M, (1793–1880)	1812–1815, T	1851
Duhamel, J.-M.-C. (1779–1872)	1813, 1814–1816, T, E, D	1840
Fourier, J. B. J. (1768–1830)	T	1817
Fresnel, A. J. (1788–1827)	1804–1806	1823
Germain, S. (1776–1831)		
Lagrange, J. L. (1736–1813)	T	1787
Lamé, G. (1795–1870)	1814–1817, T	1843
Laplace, P.-S. (1749–1827)	E	1773
Legendre, A.-M. (1752–1833)	E	1785
Liouville, J. (1809–1882)	1825–1827, T	1839
Malus, E. L. (1775–1812)	1794–1896, T, E, D	1812
Petit, A.-T. (1791–1820)	1807–1809, T	
Poinsot, L. (1777–1859)	1794–1897, T, E	1813
Poisson, S.-D. (1781–1840)	1798–1800, T, E	1812
Pontécoulant, P. G. le Doulcet (1795–1874)	1811–1813	
Sturm, J. C. F. (1803–1855)	T	1836

[a] The notations of Table 4 apply here also.

centralization of interest in Paris itself. Men chiefly involved in mechanics and engineering may well have to work, for substantial periods of time, on *particular* artifacts (bridges, canals, or whatever) in specific locations, which may well be outside Paris. But those whose physical requirements are basically pen, paper, and a limited amount of apparatus can pursue it more or less anywhere; and Paris is the best place, not only for its size (it had a population of 700,000 in the 1810s) but also for its facilities. Thus a young student, when choosing his topics of possible research interest, will note the difference between being associated with publishing papers in Paris and building bridges in Besançon, and naturally aspire to the former. Again, the mathematical physics/mathematical analysis group will be in town almost all of the time, while their practical colleagues could be away for periods on assignments. Factors such as these are noticeable in the career of Coulomb, who worked hard to forsake provincial engineering for Parisian mathematical physics, and achieved his ambition in 1781 (see Gillmor 1971). Among Coulomb's successors, the desire for Paris residence is especially marked in de Prony, who began to compile an enormous set of trigonometric and logarithmic tables in the 1790s because he got himself nominated director of the cadastral survey in Paris in order to avoid being posted to chief engineer to a provincial department! Nevertheless, even he was out of town for periods later.

Cultural theories are to be nurtured like flowers, not hewn into shape like paving stones. If the division that I have proposed here for the French community is regarded in the proper horticultural spirit, with qualifications such as those just given carefully borne in mind, then I think that it is a valuable indication of both the knowledge and activity of the community.

With regard to the companion philosophical aspects described in Section 7, there appear to be only partial correlations between attitudes to them and group membership—*and it is an indication of the complications of the historiography of mathematical physics that this is so*. For example, concerning the style of conception of the mathematics (aspect 7.6), rather more of the first group think geometrically than do the second group (though both groups have algebraists: for example, de Prony, Ampère, Poisson). Correlations *between* aspects are only partial, too: for example, those who think out their mathematical physics geometrically (such as Fourier) seem to be less inclined to slip into notional mathematical physics (aspect 6.3) than the algebraists (Poisson) or analysts (Cauchy). Again, concerning physical reference (aspects 7.5 and 7.6), there seems to be a move away from ontology: the positivism of A. Comte (a student at the *Ecole Polytechnique* from 1814 to 1816) was prominent from the late 1820s on, was heavily influenced by Fourier and may have been philosophically in tune with many members of both groups. By and large they seem have been uncertain about the unity of the various branches (aspect 7.3); in particular, Fourier advocated the autonomy of heat diffusion from mechanics (aspect 7.4), even though he was well aware of the versatility of his new mathematical methods, and the unity that they suggested at the mathematical level (aspect 7.10).

In some cases I am not convinced that an historical figure *does* have an attitude to some of these aspects, or that he has thought about them. For example, I cannot gather from various expositions of dynamics and statics that I have read whether the author thinks that the foundations are secure (aspect 7.1), or that the whole theory is a hypothesis (aspect 7.2), or whether his presentation of the material is employed merely for its educational efficacy. But in this context there are signs of a remarkable division. For at least some of the early teachers at the *Ecole Polytechnique* (Ampère, Fourier, Lagrange, Laplace) seem to have accepted aspect 7.1, while at least some of the students of these early teachers accepted aspect 7.2 (Cauchy, Duhamel, Fresnel, Navier). (Poinsot and Poisson may be exceptions, and Biot vacillated on the issue. Some figures seem to have regarded the foundations of a theory as secure, but subsequent details as hypothetical.)

Caneva has noted a similar generational change in German physics at the same time (1974, 1978), and Cantor in British optics (1975).

A somewhat similar move occurred around that time in the transition from algebraic calculus to mathematical analysis. For in his introduction to his *Cours d'analyse* (1821), which spearheaded the change, Cauchy stressed the deductive character of his new approach as opposed to the inductive generalizations of the old algebraic style, and this change in mathematical proof was maintained as mathematical analysis developed. The parallels with the situation in physics are marked. The old views are that the foundations of a physical theory are securely grounded in inductions gleaned from primordial facts, and that mathematics is built up by inductions based on algebraic procedures; they are replaced by philosophies which regard a physical theory as deductions from hypothetical foundations, and the mathematics as also displaying a markedly deductive character (compare aspect 7.8).

If these changes in philosophical view about both the physics and the mathematics occurred, then some very interesting questions arise. Did the teachers teach their philosophical views? If so, did the students react against them? Maybe the change was effected by the very receipt of systematic education by the students, which the teachers themselves often had not had the opportunity to obtain themselves. Or perhaps the increasing use of mathematics made clear that nevertheless the world is too complicated a place for their mathematicized theories to explain "correctly," on "true" foundations; it does seem to have encouraged some to emphasize numerical methods (aspect 7.12). In the particular case of Cauchy, the change relates closely to his strong religious commitments. In his *Sept leçons sur la physique générale* (1833) he claimed that mankind should immediately accept Divine revelation. In other words, truth resides only with God, so that mankind can only frame insecure, hypothetical physical theories. (This view is somewhat in contradiction with his emphasis on rigor in mathematics, for the major purposes of rigor include exact statements of the basic concepts: I presume that his desire to distinguish his new analysis from its algebraic predecessors was overriding.)

I am not in a position as yet to describe the profile of the French community in detail with regard to these aspects. But their historical importance is clear, both within the community and relative to other communities. For influence—whether positive or negative—seems often to be guided by them as well as by technical results. For example, it is important to note the common ground of geometrical thinking in tracing a route of positive influence from Euler to Einstein via Monge, Fourier, Kelvin, and Maxwell.

I have spoken throughout this section of 'French' mathematical physics, but of course most of the members of both groups were based in Paris. So I point out a minor but very interesting group, which I would call 'outsider algebraists'. These people were mostly outside Paris, or at least outside the *Ecole Polytechnique* and the *Académie*. A few were not even professionally involved in science or research. They did not function as a group socially, although there were personal connections between some of them. Intellectually they were much more concerned with mathematics than with physics, and then with its traditional algebraic forms; power-series solution forms to differential equations, the calculus of derivations (a general theory of power-series expansions due to L.-F.-A. Arbogast: see Tanner 1891), differential operators, functional equations, complex numbers (especially their geometrical interpretation: see S. Bachelard 1966), and the foundations of mechanics. Several also worked on the theory of equations and/or geometry. They rarely became even *correspondants* of the *Académie*, or published in its journals; much of their work appeared in the journals of the provincial French *Académies* and the provincial *Annales de mathématiques pures et appliquées* edited by Gergonne (a member of the group). Few of them went to the *Ecole Polytechnique*, or taught there. Although most of them taught elsewhere, they produced few textbooks. They display the mathematical old-fashionedness of the mechanics/engineering/calculus group and usually the physical uninterest of several of the other group, but their algebraic

TABLE 6

Principal Members of the Outsider Algebra Group[a]

Name	*Ecole Polytechnique* career	Election to *Académie* (as *correspondant*)
Arbogast, L. F. A. (1759–1803)	T	1792
Argand, J. R. (1768–1822)		
Bérard, J. B. (1763–18??)		
Bret, J. J. (1781–1819)	1800–1802	
Brisson, B. (1777–1828)	1794–1895, T	
Français, F. J. (1768–1810)		
Français, J. F. (1775–1833)	1797–1800	
Galois, E. (1811–1832)		
Gergonne, J.-D. (1771–1859)		1830
Kramp, C. (1760–1826)		1817
Parseval (-Deschènes) M.-A. (1755–1836)		
Sarrus, P. F. (17??–1861)		
Servois, F. J. (1767–1847)		
Woisard, J. L. (1798–1828)	1814–1816, T	
Wroński, J. M. Hoëné(-) (1776–1853)		

[a] As in Tables 4 and 5, the dates of studentship in the *Ecole Polytechnique* are given, and 'T' indicates that a teaching position was held there later.

experiments are extremely interesting: in any other community they would have stood out quite well. Their principal members are listed in Table 6.

9. SOME REMARKS ON THE SOCIAL HISTORY OF SCIENCE

In the last section I did not discuss class background, religious views, or other factors of a nonscientific character. I would not deny that there may be some correlations between these factors and those discussed there, although I have not noticed any of significance so far. (Even the correlation between practical concerns and Saint-Simonism (the socialist movement of the day) is not clearcut, for it, and the succeeding Comtism, influenced also the more theoretically inclined.) But I find exaggerated the importance that is nowadays often assigned to social factors. Indeed, there seems to be a contradiction between the advocacy of the need to be 'history-minded' in the history of science, and the extolling of its social above its scientific content. *For the historical figures were predominately concerned with the scientific content*; hence the historians should be so concerned also in aspiring to their cause of history-mindedness. Furthermore, the relevant scientific content may be greater than it appears. When we read a mathematical physicist attempting to develop a new technique, say, we must bear in mind not only its own technical content but also other kinds of technical feature (for example, the problems which he hopes that the technique will help him solve, the other techniques which have failed so far, and so on). In order to approach closely the problem situation of the historical figure and thus achieve history-mindedness, the social historian must become familiar with all these technical features—even if it means that he has to learn up some science in the process (which is what other kinds of historian have to do all the time). But the rewards of bringing these technical features to bear on the social aspects would be enormous; for the more that they are involved, the more possibilities are laid open to establish connections with social factors. *The social historian of science makes his work harder, not easier, by demoting the technical features*. Not only does he distance himself from his historical figures, but he also renders the task of establishing connections with social factors unnecessarily difficult by discarding (most of) the scientific objects with which connection might be established.

I think that this matter requires special stress here, because 19th-century mathematical physics, and its forebear mechanics and calculus, have been sources for social thought from that time onwards. For example, much of the Marxist movement was (and is) enchanted with the language and conceptions of elementary Newtonian mechanics. Again, there are other analogies between variational principles describing systems of mass points acted on by forces and teleologically returning to equilibria states after disturbance, and social theories describing communities of individuals living by common rules and inevitably returning to stability after suffering social change. It is worth recalling from Section 5 that variational principles fell markedly in status during the great development in French mathematical physics in the Napoleonic period and later (they were revived in the 1830s by figures such as W. R. Hamilton and C. G. J. Jacobi), and that one reason for their fall in status was the difficulty in using them when the answers to the problems were not already known. Their teleological assumptions (that they apply over finite quantities of space–time, and that—in a rather deterministic way—the final situation inevitably follows) are also philosophically contentious. Now analogies with social developments in France from 1800 to 1840 leap to the eye here; but I am sure that possible connections, especially *from* society *to* science, require quite detailed study of the scientific contexts before being assigned to a suitable position in historical "explanation."

A further reservation concerns the claims made for social explanations based on biographical and/or prosopographical techniques. It may indeed be true that one or several historical figures have the social profile P *and* work in a science S in a certain way; but an *explanation* requires 'because' rather than 'and', and to claim that these figures do S *because* they have P is a logical move whose achievement requires more information than biography may be able to provide on its own. *The distinction between description and explanation is a distinction in logic*, between $p\&q$ and $p; p \rightarrow q; q$. (It is not a distinction *of* logic; logic *itself* does not supply a description or explanation.) It is *not* to be drawn on the question of whether or not social aspects are being used, as seems on occasion to be thought; there may be (purported) social descriptions or explanations of social and/or technical developments, technical descriptions or explanations or social and/or technical developments, and combinations of sociology and technicalities in description and explanation.

Technical explanations of social developments are underrated by social historians of science, who seem to assume that social explanations of technical developments are the *only* kind of relationships between sociology and technicalities that may apply. Most of this paper is descriptive, but where explanations are suggested, they are usually of social developments in terms of technical aspects. The next section contains several further examples.

10. ON THE DECLINE OF FRENCH MATHEMATICAL PHYSICS

It has often been noted that the French community declined in quality from about 1830 onward. I do not dispute that a decline took place, but I disagree about its nature. The grounds of my disagreement are largely related to consideration of the mathematical part of mathematical physics, which, for some reason, historians of physics do not regard as part of their historical investigations. To me the decline is marked only from about 1840 onward (this is one reason why I chose the period to end then), and even after that may not be as substantial as is thought.

In more detail, here are eight comments on the decline:

10.1. There is a certain "glamour" for us historians in observing the *very basic* breakthrough achieved by Fourier in heat diffusion and Fourier analysis, Cauchy in complex variables, Fresnel in optics, Ampère in electrodynamics, and so on. By its nature, the later exegesis does not have the same glamour; but it does not follow that it was intellectually any less demanding. Put another way, Ampère, Fourier, and the others were not only brilliant; they were also *lucky* in the state of problem formation that applied at the time when they were active. In a similar vein, the decline of the French community is cast against the great rise in quality of the work done in other countries from 1830 onward. Without doubt the difference in quality between the French and the other communities decreases greatly; but the French themselves do not decrease by the same amount.

10.2. Our conception of mathematical physics normally excludes engineering; but it is a notable feature of the French community that they regarded it as part of their concerns. The first group were heavily involved in in it; and the second, while not usually interested in it, are not necessarily to be construed as contemptuous of it (as were some of the British, and British society in general, later) or hostile to it (as were the Germans later). Thus our assessment of the decline should be tempered by an appropriate consideration of the engineering component of the French achievement. Such an approach would have improved Ben-David 1970.

Indeed, we may point to a fillip in engineering and its teaching in the late 1820s. During that decade several of the French engineering mathematicians at last made clear the concept of work (force multiplied by distance), after a multitude of confusions concerning energy, force, momentum and the concept of a hard body (see Scott 1970). The importance of the concept of work was also emphasized, especially its conversion to and from energy and its use in connection with the performance and efficiency of machines.

Such ideas were soon widely taught, including in some new, or revived, institutions: by Poncelet at his *Ecole d'Application de l'Artillerie et du Génie* at Metz, for example, and later by Coriolis at the *Ecole Centrale des Arts et Manufactures*. This latter *Ecole* opened in 1829 as a self-financing institution for the training of scientists for industry (see Comberousse 1879). Descriptive geometry was also well represented, for one of the *Ecole*'s four founders was T. Olivier, the most enthusiastic of Monge's followers. The next year, shortly after the 1830 revolution, several polytechnicians founded the *Association Polytechnique*, which provided education for the "labouring classes." Poncelet's *Ecole* in Metz was again an inspiration, and descriptive geometry as well as the theory of machines were in the curricula: "work for the workers," indeed.

Much of this activity carried with it a positivist philosophy, and connections with Saint-Simonism and Comtism. In fact, Comte was both prominent in the *Association Polytechnique*, and was also a collaborator in the new engineering *Journal du génie civil* (1828–1848), along with Dupin, Girard, and others who helped make the concept of work clear and important.

10.3. Two events of 1830 struck at the mathematical effort: the death of Fourier, and Cauchy's self-exile after the fall of the Bourbons. (Cauchy returned to France in the late 1830s, but his contributions to mathematical physics thereafter are far less important than those of the years up to 1830.) But on the whole the mathematical side of mathematical physics kept up well in the 1830s (in contrast with the physics, which *did* decline noticeably in this decade). In particular, followers of Fourier—Duhamel, Lamé, Liouville, and Sturm—did very important work in differential equations and some of their physical interpretations. Further, the mathematical physics/mathematical analysis group gained several valuable new recruits from the *Ecole Polytechnique* during the decade 1835–1845: J. L. Bertrand (a student at the *Ecole*, 1839–1841), P. O. Bonnet (1838–1840), A. Bravais (1829–1831), E. C. Catalan (1833–1835), C. E. Delauney (1834–1836), C. Hermite (1842–1844), U. J. J. Leverrier (1831–1833), and J. A. Serret (1838–1840). Significantly, the other group received only two notable new members, and they were both late starters: F. M. G. Pambour (1813–1815) and A. J. C. Barré de Saint-Venant (1813–1816). It is interesting to note that in the late 1830s the usefulness of the *Ecole Polytechnique* and the content of its teaching were debated [compare de Chambray (1801–1803) 1836 with Bugnot (1816–1818) 1837]. An event of 1840 to mention is the death of Poisson, although in view of his wizened mathematical style it is doubtful if this loss contributed much to the decline.

10.4. The trend away from engineering and practical applications may also be observed by following the development of the *Ecole Normale*. A lengthy study of the role of science in this institution would be most valuable; Zwerling 1976 outlines its development, much data is available from its 1884 and 1895, and a survey of its recruitment is provided by Karady 1979.

Unlike the *Ecole Polytechnique*, the *Ecole Normale* was largely devoted to the teaching of humanities subjects. It ran briefly at the start of the Revolutionary period to train school teachers, and was reformed in an 1808 decree, for the same main purpose, as part of the founding of the *Université de Paris*.

Promising students joined it, although they could take courses elsewhere; they stayed on for an extra third year, and taught the students in the lower years.

The *Ecole* was suppressed from 1822 to 1826 because of its dangerously liberal tendencies, and was reopened as an *école préparatoire*. But it was reconstituted fully after the 1830 Revolution, and encouraged to grow. The philosopher Victor Cousin was crucial for this phase (and also for the revival of the *Sorbonne* in the late 1820s). A graduate of the founding promotion of 1810, he was closely involved in its development throughout the 1830s and was Director from 1835 to 1840. He was sympathetic to scientific interests (Fourier was one of his close friends, for example), and during his time the proportion of science graduates increased from (around) 20% to 40%: the total number of graduates settled down then to about 30 annually.

Cousin was succeeded for a decade as Director by the journalist P.-F. Dubois (1812 promotion), who reformed the teaching syllabi in 1842 to bring them closer to those at the *Ecole Polytechnique*: even descriptive geometry, usually associated with practical matters, took a more prominent place. In the same year was started the *Nouvelles annales de mathématiques*, a journal pitched at a lower level of difficulty than those hitherto available and subtitled 'Journal des candidats aux Ecoles Polytechnique et Normale'. In 1845 the school was renamed the *Ecole Normale Supérieure*, to distinguish it from the *Ecoles normales secondaires*.

Prior to the late 1830s, the only *normaliens* to capture our attention are A. A. Cournot (1821 promotion) who worked on mathematical analysis and then went into economics; and E. Galois (1829), who did not complete his course for well-known reasons. But, as would be expected, from the late 1830s onwards we find significant research mathematical physicist *normaliens* appearing: J.-C. Bouquet (1839), C. A. Briot (1838), A. H. Desboves (1839), J.-F. Frenet (1840), J. Houël (1843), J. A. Lissajous (1841), H. Molins (1832), V. Puiseux (1837), and M.-E. Verdet (1842). Further, they were all of the mathematical physics/mathematical analysis type; indeed, some were only concerned with "pure" mathematics (modes 6.1, 6.2). In the usual incestuous way of educational institutions, several of them taught at the *Ecole* later, as did also some of the ex-*polytechniciens* named in Subsection 10.3 above. (Cauchy-style real and complex analysis became the dominant theme from the 1860s.) As the century went on, the *Ecole* continued to grow in scientific status, which took physical form in 1864 with the founding of its *Annale scientifiques*. The chief inspiration for this venture was then the director of scientific studies: L. Pasteur (1843).

10.5. The division of the community into a group whose members mostly are of limited creative mathematical ability and a group most of whose members have limited experimental skill or even interest, suggests that the community lacked mathematical physicists who were of high caliber in both the mathematical and the experimental sides; probably Ampère, Fresnel and Lamé were the most capable in this regard. It is men of this kind that we see emerging in other countries—W. F. Bessel, W. E. Weber, C. G. Neumann in Germany, for example, and Kelvin (W. Thomson) and W. J. M. Rankine in Scotland—thus making their lack in France all the more marked.

10.6. Fox 1973 provides evidence that state and individual support for experimental science and engineering declined in France during the first half of the nineteenth century (see Bradley 1979 on the Revolutionary period). This process is easily reconcilable with the changes in mathematical physics indicated earlier in this section; for it fits in with the fact that the experimental sides of the subject declined much more than did the mathematical sides (which required only pens, paper, and a few books and journals for their furtherance). There was a similar decline in the prestige of science in French society, which may have nudged the mathematical physicists still further away from pursuing those sides of their subject which could have the most substantial social impact.

The swing away from practical science can probably be seen as reaction against the policies of the Revolutionary period. The Revolution extolled practical science; the Revolution was a Bad Thing; therefore practical science is also a Bad Thing. Such sentiments surely underlay the elimination of the course in 'art militaire' and the reduction in the enrollment to the *Ecole Polytechnique* after the 1816 *restauration*; these sentiments increased in the early 1820s, when Louis XVIII and his government brought in strongly antiliberal policies (the closing of the *Ecole Normale*, mentioned in Subsection 10.4, is another manifestation). Poisson was particularly active at that time in defending the interests of science.

10.7. Of all the areas of physics listed in Table 1 in Section 2, the one containing by far the richest sources of problems is electricity and magnetism. It is notable that three of the non-French mathematical physicists that I named in Subsection 10.5 began to make their mark substantially in these areas (as did G. S. Ohm in Germany and G. Green in England). But for some reason which I have not yet divined, the French did not pay much attention to this area, preferring instead certain parts of mechanics, optics, and heat diffusion. Indeed, those French who did contribute to it (see Tricker 1965) were rather outside the community in one way or another: Ampère (always rather outside the community, for his general philosophical position among other reasons), Biot, and Poisson (both moving toward the periphery around 1820 because of their commitment to Laplacian physics and Lagrangian mathematics). By this low degree of interest, especially from the 1820s onwards (see Brown 1969), *the French handicapped themselves intellectually*.

Another area where French lack of interest is striking is thermodynamics. The small degree of attention paid to S. Carnot's 1824 *Réflexions* may in part be due to Lazare having been his father; but it still seems difficult to reconcile with the great interest of the mathematical physics/mathematical analysis group in heat diffusion (see G. Bachelard 1928: perhaps their interest was *too* great) and of the mechanics/engineering/calculus group in steam engines (where the interest may have been in *safety* more than in *efficiency*). The only notable French writing on Carnot's work in this period is by Clapeyron, who, as was mentioned in Section 8, had been an outsider even from the geographical point of view; thermodynamics became of major importance only in British and German hands (see Brush 1976 *passim*).

The French responded much more vigorously to new developments in mathematics. In complex variable theory, progress was rather slow, but largely because Cauchy himself was curiously intermittent in preparing his basic papers on the subject.

10.8. France is said to have been very competitive during the early 19th century. But the competition was almost entirely "internal": to enter the *Ecole Polytechnique*, for example, or to be a leading student there; or again, to become a member of the *Académie*, or to compete over scientific issues there. France is *low* on "external" competition, i.e., competition *between écoles* and between societies and groups. Even the journals seem to be fairly distinct in content and/or level. This low degree of external competition may have caused some aspects of the decline in France, or at least may have done little to prevent its occurrence. The new journal and the organisations mentioned in Subsection 10.2 suggest the emergence of, and need for, some external competition in the late 1820s, although even then the intrusion is *not* of *major* significance.

Several of the comments in this section have referred to developments outside France. It is time now to look at them more closely.

11. ON MATHEMATICAL PHYSICS OUTSIDE FRANCE, 1800–1840

To sum up at once, the states of affairs elsewhere are not as moribund as is usually thought. I shall comment briefly on six principal countries: Germany, England, Scotland, Ireland, Russia, and Italy. (There was a modicum of activity in other countries; surprisingly little in Holland.) As throughout this paper, I am thinking of communities *at work* in a country rather than groups of people who happened to have been born in it but spent their careers elsewhere.

11.1. Germany The rise of the Germans occurred around the late 1820s, as a result largely of specific governmental policies. Young Germans were encouraged to look at French mathematical physics and do likewise. In mathematics A. L. Crelle was particularly influential (see Eccarius 1977), although he was intellectually out of sympathy with mathematical developments. In terms of the division of the French community presented in Section 8, he was a mechanics/engineering/calculus type, being one of the few explicitly to affirm the form of the calculus espoused by Lagrange (whose major books he translated into German). But the principal younger contributors were mathematical physicists/mathematical analysts: Weber, Neumann, and Ohm mentioned in Subsections 10.5 and 10.7 above, and also J. P. G. Lejeune-Dirichlet, Jacobi, and A. F. Möbius. Some of the older German astronomers made a contribution of a rather different kind; for Bessel, K. F. Gauss, and H. W. F. Olbers used interpolative techniques to describe approximately the motions of the heavenly bodies (aspect 7.12) rather than continue in Laplace's tradition of seeking "exact" descriptions. (Despite his great genius, Gauss is an isolated figure, usually formidable to read: the extent of his influence is hard to judge, and in any case much of his main work in mathematical physics was published from 1840 onward.) Around 1800 German mathematics was largely centered on a 'combinatorial' school around C. F. Hindenburg, whose work bears some resemblance to that of the 'outsider algebra' French group listed in Table 6 at the end of Section 8. (In fact, the Strasbourgian C. Kramp belonged to both groups.) Hindenburg's *Archiv der reinen und angewandten Mathematik* (1795–1800) was the precursor for specialist mathematical journals in Germany of Crelle's *Journal für die reine und angewandte Mathematik*, which was of major importance from its inception in 1826. At that time German interest in educational questions in mathematics and physics began to emerge (with the seminar at Königsberg, for example), and from 1841 took the form of the *Archiv der Mathematik und Physik*, a lower-level mathematics journal similar to the contemporaneous *Nouvelles annales de mathématiques* mentioned in Subsection 10.4. As Biermann 1973 shows so well, Berlin did not begin to take its prominent position until around the 1850s.

The Germans seem to have been more inclined than the French to seek means of unifying the sciences (aspect 7.3), and may also show the generation change in moving from regarding physical theory as based on true foundations to regarding it as a hypothesis (aspects 7.1, 7.2). They seem often to be geometrical in their thought (aspect 7.7), which is one source of Fourier's substantial influence in Germany.

There are surviving *Nachlässe* for many major figures. Compact but invaluable guides to information on German *Nachlässe* are provided by Denecke 1969 and Mommsen 1971; similar sources on *Nachlässe* in other countries would be most desirable.

11.2. Britain The British have certain features in common across the constituent nations. They were much less professionalized than the French (in as much as there was a notable proportion of self-supporting scientists among their ranks), and a proportion of those in professional science held governmental rather than university appointments. They were also remarkably active in contributing to many large-scale general encyclopaedias, in which, as a result, the treatment of science and engineering took up a much greater proportion than is usual in modern encyclopaedias.

This effort was part of the general move in Britain to educate the populace on topics which were held to be useful to them. The subtitles of the encyclopaedias sometimes make this point explicitly (for example, *Nicholson's Encyclopaedia* provides 'an accurate and popular view of the present improved state of human knowledge'), as does the name of 'The Society for the Diffusion of Useful Knowledge', which was involved in *The Penny Cyclopaedia* (and also published an interesting *Quarterly Journal of Education* between 1831 and 1835). There was, of course, considerable commercial potential associated with this noble sentiment—indeed, it is a manifestation of the twin utilitarian ideals of spreading enlightenment and making money—and publishers were in competition with each other to secure good editors and contributors. Another manifestation of commercial enterprise is the "popular" journals such as *The Ladies' Diary*, *The Gentlemen's Diary*, *The Mechanics' Magazine*, and *The Penny Magazine* (from the same Society as *The Penny Cyclopaedia*), where mathematics and physics would appear in suitably watered-down forms.

However, some of the articles venture *far* beyond the level of profitable popular exposition: they are substantial treatises on branches of science (with parts of mathematical physics well represented). These articles are of considerable importance to the historian of science, not only for the information they contain but also for the insight that they provide us into *how* a branch of science was viewed at that time. Thus, bearing in mind their usefulness, I have listed in Table 7 the most important ones that were published in the period 1800–1840. The table excludes reprints of encyclopaedias that were published in the 18th century. Unfortunately it is not often possible to specify the date of preparation or publication of an article, and the dates on the title pages of volumes are usually nominal (especially when the *same* date appears on *every* volume). I have provided where possible the period over which an encyclopaedia was published, and have indicated my reliance on the nominal date by placing it in brackets.

TABLE 7

Principal British and French General Encyclopaedias with Relevance to Mathematical Physics, 1800–1840[a]

Short title	Volumes	Place and date of publication	Principal editor(s)	Comments
The British Encyclopaedia	6	London, (1809)	W. Nicholson	Often called '*Nicholson's encyclopaedia*'
The Cabinet Encyclopaedia	133	London, 1829–1846	D. Lardner	61 titles in all; some appeared in 2nd eds. 12 titles for the cabinet on 'natural philosophy'
The Cyclopaedia	45	London, 1802–1820	A. Rees	Edited by the revisor of E. Chambers's *Cyclopaedia*. Sometimes called '*The new cyclopaedia*'
The Edinburgh Encyclopaedia	18	Edinburgh, 1808–1830	D. Brewster	Contains several important articles for mathematical physics
The Encyclopaedia Britannica:				
4th edition	20	Edinburgh, 1801–1810	J. Millar	Scots responsible for organising articles on science (J. Playfair, J. Robison, D. Stewart, T. Thomson)
5th edition	20	Edinburgh, 1810–1817	J. Millar	First five volumes reprinted from 4th ed.
6th edition	20	Edinburgh, (1823)	C. MacLaren	Not a substantial revision of the 4th ed.
Supplement to the 4th, 5th and 6th editions	6	Edinburgh, 1815–1824	Mac. Napier	Several lengthy articles relating to mathematical physics
7th edition	21	Edinburgh, 1830–1842	Mac. Napier	A substantial revision of the 6th ed., using material from the supplement
Encyclopaedia Edinensis	6	Edinburgh, (1827)	J. Millar	A more popular work, though still scholarly
Encyclopaedia Mancuniensis	2	Manchester, 1813		Principally concerned with science and engineering

The Encyclopaedia Metropolitana	29	London, 1817–1845	T. Curtis; E. Smedley, H. J. Rose, H. J. Rose	Contains several important articles for mathematical physics. 2nd ed. 1848–1858 (40 volumes)
Encyclopaedia Perthensis	23	Perth, 1796–1806	A. Aitchison	2nd ed. 1807–1816 (24 volumes)
The Oxford Encyclopaedia	7	Oxford, 1828–1831	Various	
Pantologia	12	London, (1813)	J. M. Good, O. Gregory, N. Bosworth	
The Penny Encyclopaedia	29	London, 1833–1856	E. G. Long	Associated with The Society for the Diffusion of Useful Knowledge
The Popular Encyclopaedia	7	Glasgow, (1841)	A. Whitelaw	Includes a survey of science by T. Thomson
Dictionnaire de l'industrie	10	Paris, 1833–1841	Various, including Coriolis	Oriented towards manufacturing industries, commerce, and agriculture
Dictionnaire technologique	22	Paris, 1822–1835	Various, including Francoeur	Oriented towards technology, commerce, and industry
Encyclopédie methodique	196	Paris, 1782–1832	Various	An incomplete partial revision of the D'Alembert–Diderot *Encyclopédie*
Encyclopédie moderne	26	Paris, 1823–1832	Courtin	A general 'dictionaire abrégé'
Encyclopédie portative	51	Paris, 1825–1830	Various, including Dupin	A series of small pocket volumes on many subjects. Not much on mathematical physics

[a] The number of volumes includes all index, plates and supplementary volumes, and the dates of publication cover them also. For much information on the history of encyclopaedias see Collison 1964.

For completeness of this paper, I have appended the corresponding French encyclopaedias; but their efforts in this direction were *much* slighter. Relative to mathematical physics, they were mainly oriented to engineering. The main French contributions to this sort of literature were made in British encyclopaedias (for mathematical physics, Arago and Biot wrote articles on branches of the subject—as did Oersted, by the way). The apparent indifference of the French to this kind of writing (or perhaps the lack of entrepreneurs to make such ventures commercially viable) would be worth studying, especially as it came after the great d'Alembert–Diderot *Encyclopédie* of the 18th century (and its revisions in the *Encyclopédie méthodique*, which were largely completed for mathematical physics and engineering *before* 1800). The lack of interest accords well with their growing lack of concern with the social and practical aspects of the subject (Section 8, Subsection 10.6) and the limitations of educational literature (Section 4).

Many of the points made above about the British are exemplified also by the British Association for the Advancement of Science, founded in 1831. In particular, the accounts of 'recent progress' in various branches of science that grace its yearly *Reports* are often substantial and important accounts similar to the lengthy encyclopaedia articles mentioned above: indeed, they are often written by the same people (whose names appear in the next three subsections). Once again mathematical physics is well represented. The writing of these reports was suggested by W. Whewell, and one of his models was the yearly reports of the *secrétaires perpétuels* of the *Académie*.

Apart from these common factors, it is inadvisable to treat Britain as a single country in mathematical physics. (Crosland and Smith do so in their otherwise excellent 1978.) For the communities in England, Scotland, and Ireland differed in their intellectual concerns and philosophical views, although there were many contacts and movements between these countries. (Little or nothing seems to have been done in Wales.) I comment on these three countries in turn.

11.3. England The English are supposed to have been dragged out of the Newtonian fluxional doldrums into the 19th century of Leibnizian calculus by the Analytical Society in the 1810s. This view is incomplete and potentially misleading (see my 1979). First, the Analytical Society brought in the differential calculus *in its algebraic, Lagrangian form* outlined in Section 2: some of their work is close to that of the 'outsider algebra' group described at the end of Section 8. The Society took its name from the titles of Lagrange's main books: *Méchanique analitique* and *Théorie des fonctions analytiques*. But these books are totally *algebraic* in style (aspect 7.7), not 'analytical' as *we* now understand that term. This understanding, in fact, comes from Cauchy's approach introduced just after the Analytical Society's publication of its views; and Cauchy's calculus vindicates Newton's fluxional calculus, for its use of limits clarify what underlay—albeit in an incoherent form—many of Newton's procedures. (On other aspects of analysis, especially concerning infinite series, Cauchy's treatment is quite different from Newton's.) The Analytical Society converted England from Newton to Lagrange; the further transition from Lagrange to Cauchy came later, in the 1830s, with Whewell, A. De Morgan, and others.

With regard to figures, we find C. Babbage, P. Barlow, J. F. W. Herschel, J. Ivory, and Whewell active by the 1820s (or earlier), and G. B. Airy, J. Challis, De Morgan, D. Lardner, J. Lubbock, and Green emerging around 1830. Their work is more impressive than is normally allowed for, with particular interest shown in various branches of mechanics (including theoretical astronomy), optics, difference equations and differential operators, and some special functions. Several of them are algebraic in mathematical approach (aspect 7.7), an attitude which prevailed for several decades. After 1840 the emergence of G. G. Stokes, G. Boole, and D. F. Gregory (English in community terms) is especially to be noted. On science in England in general, see Cardwell 1972.

11.4. Ireland The Irish are distinguished by a desire to render geometrical conceptions in algebraic form. Hamilton is the best-known figure, but similar tendencies can be traced in J. Brinkley, H. L. Lloyd, R. Murphy, and (after 1840) especially in M. O'Brien. By and large, their strength lay more in mathematics than in physics, although J. MacCullagh was a significant mathematical physicist. Their common interest with the English in algebraic techniques is both reflected in the title, and exhibited in the contents of the *Cambridge and Dublin Mathematical Journal*, which in 1846 succeeded the *Cambridge Mathematical Journal* (founded in 1839). On the community, see McConnell 1945.

11.5. Scotland Scotland has D. Brewster, J. D. Forbes, P. Kelland, J. Leslie, J. Playfair, and J. Robison as principal figures. In tune with Scottish Common Sense philosophy, they are usually geometrical in style (aspect 7.7), which continues after 1840 with the emergence of Kelvin and J. C. Maxwell (on whom, as on the Germans, the influence of Fourier is marked). However, there were extensive disputes in Scotland about the role of mathematics in the physical sciences. The importance of Scotland in connection with encyclopaedias may be realized from Table 7.

11.6. Russia The chief representative of mathematical physics and mathematical analysis in this period was M. Ostrogradsky, who spent some time in Paris in the 1820s and continued the French style back in St. Petersburg. But of especial interest is the connection between France and Russia in engineering mathematics. Lamé and Clapeyron, students of noticeably liberal tendencies at the post-Restoration *Ecole Polytechnique* in the late 1810s, were allowed by their country to spend the 1820s in Russia on teaching, and on engineering projects. They became members of the St. Petersburg *Corps des voies des communications*, and contributed to its *Journal* when it started in 1826. They were also collaborators on the Paris *Journal du génie civil* on its commencement in 1828 (compare Subsection 10.2), and several of their St. Petersburg articles, and those of others, were reprinted there (see Bradley 1981).

11.7. Italy According to the usual history of mathematical physics, the French were very good until the British and Germans were very good instead. But there is an important and unknown qualification to make: the Italian community. Already in the late 18th century there was a noticeable amount of activity in mathematical physics in Italy, and it continued right through our period. Nearly 40 workers can be named who contributed to the subject up to 1840, often to both the mathematical and the physical sides. As far as I can tell, their work was not of fundamental significance, but it was frequently abreast of current developments and sometimes useful enough to be noticed by the French. (I began to discover the Italians by seeing them cited in the French literature.) Of particular interest is the fact that, in addition to publishing in the journals of their own *Accademie*, they also had an institutional base in the form of a *Società italiana*. It published a substantial volume of *Memorie di matematica e di fisica* every two years, much of which was concerned with mathematical physics (for a partial bibliography, see Marcolongo 1901).

Until the rise of the British and the Germans in the 1830s, Italy was the second community in mathematical physics. Yet we know next to nothing about the contributions of, for example, G. Plana, the most wide-ranging member of the community in his interests (and a student at the *Ecole Polytechnique* from 1800 to 1803: as a Piedmontese, he returned to Italy in 1814); or of other interesting figures such as G. Bellavitis, G. Bidone, A. Cagnoli, T. V. Caluso, G. Fontana, G. Frullani, G. F. G. Malfatti, O. F. Mossotti, G. Piola, P. Ruffini, and B. Tortolini. If I wanted to draw a conclusion in this paper, then this final paragraph would exemplify it to perfection.

12. BIBLIOGRAPHICAL STATEMENT

It is obviously impossible to append an extensive bibliography to this paper. I shall confine myself largely to the French community.

12.1. Primary Literature Various bibliographies are available. Poggendorff's *Biographisches-literarisches Handwörterbuch* cannot be matched as a concise source of reference to all kinds of literature. However, for papers by far the best source for the 19th century is the Royal Society's *Catalogue of Scientific Papers*, including its remarkable index volumes. For books (including editions of works and translations) national catalogues like that of the *Bibliothèque Nationale* provide the most detailed information. None of these sources is much use for finding reviews, which can often be of considerable historical value.

The most convenient source for the writings of many of the major figures is the editions of their works. I must warn, however, that the editions for the major classical French figures—Lagrange, Laplace, Cauchy, and Fourier—are poor in quality, suffering from important omissions, giving only vague indications of the original publication, sometimes anachronistically edited (especially with regard to mathematical notation), and with little or no guidance as to the significance of the works or attention to mistakes in passage work. (The edition for Fresnel is much better.) Several other major figures have no edition as yet: of particular value would be collections for Ampère (whose fascinating mathematical papers are totally ignored), Lamé, and Sturm, and selections for Biot, Legendre, Liouville, Monge, Poisson, Poncelet, and de Prony. By and large, editions for figures of other nationalities were prepared with greater care.

12.2. Manuscript Collections See especially the *Bibliothèque de l'Institut*, *Académie des Sciences*, *Ecole Polytechnique* and some other Paris schools, and the *Archives Nationales*. There are also *Nachlässe* (not always in one place, and not always substantial) for several of the major figures discussed in Section 8. My enquiries have revealed so far the following information:

Nachlass		Apparently no *Nachlass*
Ampère	Liouville	Biot
Carnot, L.	Malus	Cauchy
Delambre	Monge	Laplace
Fourier	Navier	Poisson
Germain	Poinsot	Sturm
Lagrange	Poncelet	
de Prony		

In addition, the D. E. Smith Manuscript Collection at Columbia University, New York, and the Sammlung Darmstaedter (Staatsbibliothek, Berlin (West): see Darmstaedter 1909) have extensive holding of letters and documents for many figures in French science of this period; and the American

Philosophical Society, Philadelphia possesses a similar, though smaller, collection.

12.3. Secondary Literature Among the secondary literature the appropriate parts of the *Encyklopädie der mathematischen Wissenschaften* (1898–1935, Leipzig) still offers the best guide to the technical details; indeed, in some cases the articles are not likely to be surpassed. However, in terms of historiography, they leave much to be desired; they take the form of catalogues of results often anachronistically presented, and the questions raised in Sections 6 and 7, for example, are usually not even asked. The same points apply to the analogous and very incomplete French *Encyclopédie des sciences mathématiques* (1904–1914, Paris), and to individual histories, such as those by I. Todhunter. The histories of E. Mach are more useful, although one must take note of the presence of Mach's own philosophical views.

Among more modern large-scale works, the *Dictionary of Scientific Biography* (1970–1980, New York) needs no recommendation from me as a constant source of information, of both a general and a detailed kind, about major and even relatively minor figures in the history of science. My only reservation with regard to the figures discussed in this paper is that for some of them the discussion of the mathematical aspects of their achievements is perfunctory or even nonexistent. Whitrow 1971– and May 1973 are useful, concise sources of references. Among older publications, the *Biographie universelle ancienne et moderne* contains some extremely valuable articles on French historical figures (especially those written by V. Parisot) both in the original edition (1811–1853) and then in the revised Paris/Leipzig edition which began to appear as the original one was finishing.

At the end of this paper I append a list of works, which have been of especial value to me in its preparation and which do not appear in the collections mentioned earlier in this section. Several of them possess extensive bibliographies.

ACKNOWLEDGMENTS

This paper is based on lectures given at the University of Western Australia, Monash University (Melbourne), the University of Sydney, the New York Academy of Sciences, Princeton University, the University of Michigan, the University of Maryland, and Harvard University. I am most indebted to these Australian and American audiences for their encouragement; several of the questions that were asked afterwards led me to important insights. Much of the research for the paper was done as a Visiting Member of the Institute for Advanced Study, Princeton, when my wife helped in the classification of the primary literature.

A draft of the paper was circulated to participants in the Parex Social History of Mathematics workshop held in Berlin early in July 1979, and was discussed at that meeting. For comments on the draft I am indebted to G. Cantor, J. R. Ravetz, and C. Smith. For advice on Table 7 and its surrounding text, I profited from the knowledge of J. Morrell and D. Orange. For a copy of the Aarhus University Library copy of the first sheet of Biot 1812 I am indebted to Kirsti Møller-Pedersen.

BIBLIOGRAPHY

Académie des Sciences. 1910–1922. *Procès-verbaux des séances de l'Académie [des Sciences] tenues depuis la fondation jusqu'au mois d'août, 1835*, 10 vols. (Paris).

———. 1979. *Index biographique de l'Académie des Sciences du 22 décembre 1666 au 1er octobre 1978* (Paris).

Arago, D. F. J. 1811. "Mémoire sur une modification qu'éprouvent les rayons lumineux . . . ," *Mem. cl. sci. math. phys. Inst. France*, *12*, 93–134.

Arnold, D. H. 1977. "The mécanique physique of Siméon Denis Poisson . . . ," (Toronto, Canada: Toronto University Ph.D.).

Bachelard, G. 1928. *Etude sur l'évolution d'une problème de physique. La propagation thermique dans les solides* (Paris; Repr. 1973, Paris).

Bachelard, S. 1966. *La représentation géométrique des quantités imaginaires au debut du XIX siècle* (Paris).

Ben-David, J. 1970. "The rise and decline of France as a scientific centre," *Minerva*, *8*, 160–179.

Biermann, K.-R. 1973. *Die Mathematik und ihre Dozenten an der Berliner Universität 1810-1920* (Berlin).

Biot, J. B. 1812. "Mémoire sur le nouveaux rapports qui existent entre la réflexion et la polarisation . . . ," *Mem. cl. sci. math. phys. Inst. France*, *12*, 135–280.

Bradley, M. 1979. "The financial basis of French scientific education and scientific institutions, 1790–1815," *Annals of Science*, *36*, 451–491.

———. 1981. "Franco-Russian scientific links: the careers of Lamé and Clapeyron, 1820–1830," *Annals of Science*, *38*, 291–312.

Brown, T. M. 1969. "The electric current in early nineteenth-century French physics," *Historical Studies in the Physical Sciences*, *1*, 61–103.

Brush, S. G. 1976. *The kind of motion we call heat* . . . (2 vols., Amsterdam).

Bugnot, Y. D. E. 1837. *De l'Ecole Polytechnique* (Paris).

Burkhardt, H. F. K. L. 1901–1908. "Entwicklungen nach oscillierenden Funktionen und Integration der Differentialgleichungen der mathematischen Physik," *Jahresbericht der Deutschen Mathematiker-Vereinigung*, *10* (2).

Caneva, K. L. 1974. "Conceptual and generational change in German physics: the case of electricity, 1800–1846," (Princeton, New Jersey: Princeton University Ph.D.).

———. 1978. "From Galvanism to electrodynamics: the transformation of German physics and its social context," *Historical Studies in the Physical Sciences*, *9*, 63–159.

Cantor, G. 1975. "The reception of the wave theory of light in Britain . . . ," *Historical Studies in the Physical Sciences*, *6*, 109–132.

Cardwell, D. S. L. 1972. *The organisation of science in England* (revised edition, London).

Chambray, G. de 1836. *De l'Ecole Polytechnique* (Paris).

Chappert, A. 1977. *Etienne Louis Malus (1775–1812) et la théorie corpusculaire de la lumière* . . . (Paris).

Cohen, H. 1883. *Das Princip der Infinitesimal-methode und seine Geschichte* . . . (Berlin).

Collison, R. W. L. 1964. *Encyclopaedias. Their history throughout the ages* (New York; 2nd ed., 1966).

Comberousse, C. de 1879. *Histoire de l'Ecole Centrale des Arts et Manufactures . . .* (Paris).

Crosland, M. P. 1967. *The Society of Arcueil* (London).

——— and Smith, C. 1978. "The transmission of physics from France to Britain: 1800–1840," *Historical Studies in the Physical Sciences*, *9*, 1–61.

Darmstaedter, L. 1909. *Verzeichnis der Autographensammlung von Professor Dr. Ludwig Darmstaedter* [to 1908], (Berlin).

Denecke, L. 1969. *Die Nachlässe in den Bibliotheken der Bundesrepublik Deutschland* (Boppard am Rhein).

Dickstein, S. 1899. "Zur Geschichte der Prinzipien der Infinitesimalmechnung. Die Kritiker der 'Théorie des Fonctions Analytiques' von Lagrange," *Abh. zur Gesch. der Math.*, *9*, 65–79.

Dugas, R. 1950. *Histoire de la mécanique* (Paris).

Duhem, P. 1954. *The aim and structure of physical theory* (trans. P. P. Wiener, Princeton).

Eccarius, W. 1977. "August Leopold Crelle als Förderer bedeutender Mathematiker," *Jahresbericht der Deutschen Mathematiker-Vereinigung*, *79*, 137–174.

Ecole Normale. 1884. *Ecole Normale* (*1810–1883*) (Paris).

———. 1895. *Le centenaire de l'Ecole Normale 1795–1895* (Paris).

Ecole Polytechnique. 1895–1897. *Ecole Polytechnique. Livre du centenaire 1794–1894* (3 vols., ed. G. Pinet, Paris).

Fourcy, A. 1828. *Histoire de l'Ecole Polytechnique* (Paris).

Fox, R. 1973. "Scientific enterprise and the patronage of research in France 1800–70," *Minerva*, *11*, 442–473; also in *The patronage of science in the nineteenth century* (ed. G. l'E. Turner, Leyden, 1976), 9–51.

———. 1974. "The rise and fall of Laplacian physics," *Historical Studies in the Physical Sciences*, *4*, 81–136.

Frankel, E. 1972. "J. B. Biot, the career of a physicist in early nineteenth century France," (Princeton, New Jersey: Princeton University Ph.D.).

———. 1976. "J. B. Biot and the mathematication of physics in early nineteenth-century France," *Historical Studies in the Physical Sciences*, *8*, 33–72.

Friedman, R. M. 1976. "The creation of a new science: Joseph Fourier's analytical theory of heat." *Historical Studies in the Physical Sciences*, *8*, 73–99.

Gillispie, C. C. (ed.) 1971. *Lazare Carnot, savant* (Princeton).

Gillmor, C. S. 1971. *Charles Augustin Coulomb: physics and engineering in eighteenth century France* (Princeton).

Goldstine, H. 1977. *A history of numerical analysis from the 16th through the 19th century* (Berlin).

Grattan-Guinness, I. 1970. *The development of the foundations of mathematical analysis from Euler to Riemann* (Cambridge, Mass.).

———. 1975. "On Joseph Fourier: the man, the mathematician and the physicist," *Annals of Science*, *32*, 503–514.

———. 1975a. "On the publication of the last volume of the works of Augustin Cauchy," *Janus*, *62*, 179–191.

———. 1979. "Babbage's mathematics in its time," *British Journal for the History of Science*, *12*, 82–88.

——— and Ravetz, J. R. 1972. *Joseph Fourier 1768–1830* (Cambridge, Mass.).

Hachette, J. N. P. 1828. "Note sur la création de l'Ecole Polytechnique," *Ephemerides universelles*, 11 pp.; *Journal gén. civil*, *2*, 251–263.

Karady, V. 1979. "Scientists and class structure: social recruitment of students at the Parisian Ecole Normale Superieure [*sic*] in the nineteenth century," *Hist. educ.*, *8*, 99–108.

Kline, M. 1972. *Mathematical thought from ancient to modern times* (New York).

Kuhn, T. 1976. "Mathematical vs. experimental traditions in the development of physical science," *Journ. interdisc. hist.*, *7*, 1–31; also in *The essential tension* . . . . (London and Chicago, 1977), 31–65.

Lindt, R. 1904. "Das Princip der virtuellen Geschwindigkeiten . . . ," *Abh. Gesch. Math.*, *18*, 145–195.

McConnell, A. J. 1945. "The Dublin mathematical school in the first half of the nineteenth century," *Proc. Roy. Irish Acad.*, *50*, 75–88.

Marcolongo, R. 1901. "Società Italiana della Scienze. (1782–1889)," *Rend. Circ. Mat. Palermo*, (1) *15*, pt.2, 1–29 [Bibliography only of mathematical papers].

Marielle, B. C. 1855. *Repertoire de l'Ecole Impériale Polytechnique ou renseignements sur les élèves qui ont fait partie de l'institution depuis l'époque de sa création en 1794 jusqu'en 1853 inclusivement* . . . (Paris).

May, K. O. 1973. *Bibliography and research manual of the history of mathematics* (Toronto).

Mommsen, W. A. 1971. *Die Nachlässe in den deutschen Archiven (mit Ergänzungen aus anderen Beständen)* (Boppard am Rhein).

Nielsen, N. 1929. *Géomètres français sous la Révolution* (Copenhagen).

Pinet, G. 1887. *Histoire de l'Ecole Polytechnique* (Paris).

Poinsot, L. 1975. *La théorie générale de l'équilibre et du mouvement des systèmes* (ed. and comm. P. Bailhache, Paris).

Rothenburg, S. 1910. "Geschichtliche Darstellung der Entwicklung der Theorie der singularen Lösungen totaler Differentialgleichungen von der ersten Ordnung mit zwei variabeln Grössen," *Abh. Gesch. Math.*, *20*, 315–404.

Schaffner, K. F. 1972. *Nineteenth-century aether theories* (New York and Oxford).

Scott, W. L. 1970. *The conflict between atomism and conservation theory 1644 to 1860* (London and New York).

Silliman, R. H. 1967. "Augustin Fresnel (1788–1827) and the establishment of the wave theory of light," (Princeton, New Jersey: Princeton University Ph.D.).

———. 1975. "Fresnel and the emergence of physics as a discipline," *Historical Studies in the Physical Sciences*, *4*, 137–162.

Sologub, V. S. 1975. *Razvitie teorii ellipticheskikh uravneni' v XVIII i XIX stoletnyakh* (Kiev).

Szabo, I. 1977. *Geschichte der mechanischen Prinzipien und ihren wichtigsten Anwendungen* (Basel).

Tanner, H. W. L. 1891. "On the history of Arbogast's rule," *Mess. maths.*, *20*, 83–101.

Taton, R. 1947. "Les mathématiques dans le Bulletin de Ferrusac," *Arch. d'hist. sci.*, *26*, 100–125.

———. 1951. *L'oeuvre scientifique de Monge* (Paris).

———. 1971. "Sur les relations scientifiques d'Augustin Cauchy et d'Evariste Galois," *Rev. d'hist. sci.*, *24*, 123–148.

______. 1978. "Repères pour une biographie intellectuelle d'Ampère," *Rev. d'hist. sci.*, *31*, 233–248.

Todhunter, I. 1876. *William Whewell* . . . (2 vols., London).

Tricker, R. A. R. 1965. *Early electrodynamics. The first law of circulation* (Oxford).

Whitrow, M. 1971– . *ISIS cumulative bibliography 1913–65* [3 volumes to date: 1 to appear] (London).

Whittaker, E. T. 1951. *History of the theories of aether and electricity. The classical theories* (London).

Wise, N. 1977. "The flow analogy to electricity and magnetism: Kelvin and Maxwell," (Princeton, New Jersey: Princeton University Ph.D.).
Yushkevich, A.-A. P. 1976. "The concept of function up to the middle of the 19th century," *Archive for History of Exact Sciences*, *16*, 37–85.
Zwerling, C. S. 1976. "The emergence of the Ecole Normale Supérieure as a center of scientific education in nineteenth-century France," (Cambridge, Mass.: Harvard University Ph.D.).

Middlesex Polytechnic at Enfield
Enfield, Middlesex, England

Symbolik und Formalismus im mathematischen Denken des 19. und beginnenden 20. Jahrhunderts*[1]

EBERHARD KNOBLOCH

In his famous *Formulario mathematico* the Italian mathematician Guiseppe Peano (1858–1932) wrote "Every advance in mathematics corresponds to the introduction of ideographic signs or symbols." Alfred North Whitehead and Bertrand Russell, using the symbolism invented by Peano in their own *Principia Mathematica*, extended the applicability of symbolic logic and argued that deductive thought could be extended to areas beyond mathematics. Similarly, David Hilbert and Wihelm Ackermann, in the introduction to their *Foundations of Theoretical Logic*, appealed to Leibniz to indicate what could be achieved in mathematics through a formalized logic. Clearly, symbolism leads to the advance of techniques and methods of mathematical proofs. Hilbert's proof theory actually can be traced back to Leibniz. This paper analyzes the roots of Logicism and Formalism, especially in the abstract algebra and vector analysis of the 19th and early 20th centuries. The paper is divided into six parts: (1) heuristics and principles of order, (2) symbolic algebra and formal mathematics, (3) Hamilton and the theory of quaternions, (4) operational calculus, (5) algorithmization, and (6) universalization.

EINLEITUNG

In seinem berühmten *Formulario mathematico*—der Band ist in Latino sine flexione verfaßt—sagt Guiseppe Peano (1858–1932) zu Beginn des 20.

* Herrn Prof. Dr. Kurt-R. Biermann zum 60. Geburtstag gewidment.

ISBN 0-12-204050-3

Jahrhunderts explizit, wenn auch zweifellos überspitzt:[2]

> Omni progressu de Mathematica responde ad introductione de signis ideographico vel symbolos.
>
> Jeder mathematische Fortschritt entspricht der Einführung von ideographischen Zeichen oder Symbolen.

Er betont, daß die Reduktion einer Theorie auf Symbole die Analyse jeder Idee, das Aussprechen jeder Hypothese, kurz eine Präzisierung erfordere. Alfred North Whitehead (1861–1947) und Bertrand Russell (1872–1970), die die Peanosche Symbolik ihren *Principia mathematica* zugrunde legten und erweiterten,[3] rechtfertigen die symbolische Form ihres Werkes demgemäß mit der Gedächtnisentlastung, der erzielten Exaktheit und Bündigkeit, *terseness*, der Darstellung, zumal der menschliche Geist ohne die Hilfe von Symbolen mit der Vorstellungskraft allein bestimmte Gedankengänge nicht durchführen kann. Ja ein Ziel ihres Werkes ist es zu zeigen, daß mittels ihres Symbolismus das deduktive Denken auf Gedankenbereiche ausgedehnt werden kann, die gewöhnlich nicht dem mathematischen Denken unterworfen sind.

Wenn David Hilbert (1862–1943) und Wilhelm Ackermann in der Einleitung ihres Werkes über die *Grundzüge der theoretischen Logik* unter Berufung auf Leibniz sagen,[4] was durch die Formelsprache in der Mathematik erreicht werde, solle auch in der Logik durch diese erzielt werden, eine exakte wissenschaftliche Behandlung ihres Gegenstandes, so wiederholen sie damit nur Äußerungen der drei genannten Autoren.

Die Symbolik führt also zur Fortbildung der Technik und Methodik des mathematischen Beweises. Der Grundgedanke von Hilberts Beweistheorie geht auf Leibniz zurück. Er verlangt, die Formalisierung des mathematischen Schlußprozesses soweit durchzubilden, daß jeder Widerspruch im Denken sich unmittelbar im Auftreten bestimmter Zeichenkonstellationen verrät. In seiner *Neubegründung der Mathematik* aus dem Jahre 1922 nennt er die seiner Ansicht nach zur Begründung der reinen Mathematik wie überhaupt zu allem wissenschaftlichen Denken, Verstehen und Mitteilen erforderliche philosophische Einstellung:[5] "Am Anfang—so heißt es hier—ist das Zeichen."

Die eigentliche Mathematik, zu der eine Metamathematik mit inhaltlichem Schließen zum Nachweis der Widerspruchsfreiheit der Axiome hinzutritt, wird zu einem Bestand an beweisbaren *Formeln*.[6] Dies sind die pointierten Standpunkte des Logizismus bzw. Formalismus, wie er für die abstrakte Algebra kennzeichnend ist, deren Wurzeln weit ins 19. Jahrhundert zurückreichen. Im Folgenden sollen einige wichtige Ideen und Theorien aufgezeigt werden, die einen fördernden oder hemmenden Einfluß auf diese Entwicklung ausgeübt haben, und zwar: 1. Heuristik und Ordnungsprinzipien, 2. Symbolische Algebra und formale Mathematik, 3. Hamilton und die Quaternionentheorie, 4. Operationenkalkül, 5. Algorithmisierung, 6. Universalisierung.

1. HEURISTIK UND ORDNUNGSPRINIZIPIEN

Es war offensichtlich ein Mitglied des berühmten englischen Dreigestirns James Joseph Sylvester (1814–1897), Arthur Cayley (1821–1895) und George Salmon (1819–1904), nämlich Sylvester, das in seinen 1884 veröffentlichten drei *Lectures on the principles of universal algebra*[7] die Ähnlichkeit zwischen zwei epochemachenden Entdeckungen der Mathematik im 19. Jahrhundert herausgestellt hat, ohne daß moderne Autoren wie Michael Crowe oder John M. Dubbey,[8] die ein gleiches tun, darauf hinweisen: die Entdeckung nichteuklidischer Geometrien durch den Russen Nikolai Iwanowitsch Lobatschewski ((1793–1856) aus Kasan, den Ungarn Janos Bolyai ((1802–1860) in Budapest und Carl Friedrich Gauss (1777–1855) in Göttingen auf der einen Seite, die Entdeckung nichtkommutativer Zahlensysteme, der Quaternionen William Rowan Hamiltons (1805–1865)[9] bzw. der Ausdehnungslehre Hermann Graßmanns (1809–1877) auf der anderen Seite.[10]

Beide Entdeckungen brachen Schranken, die seit Jahrhunderten dem mathematischen Denken gesetzt waren. Im ersten Fall emanzipierte sich die Geometrie von Euklids *empirischem Axiom*, wie Sylvester sagte, im zweiten Fall die Algebra von der Arithmetik, deren Gesetzen das neue Zahlensystem nicht gehorchte, vom *yoke of the commutative principle of multiplication.*

In der Tat weist die Geschichte dieser beiden Entdeckungen überraschende Parallelen auf. Lobatschewski wie Graßmann arbeiteten isoliert außerhalb mathematischer Schulen,[11] ohne lange Zeit nennenswerte Anerkennung zu finden. Die Mathematiker nahmen zunächst ihre Theorien nicht zur Kenntnis oder verkannten ihre Bedeutung, wie etwa Ernst Eduard Kummer (1810–1893), der in seinem Gutachten vom 12.7.1847 über Graßmanns Schriften allerdings leider richtig prophezeite, daß die Ausdehnungslehre auch ferner von den Mathematikern ignoriert werden würde, da die Mühe, sich in dieselbe einzuarbeiten, zu groß in Beziehung auf den wirklichen Gewinn an Erkenntnis erscheine, welchen man aus derselben schöpfen zu können vermute.[12]

Die Philosophen verhielten sich entsprechend. Die in Deutschland vorherrschende Kantische Philosophie lehnte nichteuklische Geometrien ab. Aber auch Graßmanns andersartige Verallgemeinerung der klassischen

Geometrie durch den Aufbau einer Geometrie in einem Raum von n Dimensionen traf auf Unverständnis, sofern überhaupt eine Reaktion zu verzeichnen war.

Der Jenenser Philosoph Ernst Friedrich Apelt (1812–1859), der bedeutendste Schüler von Jakob Friedrich Fries (1773–1843), bemängelte August Ferdinand Möbius (1790–1868) gegenüber, daß aus Graßmanns Buch der wesentliche Charakter der mathematischen Erkenntnis, die Anschaulichkeit, ganz verbannt zu sein scheint:[13] "So eine *abstrakte* Ausdehnungslehre, wie er sucht, könnte sich nur aus Begriffen entwickeln lassen. Aber die Quelle der mathematischen Erkenntnis liegt nicht in Begriffen, sondern in der Anschauung."

Die Ausschließlichkeit dieser Aussage wird durch die Art widerlegt, in der Symbolik und Formalismus bei der Ausprägung gerade der Vektoranalysis in engem geistigen Zusammenhang mit der Entstehung der formalen Algebra, der Matrizentheorie und der mathematischen Logik eine entscheidende Rolle gespielt haben. Angesichts seines lebhaften Interesses für den Formalismus geometrischer Darstellung unt Rechnungen nimmt es nicht wunder, daß Felix Klein (1849–1925) 1872 in den *Vergleichenden Betrachtungen über neuere geometrische Forschungen*, besser bekannt unter dem Namen *Erlanger Programm*, zu dieser Frage Stellung genommen und den heuristischen Wert eines Formalismus betont hat:[14]

> Der Formalismus soll sich doch mit der Begriffsbildung decken, mag man nun den Formalismus nur als präzisen und durchsichtigen Ausdruck der Begriffsbildung verwerten, oder will man ihn benutzen, um an seiner Hand in noch unerforschte Gebiete einzudringen. Für die Gruppe der Drehungen des dreidimensionalen Raumes um einen festen Punkt stellten die Quaternionen einen solchen Formalismus dar.

Unter Bezugnahme auf den damaligen Gegensatz zwischen der synthetischen und analytischen Richtung in der neueren Geometrie stellt er fest, daß die Formeln der analytischen Geometrie als präziser und durchsichtiger Ausdruck der geometrischen Beziehungen aufgefaßt werden kann. Wörtlich heißt es:[15] "Man hat auf der anderen Seite den Vorteil nicht zu unterschätzen, den ein gut angelegter Formalismus der Weiterforschung dadurch leistet, daß er gewissermaßen dem Gedanken vorauseilt."

Zwar sei an der Forderung festzuhalten, daß man einen mathematischen Gegenstand nicht als erledigt betrachten soll, solange er nicht begrifflich evident geworden ist, es sei aber das Vordringen an Hand des Formalismus ein erster und schon sehr wichtiger Schritt. Aus der Sicht des ersten Verfechters der Graßmannschen Theorie, Victor Schlegel (1843–1905), mußte dies ein wesentliches Argument für Graßmanns Ausdehnungslehre und ihrer Symbolik sein.[16] Graßmann hatte selbst in der Vorrede zur ersten Auflage seines Werkes gesagt, jeder Fortschritt von einer Formel zur anderen sei ihm unmittelbar nur als der symbolische Ausdruck einer parallel gehenden, begrifflichen Beweisführung erschienen, wie ihm denn seine Ausdehnungslehre die Vorteile der synthetischen und analytischen Geometrie zu vereinigen schien.

Dementsprechend charakterisiert Schlegel 1896 in seinem historisch-kritischen Exkurs über die Graßmannsche Ausdehnungslehre diese neue Disziplin als eine zur systematischen Auffindung geometrischer Wahrheiten geeignete direkte Methode.[17] Denn jeder Schritt der Rechnung bedeute eine sofort verständliche geometrische Transformation, und der unerschöpfliche Reichtum der möglichen Umformungen bilde eine ebenso unerschöpfliche Quelle neuer Sätze, ein Kriterium, das Kummer—zu Unrecht—in seinem Gutachten vermißt hatte, und das Gauss zunächst den *Baryzentrischen Kalkül* von Möbius wieder aus der Hand legen ließ, wie dieser in seinem Brief an Heinrich Christian Schumacher vom 15.5.1843 bekennt.[18]

Die gedanklichen Übereinstimmungen zwischen Kleins Erlanger Programm, das dieser selbst zu den Schriften rechnete, die zu Neuem anregen wollen, indem sie Vorhandenes ordnen, und Graßmann müssen um so größer sein, als Klein selbst bekannte, er sei durch Graßmann, der affiner Geometer, nicht Projektiviker gewesen sei, neben dem Studium von Möbius und Hamilton und den Pariser Eindrücken, auf die Konzeption seines späteren Programms geführt worden.[19]

In der Tat war Klein durch Moritz Stern (1807–1894) auf Graßmann aufmerksam gemacht worden, wenn er auch von dessen Ausdehnungslehre, die er später gelegentlich überhaupt mit der Geometrie gleichsetzte,[20] zuerst durch das Buch des Riemannschülers Hermann Hankel (1839–1873) *Theorie der complexen Zahlensysteme insbesondere der gemeinen imaginären Zahlen und der Hamiltonschen Quaternionen nebst ihrer geometrischen Darstellung* aus dem Jahre 1867 erfahren hatte.[21] Im Programm wird Graßmann oft herangezogen, insbesondere im achten Paragraphen, in dem Klein Methoden aufzählt, denen eine Gruppe von Punkttransformationen zugrunde liegt, und feststellt,[22] daß die "projektivische Geometrie aus der Geometrie aller Punkttransformationen ebenso durch Adjunktion der Mannigfaltigkeit der Ebenen zu gewinnen ist, wie die elementare Geometrie aus der projektivischen durch Adjunktion des unendlich fernen Kugelkreises."

Knüpfe man mit Graßmann die Erzeugung der algebraischen Gebilde an ihre lineale Konstruktion, so werde zum Beispiel folgendes besonders deutlich: wird eine Fläche als algebraisch von einer bestimmten Ordnung bezeichnet, so ist diese Bezeichnung vom Standpunkt aller Punkttransformationen als eine invariante Beziehung zur Mannigfaltigkeit der Ebenen aufzufassen.

Geistesverwandt ist schließlich eine Graßmann wie Klein gemeinsame Zielsetzung, die dem Bourbakismus des 20. Jahrhunderts zugrunde liegt: höhere Ordnungsprinzipien in eine schier unübersehbare Menge von

Einzelheiten einzuführen. Graßmanns Elemente können von vornherein in allgemeiner Bedeutung aufgefaßt und nachträglich zusammen mit den erzielten Resultaten in verschiedenem Sinn gedeutet werden. Dieselben Formeln können auf diese Weise Sätze der Geometrie, Mechanik oder Linien- und Kugelgeometrie ausdrücken, die Operationen und Methoden auf Räume beliebiger Dimensionszahl ausgedehnt werden.[23] So klassifizierte er die n-gliedrigen Pfaffschen Differentialausdrücke nach ihrem Verhalten bei beliebigen Punkttransformationen.[24] Er verfolgte damit einen Grundgedanken einer Forschungsrichtung, der vor allem englische Mathematiker nachgegangen werden, wovon noch zu sprechen ist.

Hankel, der zu den ersten bedeutenden Forschern gehörte, die Graßmanns Verdienste angemessen würdigten und während der Abfassung seines Werkes mit diesem darüber korrespondierte, sprach von einem Bedürfnis in der Mathematik, das in England stets rege gewesen sei und sich in seinen Tagen—1867—immer allgemeiner geltend mache, ein Bedürfnis, das er durch seine Darstellung zu wecken und nach Möglichkeit zu befriedigen versuche. Die Mathematik folge damit nur einem Bestreben der Naturwissenschaft in neuester Zeit, aus der Welt der empirischen Details zu den großen Prinzipien aufzusteigen, die alles Einzelne beherrschen und unter höheren Gesichtspunkten zu einem Ganzen zu vereinigen.[25]

2. SYMBOLISCHE ALGEBRA UND FORMALE MATHEMATIK

Hankel nahm keineswegs für sich in Anspruch, als erster den Gedanken geäußert zu haben,[26] die allgemeine Arithmetik und Algebra unter dem höheren Gesichtspunkt einer formalen Mathematik anzusehen, zu der das Prinzip ihrer formalen Gesetze führe. An der Geschichte der Mathematik ohnehin interessiert, wie auch aus einem Brief an einen nicht genannten Adressaten der *Sammlung Darmstaedter* vom 27.8.1870 hervorgeht,[27] verweist er ausdrücklich auf Vorgänger, insbesondere die von ihm sogenannte Cambridger Schule, zu der Charles Babbage (1792–1871), George Peacock (1791–1858)—Mathematiker, die Klein zu Unrecht in seinen *Vorlesungen zur Geschichte der Mathematik im 19. Jahrhundert* übergangen hat—Augustus de Morgan (1806–1871), der mit Hamilton befreundet war und mit diesem über Graßmanns Ausdehnungslehre korrespondierte, sowie Duncan Farquharson Gregory (1813–1844) gehörten, einem Enkel von David Gregory (1661–1710), dem Neffen des berühmten schottischen Miterfinders der Differential- und Integralrechnung, James Gregory (1638–1675).[28]

Dazuzuzählen ist auch Matthew O'Brien (? –1855), wie Hamilton und Möbius eigentlich Astronom, der in offensichtlichem Anschluß an Hamilton um 1850 selbst ein vektorielles System entwickelte, ohne Nachfolger zu finden oder von den Begründern der modernen Vektoranalysis, Josiah Willard Gibbs (1839–1903) bzw. dem Telegrapheningenieur Oliver Heaviside (1850–1925), je erwähnt zu werden.[29] In dieser Tradition stehen—teilweise unter namentlicher Bezugnahme auf Hamilton, Graßmann, Augustin Louis Cauchy (1789–1857) und Hankel—Cayley, Benjamin Peirce (1809–1880) und Sylvester bzw. die englischen Logiker wie George Boole (1815–1864), und zwar nicht nur *The laws of thought* von 1854.[30]

O'Brien veröffentlichte 1847 *Contributions Towards a System of Symbolical Geometry and Mechanics*,[31] in denen er unter Hinweis auf eine *bedeutende mathematische Kapazität*—gemeint ist wohl Peacock, mit dem er unter den englischen Algebraikern am meisten zusammenarbeitete[32]—erklärt, die Unterscheidung zwischen arithmetischer und symbolischer Algebra könne auf die meisten Wissenschaften übertragen werden, die der Hilfe der Algebra bedürfen. So könne man symbolische und arithmetische Geometrie, symbolische und arithmetische Mechanik unterscheiden, eine Unterscheidung, die sich auf den Grad der Allgemeinheit von deren Symbolisierung bezieht: in der arithmetischen Wissenschaft haben die Symbole eine rein numerische Bedeutung, in der symbolischen repräsentieren sie nicht nur abstrakte Größen, sondern alle Umstände, die die Quantität betreffen.

Tatsächlich hatte Peacock, von späteren ähnlichen Veröffentlichungen abgesehen, 1830 *A Treatise of Algebra* herausgegeben,[33] in der wohl zum ersten Mal festgestellt wird, daß es nichtarithmetische Algebren gibt, daß die Algebra eine Generalisierung, keine Symbolisierung der Arithmetik ist, keine *universelle Arithmetik*, wie Newton (1642–1727) gesagt hatte,[34] und in Abhängigkeit von ihm Lagrange (1736–1813). Sie besteht aus der Manipulation von Symbolen in einer von jeder speziellen Interpretation unabhängigen Weise.

Alle drei Gedanken, einschließlich Peacocks Prinzips der Permanenz äquivalenter Formen hatte bereits Babbage in einer unveröffentlichten Schrift *Philosophy of analysis* aus dem Jahre 1821 geäußert, die Peacock kannte: Jede Form, die algebraisch äquivalent einer anderen ist, die in allgemeinen Symbolen ausgedrückt ist, muß äquivalent bleiben, was immer diese Symbole bezeichnen.

Babbage hatte auch schon den rein operationalen Charakter der Algebra betont und eine Arbeit zum *calculus of functions*[35] verfaßt, die abstrakte Funktionen durch Funktionalgleichungen definiert und die überragende Bedeutung der mathematischen Symbolik für den Erfolg mathematischen Denkens herausstrich. Denn eine angemessene symbolische Sprache erleichtere die Verallgemeinerung künftigen Wissens. Babbage war die treibende Kraft in der *Analytical Society*, zu der Peacock und John Herschel (1792–1871) gehörten, deren Ziel es war, die Leibnizsche Differentialschreibweise statt der Newtonschen Punktschreibweise in England einzuführen.

An der mathematischen Notation brennend interessiert, behandelte Babbage dieses Problem unter Anknüpfung an die französischen Philosophen Etienne Bonnot de Condillac (1715–1780) und Joseph Marie Degérando (1772–1842) in einer längeren Abhandlung *On the Influence of Signs in Mathematical Reasoning*,[36] die die Leistungsfähigkeit und die Vorteile der algebraischen Sprache hervorhebt. Ein wenig pointiert gesagt, gab so der Prioritätsstreit zwischen Newton und Gottfried Wilhelm Leibniz (1646–1716) noch im 19. Jahrhundert mittelbar mit den Anlaß zu einem neuen Zugang zur Algebra, der zur modernen, abstrakten, strukturell orientierten Algebra des 20. Jahrhunderts führen sollte.

Die russischen Autoren Liusternik und Petrova haben jüngst auf die französische Schule des symbolischen Kalküls gegen Ende des 18. Jahrhunderts hingewiesen,[37] den Lagrange und Pierre Simon de Laplace (1749–1827) vorbereitet haben. Lagranges symbolische Formeln werden in den Studien von Francois Arbogast (1759–1803) und Jacques Frédéric Francais (1775–1833) zu einer Art von Operatorgleichungen. Barnabé Brisson (1777–1828) und Francois-Joseph Servois (1767–1847) wandten symbolische Methoden auf partielle Differentialgleichungen an, Cauchy untersuchte Darstellungsprobleme von Operatoren und die Gültigkeit der durch operationelle Methoden erhaltenen Resultate.

Wenn allerdings gesagt wird,[38] die Ergebnisse seien später unabhängig von den britischen Mathematikern wiederentdeckt worden, so widerspricht dem Gregorys ausdrückliche Bezugnahme auf Brisson,[39] Cauchy und Servois,[40] der durch seine formalistische Konzeption der Algebra in der Tat der Hauptvorläufer der englischen Schule der symbolischen Algebra wurde.

Gregory gründete 1838 mit Robert Ellis das *Cambridge Mathematical Journal*, in dem nicht nur Boole mit seiner Hilfe seine frühesten Aufsätze druckte, sondern in dessen ersten Jahrgängen er zeigte, wie mittels der Methode der *Separation der Symbole*, dem Prinzip der Trennung der Operationszeichen von denen der Größe, Sätze über lineare Differential- und Differenzengleichungen bewiesen werden können.[41] Darin sei ihm als erster Herschel[42]—ursprünglich glaubte er, dies treffe auf Brisson zu—vorangegangen.

Aus der linearen, inhomogenen Differentialgleichung n-ter Ordnung

$$\frac{d^n y}{dx^n} + A\frac{d^{n-1}y}{dx^{n-1}} + B\frac{d^{n-2}y}{dx^{n-2}} + \cdots + R\frac{dy}{dx} + Sy = x$$

mit konstanten Koeffizienten A, B, C, usf. entsteht zum Beispiel durch Trennung der Symbole

$$\left(\frac{d^n}{dx^n} + A\frac{d^{n-1}}{dx^{n-2}} + B\frac{d^{n-2}}{dx^{n-2}} + \cdots + R\frac{d}{dx} + S\right)y = x \quad \text{oder } f\left(\frac{d}{dx}\right)y = x.$$

Setzt man allgemein a, b, n als Operationen an, so benötigt man nur, daß diese distributive, kommutative Funktionen sind bzw. dem Gesetz der Indexfunktionen unterliegen:

$$c(a + b) = ca + cb, \qquad ab = ba, \qquad a^m(a^n) = a^{m+n}$$

Gregory benutzt bewußt die Bezeichnung Operation statt Quantität, um Bedeutungsbeschränkungen der Symbole zu vermeiden.[43]

In dem Aufsatz *On the Real Nature of Symbolical Algebra*[44] aus dem Jahre 1840 weist er Peacok das Verdienst zu, als einziger bis dahin versucht zu haben, ein auf allgemeine Prinzipien gegründetes System der Algebra entworfen zu haben. *Symbolische Algebra* ist für ihn die Wissenschaft, die die Kombinationen von Operationen behandelt, die nicht durch ihre Natur, das heißt was sie sind oder was sie tun, definiert sind, sondern durch die Kombinationsgesetze, denen sie unterliegen. Der Schritt von der arithmetischen zur symbolischen Algebra bestehe darin, daß wir die Existenz von Klassen unbekannter Operationen vermuten, die den gleichen Gesetzen gehorchen, indem wir die Natur der Operationen außer Acht lassen, welche die gebräuchlichen Symbole repräsentieren.

Gregory stellt fünf derartige Operationsklassen zusammen, ein Begriff, durch den er zwar nicht den Begriff des Isomorphismus antizipiert,[45] wohl aber den des Homomorphismus. Seine sogenannten distributiven Operationen charakterisiert er durch die Beziehung

$$f(a) + f(b) = f(a + b),$$

was sich wie die moderne Homomorphiebedingung liest. Sie schließen zusammen mit den kommutativen Operationen der Art

$$f_{,}f(a) = ff_{,}(a)$$

die wichtigsten Operationen in der Mathematik wie Differentiationen und Differenzbildungen ein. Durch seine operationalen Überlegungen[46] wird er auf Ausdrücke wie

$$(1 + a)^{\log}, \qquad (1 + a)^{\sin}, \qquad (1 + a)^{d/dx}, \qquad \log\left(1 + \frac{d}{dx}\right)$$

geführt. Fruchtbringender erscheinen seine zirkulären oder reproduktiven Operationen der ersten Klasse, die folgenden Gesetzen genügen müssen: $FF(a) = F(a)$, $ff(a) = F(a)$, $Ff(a) = f(a)$, $fF(a) = f(a)$. Arithmetisch können F und f durch $+$ und $-$, geometrisch durch die Bewegungen eines Punktes auf einem Voll- oder Halbkreis gedeutet werden. Gregory deckt damit, modern gesprochen, eine Strukturgleichheit zwischen arithmetischen und geometrischen Sachverhalten auf. Ein Symbol ist für ihn algebraisch definiert, wie er einige Jahre später ausführt, wenn seine Kombinationsgesetze gegeben sind.[47] Ein Symbol repräsentiert eine gegebene Operation, wenn die Kombinationsgesetze der Operationen die gleichen sind wie die des Symbols.

De Morgan erläuterte fast gleichzeitig in vier Aufsätzen *On the Foundation of Algebra* aus den Jahren 1839 bis 1844[48] seine Auffassung von Algebra, die er in einen technischen und einen logischen Zweig unterteilt. Er zog die Bezeichnung *technisch* dem Wort *symbolisch* vor, da dieser Ausdruck nicht den Gebrauch der Symbole von der Erklärung der Symbole trenne. Die technische Algebra hat es danach mit den Operationsregeln zu tun, die die Symbole definieren, die logische mit der Methode, den primären Symbolen Bedeutung zu geben und alle nachfolgenden symbolischen Resultate zu interpretieren.

Der erste Schritt zur logischen Algebra ist die Trennung der Regeln der gewöhnlichen Wissenschaft von ihren Prinzipien bzw. der Operationsregeln von der Erklärung der Symbole, mit denen umgegangen wird. Um A^B unabhängig definieren zu können, erweitert er die Idee des Logarithmus zum *Logometer*,[49] dem vollständigen algebraischen Logarithmus, eine Funktion, die er durch

$$\lambda A + \lambda B = \lambda(AB) \qquad \text{bzw.} \quad \lambda(a, \alpha) + \lambda(b, \beta) = \lambda(ab, \alpha + \beta)$$

definiert und die die Rolle des Logarithmus in einem vollständigen System der Algebra spielt. Dabei schlägt er als eine mögliche Definition von $\lambda(a, \alpha)$

$$\log a(m + n\sqrt{-1}) + \alpha(\mu + \nu\sqrt{-1}), \quad m, n, \mu, \nu \quad \text{beliebige Konstante,}$$

vor.

Der vierte Aufsatz *On Triple Algebra* ist eine der ersten wissenschaftlichen Reaktionen auf Hamiltons ein Jahr zuvor veröffentlichte Entdeckung der Quaternionen. Sein ausdrückliches Ziel war es dabei, Systeme zu finden, in denen die symbolischen Formen der gewöhnlichen Algebra richtig sind, ohne der Interpretation ein Opfer zu bringen. Durch die Nichtkommutativität wich Hamiltons Quadrupelalgebra, wie de Morgan sagte, aber in einem wesentlichen Punkt von den üblichen symbolischen Regeln ab, da Hamilton eine Interpretation auch auf Kosten von symbolischen Formen der Algebra sicherstellen wollte.

Er beginnt seine Überlegungen mit einer Algebra n-ten Charakters, zu der n verschiedene Symbole $\xi_1, \ldots, \xi_n$ gehören, ein Gedanke, den auch Hamilton im Rahmen seiner als Wissenschaft von der reinen Zeit aufgefaßten Algebra äußerte.[50] Hatte dieser doch eine klare Vorstellung von beliebigen Algebren von endlichem Rang über dem Körper der komplexen Zahlen.[51] De Morgan beschränkte sich dagegen darauf—ohne zu abschließenden Ergebnissen zu gelangen—verschiedene Formen von Tripelalgebren, d.h. Systeme von drei fundamentalen Einheiten, zu *prüfen*, wie er selbst einschränkend statt *beweisen* sagte. Er hoffte jedoch, daß die verallgemeinerten Begriffe von Interpretation, die seine Darstellung gab, auf die gewöhnliche *double algebra* anwendbar sei, womit er Hamiltons Darstellung komplexer Zahlen mittels geordneter Paare meinte, auch wenn die richtige Richtung der Verallgemeinerung noch nicht zu sehen sei. Auf jeden Fall kennzeichnet sein Zugang zur Algebra als Studium bestimmter abstrakter Systeme den modernen Zugang zur Algebra und in gewissem Maße zur gesamten modernen Mathematik.

3. HAMILTON UND DIE QUATERNIONENTHEORIE

Diese verfeinerte formale oder symbolische Auffassung von Algebra war eine der beiden geistigen Strömungen, denen Hamilton eingestandenermaßen seine Entdeckung der Quaternionen am 16.10.1843 verdankt, deren Vorgeschichte er im Vorwort zu seinen *Lectures on quaternions* von 1853 ausführlich schildert.[52] Die andere Hauptströmung waren die zahlreichen vorangegangenen Überlegungen deutscher, englischer und französischer Mathematiker zur Natur der imaginären Zahlen.

Da er zugestimmt habe, daß diese nicht im eigentlichen Sinne Quantitäten seien, habe er nach einer Bedeutung gesucht. Dies Ziel habe er erreicht, indem er die Algebra, ermutigt durch Kants *Kritik der reinen Vernunft*, nicht als bloße Kunst, Sprache oder Wissenschaft von den Quantitäten ansah, sondern als Wissenschaft der *order in progression*, eines stetigen, eindimensionalen Fortganges. Die aufeinanderfolgenden Zustände solch einer Progression könnten zwar durch Punkte einer Linie repräsentiert werden, aber besser durch die Augenblicke der Zeit, der abstrakten, idealen, reinen Zeit. Daher sei die Algebra die Wissenschaft von der reinen Zeit.

Hamilton rekapituliert damit im *Preface* die Hauptgedanken seiner ersten großen algebraischen Arbeit *Theorie konjugierter Funktionen oder algebraischer Paare mit einem vorausgehenden und elementaren Essay über Algebra als die Wissenschaft der reinen Zeit* aus den Jahren 1833/5.[53] Er begann sich damit in das Wesen des algebraischen Algorithmus zu einem Zeitpunkt zu vertiefen, als sein Ruhm durch Arbeiten zur Optik und Dynamik bereits gefestigt war. Da, wie er zugab, in dieser Sicht der Algebra mehr als in jeder anderen das Quadrat jeder Zahl positiv ist, führte er Paare von Zeitschritten ein, auf denen er eine Theorie von Zahlenpaaren begründete. So erhielt er für das Symbol $\sqrt{-1}$ die klare Interpretation (0, 1). Die Ableitungsprinzipien für die Hauptoperationen auf Zahlenpaaren hingen von der Separation der Symbole ab.

Er hatte den Traktat mit den Worten geschlossen,[54] daß scheinbar rein symbolische und uninterpretierbare Ausdrücke durch die andersartige

Sichtweise der Algebra Realität und Bedeutung erhalten hätten. Ausdrücklich weist er daraufhin, daß diese Sichtweise nicht grundsätzlich anders als die von Peacock und Martin Ohm (1792–1872) sei, wie denn die Terminologie bereits die geistige Zugehörigkeit aufdeckt.

Ohm, der Bruder des berühmten Physikers, hatte ab 1822 einen mehrteiligen, wiederholt aufgelegten *Versuch eines vollkommen consequenten Systems der Mathematik* veröffentlicht,[55] in dem er eine rein formale Darstellung der arithmetischen Operationen gegeben hatte. In den Vorreden zur zweiten und dritten Auflage heißt es:[56]

> In den verschiedensten Erscheinungen des Kalküls (Arithmetik, Algebra, Analysis etc.) erblickt der Verfasser nicht die Eigenschaft der Größen, sondern die Eigenschaften der Operationen, d.h. Verstandestätigkeiten . . . Es stellte sich heraus, daß man i.a. nur mit 'Formen' rechnet, d.h. mit angezeigten Operationen, Verstandestätigkeiten, die durch die Betrachtung der ganzen unbenannten Zahlen . . . angeregt werden.

Zwei Ideen leiteten ihn also: (1) ein Formalismus: er drückte die Eigenschaften der Operationen, etwa die von ihm erwähnte Kommutativität und Assoziativität, durch formale Gleichungen aus, die zugleich die Rolle von Axiomen spielen; (2) die Methode, die Gültigkeit der formalen Gleichungen, angefangen für die Menge der natürlichen Zahlen, auf die der rationalen, reellen und komplexen Zahlen zu erweitern. Das Wesen dieser Methode spiegelt das sogenannte Permanenzprinzip wider, wie es Peacock fast gleichzeitig formulierte und 1867 Hankel in folgender Form:[57]

> Wenn zwei in allgemeinen Zeichen der arithmetica universalis ausgedrückte Formen einander gleich sind, so sollen sie einander auch gleich bleiben, wenn die Zeichen aufhören, einfache Grössen zu bezeichnen, und daher auch die Operationen einen irgend welchen anderen Inhalt bekommen.

Wenn Ohm und Hankel von Formen sprechen, so erinnert dies an Graßmanns Formenlehre. Insofern stimmt Hamiltons Bemerkung nur bedingt, Ohm scheine mehr das Studium der Relationen zwischen den fundamentalen Operationen zu betonen, Peacock die Permanenz äquivalenter Formen.[58] Es muß jedoch im Gegensatz zu Hamilton offen bleiben, inwieweit Graßmann durch Ohm angeregt wurde.

Insbesondere aber wies Hamilton auf de Morgans Überlegungen hin, der von *double algebra* gesprochen habe.[59] Ja, er erwähnt ausdrücklich, daß de Morgans erster Aufsatz über die Grundlage der Algebra, den er vom Autor 1841 zugeschickt bekommen hatte, ihm geholfen habe, nicht aufzugeben, die Schwierigkeiten bei der Ausdehnung der Gesetze auf den Raum zu überwinden bzw. eine Tripelalgebra zu finden. Tatsächlich hatte de Morgan dort festgestellt:[60]

> Eine Erweiterung auf eine dreidimensionale Geometrie ist solange nicht praktikabel, wie wir nicht zwei Symbole Ω und ω so zuweisen können, daß aus $a + b\Omega + c\omega = a_1 + b_1\Omega + c_1\omega$ $a = a_1, b = b_1, c = c_1$ folgt. Kein definites Symbol der gewöhnlichen Algebra wird diese Bedingung erfüllen.

Dagegen weist Hamilton jeden Einfluß der *geistreichen Aufsätze* von Gregory im *Cambridge Mathematical Journal* zurück. Sein Wunsch war,[61] in neuer und nützlicher Weise Rechnung mit Geometrie durch eine unentdeckte Erweiterung auf den dreidimensionalen Raum zu verbinden. Demgemäß probierte er zahlreiche Tripelsysteme durch, wobei ihn störte, daß in bestimmten Fällen—modern gesprochen—Nullteiler auftraten und zuviel Raum für beliebige Konstantenwahl blieb.

Sein Freund John T. Graves habe zwei neue imaginäre Zahlen i, j verwendet, die den Charakter vierter Einheitswurzeln hatten, von denen i eine Rotation von 90° um die z-Achse verursachte. Bei der Multiplikation habe sich Kommutativität, nicht aber Distributivität ergeben. 1843 unternahm Hamilton den Versuch, das Distributivgesetz zu retten, und glaubte zunächst, auch das Kommutativgesetz beibehalten zu können, ohne zu wissen, was er mit Produkten der Form ij machen sollte. Zunächst hielt er es selbst für ein Tripel, dann für Null. Schließlich sah er die Tripel als unvollkommene Formen einer Quaternion an und legte die Produkte in der bekannten Weise durch $i^2 = j^2 = k^2 = -1, ij = -ji = k, jk = -kj = i, ki = -ik = j$ fest.

Darin, daß er so als erster die Produkte zweiter Stufe auf die ursprünglichen Einheiten zurückführte,[62] lag einer der drei Hauptunterschiede zu Graßmanns kombinatorischem oder äußerem Produkt, der das Produkt zweier Einheiten e_i, e_k nicht auf die Einheiten selbst zurückführte, sondern als Größen einer neuen Art ansah.[63] Er verschloß sich damit den Weg zu Systemen von höheren komplexen Zahlen, deren Begriff Hamilton 1853 in den *Lectures on Quaternions* allgemein aufgestellt hatte, konnte aber auf diese Weise alle geometrischen Grundelemente (Punkt, Gerade, Ebene usf.) gleichzeitig der Rechnung unterwerfen.[64]

So wie die Multiplikation zweier komplexer Zahlen als Drehstreckung der Ebene gedeutet werden konnte, so konnte die Quaternionenmultiplikation als Rotation des Raumes gedeutet werden. Dies war, wie erwähnt, eine der Ideen, die Hamilton bei seiner Entdeckung leitete. Da die Rotationen im Raum eine nichtabelsche Gruppe bilden, hatte er damit—modern gesprochen—das erste Beispiel für einen nichtkommutativen Körper gefunden, dem einzigen, wie Georg Frobenius (1849–1917) 1878 zeigte, der über dem Körper der reellen Zahlen konstruiert werden kann.[65] Im Nachhinein ist auch klar, daß Hamilton für diese Bewegungen viergliedrige Zahlen benötigte. Denn zur Festlegung einer Drehung um den Nullpunkt muß im Raum eine Achse festgelegt werden. Daher braucht eine Drehstreckung im Raum vier Parameter: zwei für die Richtung der Drehachse, einen für den Drehwinkel, einen für die Streckung.[66]

Es verdient hervorgehoben zu werden, daß Hamilton in einer der letzten Fußnoten der Vorrede zu den *Lectures on Quaternions* auch mit lobenden Worten auf Graßmann zu sprechen kommt, obwohl er zugleich seine geistige Unabhängigkeit reklamiert und die Grenze von Graßmanns Verdiensten aus seiner Sicht absteckt: Es sei angemessen, hier festzuhalten, daß ein Beispiel für eine nichtkommutative Multiplikation für Strecken (äußere Multiplikation) in einem sehr bemerkenswerten Werk von Graßmann auftrete, mit dem er erst Jahre nach der Erfindung der Quaternionen bekannt geworden sei. Dort werde auch $\beta - \alpha$ für die Strecke gesagt, die vom Punkt α zum Punkt β führt.—Hamilton ging genauso bei der Gleichheitsdefinition von Zeitintervallen vor.—Sonst handele es sich um völlig andere Begriffe, Methoden, Ergebnisse. Aus der Vorrede Graßmanns ergebe sich zudem, daß er nicht im Besitz der Quaternionenlehre gewesen sei.

Hamilton beruft sich dabei auf die Worte: "Hingegen ist es nicht mehr möglich, vermittels des Imaginären auch die Gesetze für den Raum abzuleiten. Auch stellen sich überhaupt der Betrachtung der Winkel im Raume Schwierigkeiten entgegen, zu deren allseitiger Lösung mir noch nicht hinreichend Musse geworden ist."[67] Graßmann teilte sich freilich mit Gauß die Ehre, daß ihm Hamilton zutraute, die Quaternionen gefunden zu haben, auch wenn er sich stets erneut darüber freute, daß dies offensichtlich nicht der Fall war. Dies geht aus den zwischen de Morgan und Hamilton über Graßmann gewechselten Briefen hervor.[68]

Hamilton folgte der Cambridger Tradition, wenn er für die Operationen mit Quaternionenfeldern symbolische Schreibweisen wählte und symbolische Operatoren aus den partiellen Differentiationen nach den Koordinaten des Feldpunktes zusammensetzte. Der wichtigste unter ihnen war der berühmte Nabla (nach einem alten jüdischen Musikinstrument benannt):

$$\nabla = i\frac{\partial}{\partial x} + j\frac{\partial}{\partial y} + k\frac{\partial}{\partial z}.$$

Formal war er wie ein Vektor zu handhaben und hieß in Peter Guthrie Taits (1831–1901), Hamiltons bedeutenden Parteigänger, Sprechweise unter Umkehrung der Buchstabenreihenfolge des Wortes Delta *Atled*.[69]

4. OPERATIONENKALKÜL

Graßmann und Hamilton verbanden von vornherein ihre Theorien mit der Lehre von den mehrgliedrigen komplexen Zahlen.[70] Welch unsichere Vorstellung über die Natur der komplexen Zahlen noch um 1821 vorherrschte, zeigt Cauchys *Analyse algébrique*, der erste Teil des *Cours d'analyse*, die Hamilton lobend erwähnt.[71] Im siebten Kapitel stellt Cauchy allgemeine Betrachtungen über imaginäre Ausdrücke an:[72]

> In der Analysis nennt man einen symbolischen Ausdruck oder ein Symbol jede Kombination algebraischer Zeichen, die nichts durch sich selbst bezeichnet oder der man einen Wert zuordnet, verschieden von dem, den sie natürlicherweise haben muß. Ebenso nennt man symbolische Gleichungen solche, die dem Wortlaut nach genommen und gemäß den üblicherweise eingeführten Konventionen interpretiert, ungenau oder sinnlos sind (*sont inexactes ou n'ont pas de sens*), aber aus denen man exakte Resultate ableiten kann.

Als Beispiel führt er die Gleichung

$$\cos(a+b) + \sqrt{-1}\sin(a+b) = (\cos a + \sqrt{-1}\sin a).(\cos b + \sqrt{-1}\sin b)$$

an, die den Buchstaben nach ungenau sei und keinen Sinn habe. Hankel nahm an diesen in der Tat fragwürdigen Ausführungen heftigen Anstoß und sprach drastisch von *Galimatias*, Unsinn.[73] Er erwähnte jedoch selbst, daß Cauchy 1847 eine andere, aber auch nicht befriedigende Theorie imaginärer Größen gab, in der er nur reelle Größen betrachtete, also auch i als allgemeine reelle Variable ansah. Einer der Nachteile seiner Methode war, daß sie überall die Entwickelbarkeit nach Potenzen von i voraussetzte.

Eine ganze Reihe von Mathematikern hatte sich seit Beginn des 19. Jahrhunderts um eine geometrische Darstellung oder Erklärung der komplexen Zahlen bemüht,[74] 1799 der Norweger Caspar Wessel (1745–1818), 1805/06 der Abbé Buée, 1806 Jean Robert Argand (1768–1822), 1813 Jacques Frédéric Francais, der sich auf den namentlich ihm unbekannten Argand bezog, was zu einem Briefwechsel zwischen Argand, Francais und Joseph Diaz Gergonne (1771–1859), dem Herausgeber der *Annales de mathématiques* führte (die Stellungnahme von Servois dazu ist von besonderem Interesse[75]), 1828 John Warren und C. F. Mourey.

Gauss' erste geometrische Darstellung komplexer Zahlen erschien erst 1831 in der Selbstanzeige der *Commentatio secunda* zur *Theoria residuorum biquadraticorum*,[76] obwohl seine diesbezüglichen Überlegungen über dreißig Jahre zurückreichten. Seine *wahre Metaphysik* der imaginären Zahlen bestand darin, daß er für $\sqrt{-1}$ eine anschauliche Bedeutung im Raum nachwies, was ihn zugleich veranlaßte, Kants Behauptung zurückzuweisen, der Raum sei nur eine Form unserer äußeren Anschauung, während dieser in Wahrheit unabhängig von unserer Anschauungsart reelle Bedeutung habe. Diese Bemerkung ist um so interessanter, als Hamilton umgekehrt, wie erwähnt, gerade durch Kant zu seiner Konzeption der Algebra angeregt wurde.

Bemerkenswert ist, daß die meisten der genannten Autoren, Wessel, Gauss, Argand, Servois, Francais, Mourey ebenso wie John T. Graves und de Morgan ohne rechten Erfolg nach einem Tripelsystem gesucht hatten, das die Methoden der Ebene auf den dreidimensionalen Raum übertragen sollte. Servois wurde von Hamilton zugestanden, der Entdeckung der Quaternionen am nächsten gekommen zu sein. Er bezog sich dabei auf den Brief von Servois an Gergonne, in dem dieser die Darstellungen von Argand und Francais zurückweist.

Argand stütze sich, wie Servois bemängelt, bei der Begründung einer ungewöhnlichen Lehre, die in einiger Hinsicht den bisherigen Prinzipien widerspricht, auf ein allein nicht zureichendes Mittel, die einfache Analogie. Insbesondere aber räume Argand selbst ein, daß man in seiner neuen Theorie nur die einfache Verwendung einer speziellen Notation sehen könne. Er selbst, Servois, bekenne, daß er in dieser Notation nur *un masque géométrique*,[77] nur eine auf analytische Formen angewandte geometrische Maske sehen könne, deren unmittelbarer Gebrauch ihm einfacher und schneller erscheine. Im Rahmen seiner formalistischen Konzeption der Algebra betonte er also den Primat der algebraischen Sprache. Am Ende des Briefes schreibt er Gergonne die Idee zu, was dieser bescheiden zurückweist, zur Ausdehnung auf den dreidimensinalen Raum Terme trinomialer Form konzipiert zu haben:

> Aber welchen Koeffizienten hätte der dritte Term ? . . . Die Analogie schiene zu fordern, daß das Trinom die Form $p \cos \alpha + q \cos \beta + r \cos \gamma$ hat. Dabei sind α, β, γ die Winkel, die eine Gerade mit den drei senkrecht aufeinanderstehenden Achsen bildet . . . Die Werte von p, q, r wären irrational, aber wären sie imaginär, reduzierbar auf die allgemeine Form $A + B\sqrt{-1}$?

Dem *Essai sur un nouveau mode d'exposition des principes du calcul différentiel* aus den Jahren 1814/15[78] lag die Beobachtung zugrunde, daß der Differentialkalkül auf der Beibehaltung bestimmter Eigenschaften der Operationen basiert, auf die er angewandt wird. Ohne zwischen Funktionen und Operationen zu unterscheiden, worin ihm Gregory folgte, führte er die grundlegenden Begriffe der *distributiven* und *untereinander kommutativen* Funktionen ein:

$$\varphi(x + y + \cdots) = \varphi x + \varphi y + \cdots, \qquad fFz = Ffz.$$

Wir würden heute von strukturverträglichen Abbildungen und der Kommutativität der Komposition von Abbildungen sprechen. Wenn er etwa die sinus-Funktion als Gegenbeispiel zum ersten Fall anführt,[79] so erklärt er damit ausdrücklich, daß bestimmte Operationen nicht distributiv bzw. nicht kommutativ zu sein brauchen, ein durchaus revolutionärer Gedanke, den sowohl Hamilton wie auch Graßmann eingestandenermaßen nur sehr zögernd zu akzeptieren bereit waren. Graßmann sagt anläßlich der

Erklärung seines äußeren Produktes:[80]

> Auch machte mich das merkwürdige Resultat anfangs betroffen, daß für diese neue Art des Produktes zwar die übrigen Gesetze der gewöhnlichen Multiplikation und namentlich ihre Beziehungen zur Addition bestehen blieb, daß man aber die Faktoren nur vertauschen konnte, wenn man zugleich die Vorzeichen umkehrte.

John T. Graves[81] sprach Hamilton sein Unbehagen und seine Zweifel darüber aus, ob wir imaginäre Größen nach Belieben schaffen und mit übernatürlichen Eigenschaften versehen können. So sei er froh, daß Hamilton physikalische Analogien (Raumrotationen) vorweisen könne. Noch 24 Jahre später wollte Hankel bewußt wenigstens einige Zahlensysteme behandeln,[82] die nicht allen Gesetzen der *arithmetica universalis* im Sinne Newtons folgen, und wählte dazu Hamiltons Quaternionen und seine sogenannten *alternierenden Zahlen*, die bereits Graßmann in abstrakter und schwer verständlicher Form behandelt hatte.[83] Servois entwickelte so den *Operationenkalkül*, um den Differentialkalkül auf eine rigide Grundlage zu stellen. Als methodische Hilfsmittel erwähnt er ausdrücklich die *analise combinatoire*, also die Schriften der kombinatorischen Schule Karl Friedrich Hindenburgs (1739–1808) und den *calcul des dérivations* von Louis Francois Arbogast (1759–1803).[84]

Diese Betonung von Kombinations- und Operationslehre spielte bei den englischen Algebraikern und Logikern, insbesondere bei George Boole, eine entscheidende Rolle. Und es ist von größtem Interesse zu sehen, daß völlig unabhängig von dieser englischen Tradition Graßmann die gleichen Grundprinzipien hervorhebt, indem er Kombinations- und Formenlehre parallelisiert.[85]

Boole strebte über den Zweischritt Symbolik–Kalkül hinaus den Dreischritt Symbolik–Kalkül–Formalismus an. Für ihn bestand, wie er 1847 in der *Mathematical Analysis of Logic* feststellte, der bestimmende Charakter eines wahren Kalküls darin, daß er eine Methode ist, die auf der Anwendung von Symbolen beruht, deren Kombinationsgesetze bekannt und allgemein sind und deren Ergebnisse eine konsistente Interpretation zulassen.[86] Er war davon überzeugt,[87] daß die Gültigkeit des mathematischen Prozesses nicht von der Interpretation der Symbole, sondern allein von den Gesetzen ihrer Kombinationen abhängt. Er betrachtete danach die Analysis als einen universellen Kalkül mit Symbolen bzw. Logik und Analysis als spezielle Zweige eines universellen Kalküls. Seine Logik war eine mögliche Materialisierung eines universellen Kalküls mit Symbolen.

Seine Methode der Symboltrennung bestand im bewußten Anschluß an Gregory darin,[88] die Symbole von ihren Bedeutungen zu trennen und mit den bedeutungslosen Symbolen gemäß algebraischen Regeln zu operieren.

Er betrieb eine Generalisierung durch Symbolisierung und beabsichtigte, die Methode der Symboltrennung auf neue Calculi auszudehnen. Sein *Treatise on Differential Equations* von 1859 war Heaviside wohlbekannt, der früher fälschlich als Erfinder des Operatorkalküls angesehen wurde.

Der geschickte Gebrauch der Symbole und mathematischen Operationen —Boole nahm an, die Logik sei ein dem Kalkül der Differentialoperatoren analoger Kalkül—sollte bekannte mathematische Resultate vom allgemeinstmöglichen Standpunkt betrachten, die erhaltenen Verallgemeinerungen auf scheinbar ganz verschiedene Gebiete der Mathematik und Physik ausdehnen lassen. Seine Idee war, Symbolik und Regeln zu definieren, um dann fortzuschreiten, um Resultate zu erhalten, ohne jeden Zwischenschritt zu interpretieren. Also neuerlich: Formalismus im Dienst der Heuristik!

5. ALGORITHMISIERUNG

Graßmann teilte das Schicksal, auf die folgende Entwicklung der Vektoranalysis nur einen geringen unmittelbaren Einfluß ausgeübt zu haben, mit mehreren Zeitgenossen: mit Giusto Bellavitis (1803–1880), der 1835 seinen *Saggio di applicazioni di un nuovo metodo di Geometria analitica* (*calcolo delle equipollenze*) veröffentlichte,[89] mit Adhémar Barré, mit Cauchy, der 1853 seine *clefs algébriques* in Abhängigkeit von symbolischen Polynomfaktoren einführte[90] (jene fallen mit Graßmanns alternierenden Einheiten zusammen[91]) und die Theorie der imaginären Größen und Hamiltonschen Quaternionen als Spezialfall der Theorie der Schlüssel ansah, mit Matthew O'Brien, vor allem aber mit Möbius und dessen *Baryzentrischen Calcul* aus dem Jahre 1827,[92] der zu den ganz wenigen bedeutenden Mathematikern gehörte, die Graßmann frühzeitig zu schätzen wußten, zumal sein Kalkül das Fundament für Graßmanns Punktrechnung bildete.

Der bei Servois angetroffene Trend zur Algorithmisierung und Algebraisierung leitete auch Möbius—war er doch neben Julius Plücker (1801–1868) der bedeutendste deutsche Vertreter der algebraischen Geometrie jener Zeit. Er deutete bereits durch den Titel seines Werkes dessen formalistischen Charakter an. Es handelt sich um ein vektorielles System, das der Form nach den Bedürfnissen der projektiven Geometrie angepaßt war[93] und auf dem Begriff des Schwerpunktes eines Systems von Massenpunkten beruhte,[94] einem System der ebenen Dreipunkt- und räumlichen Vierpunktkoordinaten.[95]

Mit den baryzentrischen Koordinaten eines Punktes führte er die ersten homogenen Koordinaten in die Geometrie ein. Er stellte *symbolische* Formeln der Form $aA + bB + cC + dD$, den baryzentrischen Ausdruck des

Punktes P, an den Anfang, wobei a, b, c, d baryzentrische Koordinaten, A, B, C, D Fundamentalpunkte sind. Die Rechnung mit solchen abgekürzten Formeln nannte er den *baryzentrischen, das ist den aus dem Begriffe des Schwerpunkts abgeleiteten Calcul*, der sich von den gewöhnlichen Rechnungsweisen der Algebra im ganzen nicht unterscheide. Die Gleichung

$$aA + bB + cC + \cdots = pP + qQ + rR + \cdots$$

besagt, daß die Gewichte a, b, c, ... in den Punkten A, B, C, ... dieselbe Summe und denselben Schwerpunkt wie die Gewichte p, q, r, ... in den Punkten P, Q, R, ... haben.[96]

Die Koordinaten eines Punktes sind also nicht mehr Linien, sondern zum Beispiel bezüglich eines Dreieckes in einer Ebene, in der der Punkt liegt, drei Zahlen, die proportional so zu Gewichten sind, daß der gegebene Punkt Schwerpunkt des Systems ist, wenn sie an den Ecken des Dreieckes plaziert werden.[97] Nur die Verhältnisse der Gewichte haben Bedeutung. Durch die Untersuchung kollinearer Gebilde gelangt er zum abgekürzten, baryzentrischen Kalkül, der eine Verallgemeinerung des früheren ist, da A, B, C, ... nicht mehr Punkte mit derselben Masse 1, sondern mit beliebigen Massen sind.[98]

Gauss bekannte in dem oben zitierten Brief an Schumacher, daß er zunächst zweifelte, ob es der Mühe wert sei, eine recht artig ausgesonnene Rechnungsweise sich anzueignen, wenn man durch dieselbe nichts leisten könne, was sich nicht ebenso leicht ohne sie leisten lasse—der gleiche Vorwurf kehrte in Kummers Gutachten über Graßmann wieder. Er überzeugte sich jedoch davon, daß gerade der *baryzentrische Calcul* auf dem leichtesten Wege zur Auflösung aller dahin gehörenden Aufgaben führt. Wörtlich stellte er fest:[99]

> Der Vorteil ist aber der, daß, wenn ein solcher Calcul dem innersten Wesen vielfach vorkommender Bedürfnisse correspondiert, jeder, der sich ihn ganz angeeignet hat, auch ohne die gleichsam unbewußten Inspirationen des Genies, die niemand erzwingen kann, die dahin gehörenden Aufgaben lösen, ja selbst in so verwickelten Fällen gleichsam mechanisch lösen kann, wo ohne eine solche Hilfe auch das Genie ohnmächtig wird. So ist es mit der Erfindung der Buchstabenrechnung überhaupt, so mit der Differentialrechung gewesen, ... und mit Möbius' Calcul. Es werden durch solche Conceptionen unzählige Aufgaben, die sonst vereinzelt stehen, und jedesmal neue Efforts ... des Erfindungsgeistes erfordern, gleichsam zu einem organischen Reiche.

Graßmann konstruiert dagegen ein algebraisch-geometrisches Gebäude, das auf einer geometrischen oder *intrinseken*, das heißt beinahe axiomatisierten Auffassung des n-dimensionalen Vektorraumes beruht. Allerdings vergingen noch 44 Jahre, bevor Peano, einer der Schöpfer der axiomatischen Methode und einer der ersten, die Graßmanns Werk zu schätzen wußten, eine axiomatische Definition der Vektorräume über dem Körper der reellen Zahlen mit der Definition linearer Abbildungen zwischen solchen Räumen im *Calcolo geometrico secondo l'Ausdehnungslehre di H. Grassmann* gab.[100]

In der ersten Auflage seiner Ausdehnungslehre von 1844 deduzierte Graßmann zunächst aus allgemeinsten philosophischen Begriffen auf der Grundlage der Schleiermachschen Philosophie ohne irgendwelche Formeln,[101] was neben dem hohen Grad der Abstraktheit und Allgemeinheit die Aufnahme seines Werkes entscheidend erschwerte. Sein Untersuchungsgegenstand war ein Kontinuum von n Variablen, also—modern gesprochen—ein R^n, wodurch die Ausdehnungslehre als Disziplin umfassender und weitreichender als die Quaternionentheorie wurde. Er baute zunächst nur eine affine n-dimensionale Geometrie auf, zu der in der Neubearbeitung von 1862 eine Metrik hinzukam.[102]

Methodisch trennte er zwischen reiner Mathematik oder Formenlehre[103] und Ausdehnungslehre. In seiner Übersicht über die allgemeine Formenlehre heißt es:[104]

> Unter der allgemeinen Formenlehre verstehen wir diejenige Reihe von Wahrheiten, welche sich auf alle Zweige der Mathematik auf gleiche Weise beziehen, und daher nur die allgemeinen Begriffe der Gleichheit und Verschiedenheit, der Verknüpfung und Sonderung, voraussetzen. Es müßte daher die allgemeine Formenlehre allen speziellen Zweigen der Mathematik vorangehen.

Mangels ihm bekannter Vorbilder sieht er sich gezwungen, diese Formenlehre selbst aufzubauen. Diese rein *formale* Mathematik hat sich, wie Hankel[105] richtig bemerkte, als eminent fruchtbar für den ganzen Organismus der Mathematik erwiesen. Graßmann hat so den abstrakten Begriff der Verknüpfung systematisch behandelt,[106] bevor er auf der kombinatorisch-stetigen oder extensiven Größe, wie er sagt, seine sogenannte Ausdehnungslehre aufbaut. Es ist kein Zufall, daß Hilbert denselben Ausdruck in seiner *Neubegründung der Mathematik* verwendet, als er von der Begründung der Theorie des Kontinuums spricht.[107] Die extensive Größe hat mit höheren komplexen Zahlen die Eigenschaft der linearen Ableitbarkeit—Graßmann sagt *numerische Ableitbarkeit*—aus gewissen Systemen von Einheiten gemein:[108]

$$\sum \alpha_r e_r, \qquad e_r \text{ Einheit}, \quad \alpha_r \text{ (reelle) Zahl}$$

Unter den Produktbildungen ragen die äußere oder kombinatorische Multiplikation, die Gibbs als das vielleicht größte Denkmal des Genius des Autors bezeichnete,[109] und die innere an Bedeutung hervor.

6. UNIVERSALISIERUNG

Ohne anzuerkennen, daß eine allumfassende geometrische Symbolik nicht gefunden werden könne, waren Graßmann und Hamilton von der Leistungsfähigkeit ihrer Symbolismen und Kalküle glühend überzeugt. In beiden Fällen bildete sich eine nahezu fanatische Anhängerschaft, die im Falle Hamiltons 1907 zur Gründung des *Weltbundes zur Förderung der Quaternionenlehre* führte, der nach dem ersten Weltkrieg wieder verschwand. Um mit Felix Kleins Worten zu sprechen: "Ein mit Scheu und Verehrung gehandhabter Formalismus, dem die symbolische Schreibweise reichlich Nahrung gab, förderte die Entwicklung der grenzenlosen Begeisterung der Quaternionisten."[110]

Während Carl Neumann in der Festschrift für Ludwig Boltzmann abfällig von der Quaternionenstenographie sprach,[111] verehrte Erwin Schrödinger Hamilton in geradezu enthusiastischer Weise. Und doch wurde Schrödinger an einer entscheidenden Stelle in der theoretischen Physik vom Mangel an Anschaulichkeit—man fühlt sich an Apelts Kritik an Graßmann erinnert—abgestoßen.

Werner Heisenberg, Max Born und Pascual Jordan hatten 1925 in drei Aufsätzen[112] der Quantenmechanik eine matrizentheoretische Grundlage gegeben, wonach eine quantentheoretische Größe eine unendliche Matrix ist und die Ausgestaltung des Heisenbergschen Formalismus auf der Matrizenmultiplikation beruht. Die unendliche Matrix ist der Repräsentant einer physikalischen Größe, die in der klassischen Theorie als Funktion der Zeit angegeben wird. Damit knüpft die Heisenbergsche Theorie die Lösung eines Problems der Quantenmechanik an die Auflösung eines Systems von unendlich vielen algebraischen Gleichungen, deren Unbekannte—unendliche Matrizen—den klassischen Lage- und Impulskoordinaten des mechanischen Systems und Funktionen derselben zugeordnet sind *und eigenartige Rechengesetze befolgen.*

Schrödinger hatte, wie er ausdrücklich bekennt,[113] von Heisenbergs Theorie natürlich Kenntnis, fühlte sich aber durch die ihm sehr schwierig scheinenden Methoden der transzendenten Algebra und durch den Mangel an Anschaulichkeit abgeschreckt, um nicht zu sagen abgestoßen. Durch seine Undulationsmechanik, die mit der Heisenbergschen Quantenmechanik vom mathematischen Standpunkt aus äquivalent ist, führte er die Auflösung des ganzen Systems der Matrizengleichungen auf das Randwertproblem einer linearen, partiellen Differentialgleichung zurück, nach dessen Lösung jedes Matrizenelement durch Differentiationen und Integrationen ausgerechnet werden kann.

Den fruchtbaren Gedanken zur richtigen Einordnung der Quaternionen lieferte Cayley. Er hatte bereits 1846 gesagt,[114] wie man hinsichtlich der

Möglichkeit des vierdimensionalen Raumes schlußfolgern kann, ohne, wie er betont, auf irgendeinen metaphysischen Begriff zurückzugehen. Aber auch Gauß vor ihm und Hankel nach ihm verzichteten, wie wir sahen, nicht auf Metaphysik. Zwölf Jahre später begründete er mit seiner berühmten Abhandlung *On the Theory of Matrices* den Matrizenkalkül,[115] dessen Multiplikation er mit Hilfe der Theorie linearer Substitutionen erklärte.

In seinen witzig-spritzig geschriebenen, anfangs zitierten *Lectures on the Principles of Universal Algebra*, die Gibbs nachdrücklich hervorhebt, ließ Sylvester mit dieser Arbeit seines Freundes das Reich von Algebra II beginnen.[116] Er selbst habe unabhängig von Cayley Entdeckungen gemacht, indem er die Matrizen als Größen betrachtete. Denn der Begriff der Matrizenaddition sei der Substitutionsidee völlig fremd.

Anders als Cayley weist er den Quaternionen in seiner Matrizentheorie einen, den richtigen Platz zu: es sind unter einem bestimmten Aspekt Matrizen zweiter Ordnung. Er habe eine unabhängige, algebraische Begründung für die Einführung und den Gebrauch der Symbole entdeckt, die in Hamiltons Theorie angewendet werden.[117] Er weist aber daraufhin, daß die beiden Peirces, Benjamin und sein Sohn Charles Santiago Saunders, die Universalisierung der Hamiltonschen Theorie dargestellt und vermutet hatten, daß alle Systeme algebraischer Symbole mit assoziativer Multiplikation eventuell identisch mit linearen Transformationsschemata seien, die matrizenförmiger Darstellung fähig sind.[118]

Benjamin Peirce hatte in der Tat 1870 die Arbeit *Linear Associative Algebra* vor der Nationalen Akademie der Wissenschaften in Washington verlesen, die allerdings erst 1881 postum erschien und der beim Abdruck beinahe der Rang der *Principia* des philosophischen Studiums der Gesetze der algebraischen Operationen verliehen und als einer der ersten originalen Beiträge der USA zur Mathematik gewertet wurde.[119]

Peirce definiert unter bewußter Anknüpfung an de Morgan und Hamilton Algebra als formale Mathematik, deren Symbole mit den Gesetzen der Kombinationen ihre Sprache konstituieren. Die Methode, ihre Svmbole beim Schlußfolgern zu gebrauchen, sei ihre Kunst, ihre Interpretationen ihre wissenschaftliche Anwendung. Die von ihm untersuchten und klassifizierten quadratischen linearen assoziativen Algebren erster bis sechster Ordnung sind, wie sein Sohn Charles erkannte, mit Matrizen identisch.[120]

ANMERKUNGEN

1. Diese Arbeit ist eine Fortsetzung und Spezialisierung der Untersuchung von Knobloch, E. 1980. "Einfluß der Symbolik und des Formalismus auf die Entwicklung des mathematischen Denkens," *Berichte zur Wissenschaftsgeschichte*, *3*, 77–94.

2. Peano, G. 1908. *Formulario mathematico, Editio V (Tomo V de Formulario completo)* (Turin; riproduzione con introduzione e note di Ugo Cassina e col contributo del Conume di Cuneo: Rom, 1960), V.
3. Whitehead, A. N. und B. Russell. 1927. *Principia mathematica* (2. Auflage, 3 Bde., Cambridge; Nachdruck, 1973), hier Bd. *1*, VII–VIII und 1–4.
4. Hilbert, D. und W. Ackermann. 1959. *Grundzüge der theoretischen Logik* (4. Auflage, Berlin-Göttingen-Heidelberg), 1.
5. Hilbert, D. 1922. "Neubegründung der Mathematik. Erste Mitteilung," *Abhandlungen aus dem Mathematischen Seminar der Hamburger Universität, 1*, 157–177 = Hilbert, D. 1932–1935. *Gesammelte Abhandlungen* (3 Bde., Berlin), hier Bd. 3, 157–177, insbesondere 163.
6. Hilbert, D. 1932–1935. *Gesammelte Abhandlungen* (wie Anm. 5), 174.
7. Sylvester, J. J. 1884. "Lectures on the principles of universal algebra," *American Journal of Mathematics, 6*, 270–286 = *The collected mathematical papers of J. J. Sylvester* (4 Bde., Cambridge, 1904–1912; Nachdruck New York, 1973), hier Bd. *4*, 208–224, insbesondere 209.
8. Crowe, M. J. 1967. *A history of vector analysis, The evolution of the idea of a vectorial system* (Notre Dame—London), 30; Dubbey, J. M. 1977. "Babbage, Peacock and modern algebra," *Historia Mathematica, 4*, 295–302; hier 302.
9. Eine mathematische Analyse dieser Entdeckung findet man bei van der Waerden, B. L. 1973. *Hamiltons Entdeckung der Quaternionen* (Göttingen, 14 S.).
10. Graßmann, H. 1844. *Die lineale Ausdehnungslehre, Ein neuer Zweig der Mathematik, dargestellt und durch Anwendungen auf die übrigen Zweige der Mathematik, wie auch auf die Statik, Mechanik, die Lehre vom Magnetismus und die Krystallonomie erläutert* (Leipzig; 2. Auflage, 1878) = Graßmann H. 1894–1911. *Gesammelte mathematische und physikalische Werke*, hrsg. v. F. Engel, J. Graßmann, H. Graßmann jr., J. Lüroth, G. Scheffers, E. Study (3 Bde. in 6, Leipzig; Nachdruck New York–London, 1972), hier Bd. *1*(1), 1–319.
11. Struik, D. J. 1976. *Abriß der Geschichte der Mathematik* (6. Auflage, Berlin), 177.
12. Engel, F. 1911. "Grassmanns Leben," in Graßmann, H. 1894–1911. *Gesammelte mathematische und physikalische Werke* (Wie. Anm. 10), hier Bd. *3*(2), 127.
13. Engel, F. 1911. "Grassmanns Leben," (Wie Anm. 12), *3*(2), 101.
14. Klein, F. 1893. "Vergleichende Betrachtungen über neuere geometrische Forschungen (Programm zum Eintritt in die philosophische Fakultät und den Senat der k. Friedrich-Alexanders-Universität zu Erlangen 1872)," *Mathematische Annalen, 43*, 63–100 = Klein, F. 1921–1923. *Gesammelte mathematische Abhandlungen*, hrsg. v. R. Fricke, A. Ostrowski, H. Vermeil, E. Bessel-Hagen (3 Bde., Berlin), hier Bd. *1*, 460–497; wiederabgedruckt in *Geometrie*, hrsg. v. K. Strubecker (Darmstadt, 1972), 118–155; nochmals wiederabgedruckt in *Mathematical Intelligencer, 0*, (1977), 23–30; hier wird nach der Werkausgabe zitiert, nämlich Bd. *1*, 488.
15. Schlegel, V. 1896. "Die Grassmann'sche Ausdehnungslehre, Ein Beitrag zur Geschichte der Mathematik in den letzten fünfzig Jahren," *Zeitschrift für Mathematik und Physik, 41*, 1–21 und 41–59.
16. Graßmann, H. 1844. *Die lineale Ausdehnungslehre* (Wie Anm. 10), Bd. *1*(1), 9.
17. Schlegel, V. 1896. (Wie Anm. 15), 4.
18. Gauß, C. F. 1863–1933. *Werke*, hrsg. von der Gesellschaft der Wissenschaften zu Göttingen (12 Bde. in 14, Göttingen; Nachdruck Hildesheim, 1973), hier Bd. *8*, 297–298.
19. Engel, F. 1911. "Grassmanns Leben," (Wie Anm. 12), *3*(2), 312.
20. Klein, F. 1926–1927. *Vorlesungen über die Entwicklung der Mathematik im 19. Jahrhundert* (2 Bde., Berlin; Nachdruck New York, 1967), hier Bd. *2*, 28.

21. Hankel, H. 1867. *Vorlesungen über die complexen Zahlen und ihre Functionen in zwei Theilen, I. Theil: Theorie der complexen Zahlensysteme inbesondere der gemeinen imaginären Zahlen und der Hamilton'schen Quaternionen nebst ihrer geometrischen Darstellung* (Leipzig).
22. Klein, F. 1893. (Wie Anm. 14), Bd. *1*, 483.
23. Schlegel, V. 1896. (Wie Anm. 15), 4.
24. Klein, F. 1926–1927. (Wie Anm. 20), Bd. *2*, 42.
25. Hankel, H. 1867. (Wie Anm. 21), IX–X.
26. Hankel, H. 1867. (Wie Anm. 21), 17.
27. Berlin, Staatsbibliothek Preußischer Kulturbesitz, Handschriftenabteilung, Sammlung Darmstaedter H 1864 (8): H. Hankel.
28. Knappe Bemerkungen hierzu sowie ein genaueres Eingehen auf Graßmann findet man in der Dissertation von Jahnke, H. N. 1978. "Zum Verhältnis von Wissensentwicklung und Begründung in der Mathematik—Beweisen als didaktisches Problem," (Bielefeld) 38–54.
29. Crowe, M. J. 1967. (Wie Anm. 8), 100.
30. Struik, D. J. 1976. (Wie Anm. 11), 181.
31. O'Brien, M. 1847. "Contributions towards a system of symbolical geometry and mechanics," *Transactions of the Cambridge philosophical society*, *8*, 497–507.
32. Crowe, M. J. 1967. (Wie Anm. 8), 108.
33. Genaueres dazu bei Dubbey, J. M. 1977. "Babbage, Peacock and modern algebra," *Historia Mathematica*, *4*, 295–302.
34. Whiteside, D. T. ed. 1967–1980. *The mathematical papers of Isaac Newton* (8 Bde., Cambridge), hier Bd. *5*, 539.
35. Babbage, C. 1821. "Observations on the notation employed in the calculus of functions," *Transactions of the Cambridge Philosophical Society*, *1*, 63–76.
36. Babbage, C. 1827. "On the influence of signs in mathematical reasoning," *Transactions of the Cambridge Philosophical Society*, *2*, 325–377.
37. Lusternik, L. A. und S. S. Petrova. 1977. "Iz istorii simbolicheskogo ischisleniia (*Aus der Greschichte der Symbolrechnung*)," *Istoriko-matematičeskie Issledovanija*, *Vypusk 22*, 85–101.
38. Chogoshvili, G. 1979. Rezension des Aufsatzes von L. A. Liusternik und S. S. Petrova (Wie Anm. 37): *Zentralblatt für Mathematik und ihre Grenzgebiete*, *395*, Nr. 01002.
39. Gregory, D. F. 1839. "On the solution of linear differential equations with constant coefficients," *The Cambridge Mathematical Journal*, *1*, 22–32, hier 22.
40. Gregory, D. F. 1840. "On the real nature of symbolical algebra," *Transactions of the Royal Society of Edinburgh*, *14*, 208–216, hier 211f.
41. Gregory, D. F. 1839a. "On the solution of linear equations of finite and mixed differences," *The Cambridge Mathematical Journal*, *1*, 54–61; 1839b. "Demonstrations of theorems in the differential calculus and calculus of finite differences," *The Cambridge Mathematical Journal*, *1*, 212–222; siehe auch Anm. 39.
42. Gregory, D. F. 1839b. (Wie Anm. 41), 222.
43. Gregory, D. F. 1841. "On the elementary principles of the application of algebraical symbols to geometry," (November 1839) *The Cambridge Mathematical Journal*, *2*, 1–9, hier 1.
44. Gregory, D. F. 1840. Wie Anm. 40.
45. Novy, Lubos. 1973. *Origins of modern algebra* (Prag), 195.
46. Eine gründliche Erörterung des operationalen Aspekts gibt die Dissertation von Koppelman, Elaine. 1971/2. "The calculus of operations and the rise of abstract algebra," *Archive for History of Exact Sciences*, *8*, 155–242, die auch die Entwicklungslinie von

den französischen Mathematikern über die Mitglieder der Analytical Society zur Formulierung eines neuen Begriffes der abstrakten Algebra in England herausstellt.

47. Gregory, D. F. 1843. "On a difficulty in the theory of algebra," (November 1842) *The Cambridge Mathematical Journal*, *3*, 153–159, hier 153.
48. de Morgan, A. 1841. "On the foundation of algebra," *Transactions of the Cambridge Philosophical Society*, *7*, 173–187; 1842. "On the foundation of algebra, No. II," *Transactions of the Cambridge Philosophical Society*, *7*, 287–300; 1844. "On the foundation of algebra, No. III," *Transactions of the Cambridge Philosophical Society*, *8*, 139–142; 1847. "On the foundation of algebra, No. IV., on triple algebra," *Transactions of the Cambridge Philosophical Society*, *8*, 241–254.
49. de Morgan, A. 1842. (Wie Anm. 48), 292 bzw. de Morgan 1844. (Wie Anm. 48), 139f.
50. Hamilton, W. R. 1853. *Lectures on quaternions* (Dublin), 1–64. Preface = *W. R. Hamilton: The mathematical papers*, hrsg. v. A. W. Conway, J. L. Synge, J. Mc Conwell, H. Halberstam, R. E. Ingram (3. Bde., Cambridge, 1931–1967), hier Bd. *3*, 117–155, insbesondere 133.
51. Bourbaki, N. 1971. *Elemente der Mathematikgeschichte* (Göttingen), 140.
52. Siehe Anm. 50.
53. Hamilton, W. R. 1837. "Theory of conjugate functions, or algebraic couples; with a preliminary and elementary essay on algebra as the science of pure time," *Transactions of the Royal Irish Academy*, *17*, 293–422 = *W. R. Hamilton: The mathematical papers* (Wie Anm. 50), Bd. *3*, 3–96.
54. *W. R. Hamilton: The mathematical papers* (Wie Anm. 50), 96 bzw. 124f.
55. Ohm, M. 1822–1852. *Versuch eines vollkommen consequenten Systems der Mathematik* (9 Teile, Berlin; 3. Auflage der Bde. 1–3: Berlin, 1853–1854).
56. Ohm, M. 1853–1854. (Wie Anm. 55), Bd. *1* (3. Auflage), VII bzw. XIV.
57. Hankel, H. 1867. (Wie Anm. 21), 11.
58. *W. R. Hamilton: The mathematical papers* (Wie Anm. 50), 125.
59. *W. R. Hamilton: The mathematical papers* (Wie Anm. 50), 136.
60. de Morgan, A. 1841. (Wie Anm. 48), 177.
61. *W. R. Hamilton: The mathematical papers* (Wie Anm. 50), 134.
62. Graßmann, H. 1844. *Gesammelte mathematische und physikalische Werke* (Wie Anm. 10), Bd. *2*(1), 434.
63. Klein, F. 1926–1927. (Wie Anm. 20), Bd. *1*, 187.
64. Graßmann, H. 1844. *Gesammelte mathematische und physikalische Werke* (Wie Anm. 10), Bd. *1*(2), 399, bzw. Bd. *3*(2), 197; Lotze, A. 1914–1931. "Die Graßmannsche Ausdehnungslehre," *Encyklopädie der mathematischen Wissenschaften mit Einschluss ihrer Anwendungen*, Bd. *3*(1, 2), (Berlin), 1426–1550, hier 1428.
65. Bourbaki, N. 1971. (Wie Anm. 51), 80 bzw. 142; Lotze, A. 1914–1931. (Wie Anm. 64), 1307.
66. Klein, F. 1926–1927. (Wie Anm. 20), Bd. *1*, 185.
67. Graßmann, H. 1844. (Wie Anm. 10), Bd. *1*(1), 14.
68. Engel, F. 1911. "Graßmanns Leben," (Wie Anm. 12), *3*(2), 204–208.
69. Klein, F. 1926–1927. (Wie Anm. 20), Bd. *1*, 187.
70. Klein, F. 1926–1927. (Wie Anm. 20), Bd. *2*, 36.
71. *W. R. Hamilton: The mathematical papers* (Wie Anm. 50), 123.
72. Cauchy, A. L. 1821. *Cours d'analyse de l'Ecole Royale Polytechnique* (Paris) = Cauchy, A. L. 1882ff. *Oeuvres complètes* (éd. l'Académie des sciences, Paris). 2 Serien, hier Bd. *2*(3), *Cours d'analyse de l'Ecole Royale Polytechnique, 1re partie Analyse algébrique*, 153f.
73. Hankel, H. 1867. (Wie Anm. 21), 14 bzw. 73.

74. Crowe, M. J. 1967. (Wie Anm. 8), 6–12.
75. Francais, J. F. 1813/14. "Nouveaux principes de géometrie de position, et interprétation géométrique des symboles imaginaires," *Annales de mathématiques pures et appliquées*, *4*, 61–71; Argand, J. R. 1813/14. "Essai sur une manière de représenter les quantités imaginaires dans les constructions géométriques," *Annales de mathématiques etc.*, *4*, 133–147; Lettre de Francais. 1813/14. *Annales de mathématiques etc.*, *4*, 222–227; Lettre de M. Servois. 1813/14. *Annales de mathématiques etc.*, *4*, 228–235.
76. Gauß, C. F. 1831. "Theoria residuorum biquadraticorum, commentatio secunda (Selbstanzeige)," *Göttingische Gelehrte Anzeigen* = Gauß, C. F. 1863–1933. *Werke* (Wie Anm. 18), Bd. *2*, 169–178.
77. Servois, F.-J. 1813/14. (Wie Anm. 75), 230.
78. Servois, F.-J. 1814/15. "Essai sur un nouveau mode d'exposition des principes du calcul différentiel," *Annales de mathématiques pures et appliquées*, *5*, 93–140.
79. Servois, F.-J. 1814/15. (Wie Anm. 78), 98.
80. Graßmann, H. 1844. (Wie Anm. 10), Bd. *1*(1), 8.
81. Crowe, M. J. 1967. (Wie Anm. 8), 34.
82. Hankel, H. 1867. (Wie Anm. 21), VI.
83. Hankel, H. 1867. (Wie Anm. 21), 140.
84. Servois, F.-J. 1814/15. (Wie Anm. 78), 140.
85. Klein, F. 1893. *Gesammelte mathematische Abhandlungen* (Wie Anm. 14), Bd. *1*, 489.
86. Boole, G. 1847. *The mathematical analysis of logic, being an essay towards a calculus of deductive reasoning* (Cambridge; Nachdruck Oxford, 1965), 4.
87. Über diese Problematik handelt ausführlich Laita, L. M. 1977. "The influence of Boole's search for a universal method in analysis on the creation of his logic," *Annals of Science*, *34*, 163–176.
88. Boole, G. 1841. "On the integration of linear differential equations with constant coefficients," *The Cambridge Mathematical Journal*, *2*, 114–119, hier 115; Laita, L. M. 1977. (Wie Anm. 87), 169.
89. Bellavitis, G. 1835. "Saggio di applicazioni di un nuovo metodo di Geometria analitica (Calcolo delle equipollenze)," *Annali delle Scienze del Regno Lombardo-Veneto*, *5*, 244–259.
90. Cauchy, A. L. 1853. "Sur les clefs algébriques," *Comptes rendus de l'Académie des Sciences*, *36*, 70–75 und 129–136 = Cauchy, A. L. 1882ff. *Oeuvres complètes* (Wie Anm. 72), Bd *1*(11), 439–445 und Bd. *1*(12), 12–20.
91. Hankel, H. 1867. (Wie Anm. 21), 140.
92. Möbius, A. F. 1827. *Der baryzentrische Calcul, ein neues Hülfsmittel zur analytischen Behandlung der Geometrie dargestellt und insbesondere auf die Bildung neuer Classen von Aufgaben und die Entwicklung mehrerer Eigenschaften der Kegelschnitte angewendet* (Leipzig) = Möbius, A. F. 1885–1887. *Gesammelte Werke*, hrsg. v. R. Baltzer, F. Klein, W. Scheibner (4 Bde., Leipzig), hier Bd. *1*, 1–388.
93. Bourbaki, N. 1971. (Wie Anm. 51), 79.
94. Rothe, H. 1914–1931. "Systeme geometrischer Analyse, 1. Teil," *Encyklopädie der mathematischen Wissenschaften mit Einschluss ihrer Anwendungen* Bd. *3*(1, 2) (Berlin), 1277–1423, hier 1289.
95. Hankel, H. 1867. (Wie Anm. 21), 118.
96. Gibbs, J. W. 1886. "On multiple algebra," *Proceedings of the American Association for the Advancement of Science*, *35*, 37–66 = Gibbs, J. W. 1906. *The Scientific papers*, ed. by H. A. Bumstead, R. Gibbs van Name (2. Bände; Nachdruck New York, 1961), hier Bd. *2*, 91–117, insbesondere 92.
97. Boyer, C. B. 1956. *History of analytic geometry* (New York), 242.
98. Rothe, H. 1914–1931. (Wie Anm. 94), 1292.

99. Siehe Anm. 18.
100. Peano, G. 1888. *Calcolo geometrico secondo l'Ausdehnungslehre di Grassmann, proceduto dalle operazioni della logica deduttiva* (Turin). Siehe Bourbaki, N. 1971. (Wie Anm. 51), 84 und Klein, F. 1926–1927. (Wie Anm. 20), Bd. *2*, 48.
101. Genaueres dazu in Lewis, A. C. 1977. "H. Grassmann's 1844 Ausdehnungslehre and Schleiermacher's Dialektik," *Annals of Science*, *34*, 103–162.
102. Graßmann, H. 1862. *Die Ausdehnungslehre vollständig und in strenger Form* (Berlin) = Graßmann, H. 1894–1911. *Gesammelte mathematische und physikalische Werke* (wie Anm. 10), *1*(2); siehe Klein, F. 1926–1927. (Wie Anm. 20), Bd. *1*, 178.
103. Diesen Begriff entwickelte Graßmann zusammen mit seinem Bruder Robert; siehe dazu die Dissertation von Mehrtens, H. 1979. *Die Entstehung der Verbandstheorie. Arbor scientiarum Reihe A*, *6* (Hildesheim), 24–28.
104. Graßmann, H. 1894–1911. (Wie Anm. 10), Bd. *1*(1), 33.
105. Hankel, H. 1867. (Wie Anm. 21), 12.
106. Lotze, A. 1914–1931. (Wie Anm. 64), 1479.
107. Hilbert, D. 1922. (Wie Anm. 5), 159.
108. Graßmann, H. 1894–1911. (Wie Anm. 102), Bd. *1*(2), 12.
109. Gibbs, J. W. 1961. (Wie Anm. 96), Bd. *2*, 94.
110. Klein, F. 1926–1927. (Wie Anm. 20), Bd. *1*, 188.
111. Neumann, C. 1904. "Über die sogenannte absolute Bewegung," *Festschrift Ludwig Boltzmann gewidmet zum sechzigsten Geburtstage, 20. Februar 1904* (Leipzig), 252–259, hier 259.
112. Heisenberg, W. 1925. "Über quantentheoretische Umdeutung kinematischer und mechanischer Beziehungen," *Zeitschrift für Physik*, *33*, 879–893; Born, M. und P. Jordan. 1925. "Zur Quantenmechanik," *Zeischrift für Physik*, *34*, 858–888; Born, M. W. Heisenberg und P. Jordan. 1926. "Zur Quantenmechanik II," *Zeitschrift für Physik*, *35*, 557–615.
113. Schrödinger, E. 1926. "Über das Verhältnis der Heisenberg-Born-Jordanschen Quantenmechanik zu der meinen," *Annalen der Physik*, 4. Folge, *79*, 734–756, hier 735, Fußnote 2.
114. Cayley, A. 1846. "Sur quelques théorèmes de la géométrie de position," *Journal für die reine und angewandte Mathematik*, *31*, 213–227 = Cayley, A. 1889–1898. *The collected mathematical papers* (14 Bde., Cambridge; Nachdruck New York, 1963/4), hier Bd. *1*, 317–328, insbesondere 321.
115. Cayley, A. 1858. "A memoir on the theory of matrices," *Philosophical Transactions of the Royal Society of London*, *148*, 17–37 = Cayley, A. 1889–1898. *The collected mathematical papers* (wie Anm. 114), Bd. *2*, 475–496.
116. Sylvester, J. J. 1884. (Wie Anm. 7), Bd. *4*, 209; siehe Gibbs, J. W. 1961. (Wie Anm. 96), Bd. *2*, 97.
117. Sylvester, J. J. 1884. (Wie Anm. 7), Bd. *4*, 224.
118. Sylvester, J. J. 1884. (Wie Anm. 7), Bd. *4*, 210.
119. Peirce, B. 1881. "Linear associative algebra," *American Journal of Mathematics*, *4*, 97–229; siehe Struik, D. J. 1962. *Yankee science in the making* (New York), 415. Und Pycior, H. M. 1979. "Benjamin Peirce's Linear associative algebra," *Isis*, *70*, 537–551.
120. Bumstead, H. A. 1961. "Josiah Willard Gibbs," in *J. W. Gibbs*: *The scientific papers* (wie Anm. 96), Bd. *1*, XI–XXVI, hier XVIII.

Institut für Philosophie, Wissenschaftstheorie
Wissenschafts- und Technikgeschichte
Technische Universität Berlin
Berlin, Bundesrepublik Deutschland

An Early Version of Gauss's *Disquisitiones Arithmeticae**

UTA C. MERZBACH

In 1975, while working at the Staatsbibliothek Preussischer Kulturbesitz in Berlin, the author found two sheets in the papers of G. P. L. Dirichlet which seemed very much like the first part of Section 3 of Gauss's *Disquisitiones arithmeticae*. Comparison of the handwriting and paper showed that the pages indeed matched parts of a manuscript by Gauss, his *De Analysis Residuorum* in the Niedersächsische Staats- und Universitätsbibliothek in Göttingen. A year later, another unidentified manuscript was discovered in the Dirichlet Nachlass in the archives of the Akademie der Wissenschaften der DDR. Together, these two finds provide an early version of Gauss's *Disquitiones arithmeticae*. This paper discusses the significance of these documents which represent the only cohesive sources illustrating Gauss's work prior to 1799.

In September 1975, while studying the Nachlass Dirichlet in the manuscript division of the Staatsbibliothek Preussischer Kulturbesitz in West Berlin, I encountered among some unidentified students' papers two sheets (8 pages) entitled "Caput tertium. De Residuis functionum exponentialium." The manuscript in question read like a paraphrase of the first part of Section 3 of Gauss's *Disquisitiones arithmeticae* (*DA*). Comparison of handwriting

* Dedicated to Professor Dr. Kurt-R. Biermann on the occasion of his 60th Birthday.

ISBN 0-12-204050-3

and paper revealed that the manuscript fragment matched the portions of the Gauss manuscript in the Niedersächsische Staats- und Universitätsbibliothek Göttingen known as *De Analysis Residuorum* (*AR*) and partly reproduced in Gauss [1863b, 199–240]. Aside from confirming that the *AR* was an earlier version of the *DA*, the newly found manuscript appeared to have little more than antiquarian interest.

On 3 June 1976, while working on the Dirichlet Nachlass in the Archive of the Akademie der Wissenschaften der DDR, I found another unidentified manuscript headed "Elementa doctrinae Residuorum./Caput Primum" [Archiv-Signatur: Nachlass Dirichlet No. 55]. The manuscript contained most of four chapters corresponding to the first four sections of the *DA*; the pagination of the two sheets in West Berlin exactly matched the beginning of the third chapter missing in this manuscript.

Together with the previously known Chapters 6–8 of the *AR* in Göttingen, these two finds provide us with an early version of Gauss's *DA*, complete except for that portion of Chapter 4 starting with the statement and proof of the quadratic reciprocity law, the important Chapter 5, and some isolated pages. As it is the only cohesive source document known to us that illustrates Gauss's work prior to 1799, it merits some consideration.

PROVENANCE

The existence of the Göttingen manuscript has been known to Gauss scholars since 1863 when part of Chapter 6 and Chapter 8 were published with editorial notes by Dedekind [Gauss 1863b, 199–240]. From that time until the winter of 1896–1897 no one seems to have been aware of the existence of any part of the *AR* other than Chapters 6–8. In a letter of 10 June 1885, Ernst Schering, who edited Volumes 1–6 of Gauss's *Werke* (published under the auspices of the Gesellschaft der Wissenschaften in Göttingen), mentioned that he had looked for Gauss's manuscript of the *DA* in Dirichlet's house. Writing to Kronecker, who was preparing the edition of Dirichlet's *Mathematische Werke*, Schering stated that when he sorted and catalogued Dirichlet's library after his death, he searched diligently for the *DA* manuscript because Dirichlet had once told him

> that when he was visiting Gauss, Gauss made the *fidibus* with which he lit his pipe from the old manuscript of the Disquiss. Arthm. Dirichlet expressed his great surprise at this, and, upon his request, received from Gauss the remaining portion of the manuscript of the diss. arith. [Translated from Akademie der Wissenschaften der DDR—Archiv-Signatur: Nachlass Dirichlet No. 67/3]

Schering may have missed the *AR* manuscript because he did not have access to all of Dirichlet's papers; or he may have confined his diligence to searching for a manuscript that corresponded to the published version of the *DA*. Whatever the circumstances, a manuscript surfaced and was identified by Dedekind in the winter of 1896–1897. On 13 February 1897, Dedekind, who was advising L. Fuchs on questions arising during the preparation of Volume 2 of Dirichlet's *Werke*, wrote to him:

> My thoughts are particularly occupied with the quarto folder of antique appearance which you showed me and in which I thought to recognize the handwriting of Gauss, presumably the first treatment of the beginning of the Disquisitiones Arithmeticae. Unfortunately so little time existed for closer examination that it now appears to me like a dream! [Translated from Akademie der Wissenschaften der DDR—Archiv-Signatur: Nachlass Dirichlet No. 76/9]

Dedekind continued the letter by suggesting that, since upon his appointment to Göttingen in 1855 Dirichlet was charged with sorting the Gauss Nachlass preparatory to publication of Gauss's *Werke*, it was likely that the Gauss material had become mixed up with his own papers and so was subsequently sent to Berlin as part of the Dirichlet Nachlass. While registering surprise that those charged with reviewing Dirichlet's Nachlass had not identified the manuscript, he encouraged Fuchs to pursue the matter further, particularly because he felt that Göttingen should lay claim to the Gauss material.

When Fuchs informed Schering in Göttingen of the find and of Dedekind's supposition, Schering repeated the story he had written Kronecker 12 years earlier:

> The piece of Gauss's Disquisitiones Arithmeticae, which is found among Dirichlet's papers, is probably that portion which, as Dirichlet told me himself, he saved from the hand of Gauss when the latter lit his pipe with his manuscript of the disquisitiones arithmeticae on the day of his doctoral jubilee. [Translated from Akademie der Wissenschaften der DRR—Archiv-Signatur: Nachlass Dirichlet No. 81][1]

Dedekind expressed skepticism when informed of Schering's tale, reiterating his conjecture that the manuscript had accidentally found its way from Gauss's to Dirichlet's Nachlass. He noted that if Gauss had saved the manuscript for 50 years he presumably valued it, and that if the anecdote were true, Dirichlet surely would have told it to him [2]!

Further discussion, verification, or publicizing of the manuscript was prevented by the pressure on Fuchs to produce the long overdue second volume of Dirichlet's *Werke* and by the death of Schering in the same year. Thus neither the existence of the manuscript nor this related correspondence came to light until now[3].

The materials collected by Kronecker and Fuchs for the publication of Dirichlet's works, under the auspices of the Berlin Academy, make up the Dirichlet Nachlass now in the Archive of the Akademie der Wissenschaften der DDR. Aside from Dirichlet manuscripts, this contains two appendixes. The first is composed of letters and papers by others; this includes the Gauss manuscript. The other is made up of the editors' notes and correspondence concerning the Dirichlet *Werke*.

It appears that the manuscript portion now in West Berlin had become separated from the mathematical papers turned over to Dirichlet's editors and remained in the Dirichlet household, along with a group of letters addressed to Dirichlet that were retained by the family. This material was purchased by the Staatsbibliothek Preussischer Kulturbesitz in 1969 at an auction of materials from the Nachlass of the philosopher Leonard Nelson, a great-grandson of Dirichlet.

DESCRIPTION

The manuscript in all three locations is made up of sheets primarily measuring approximately 205 × 330 mm. These sheets have been folded once in sets of two, forming "quarto" sets of four leaves or eight pages. The first page of each set carries the signature on the upper right-hand corner: A, . . . P, . . . LL, . . . TT, . . . $\circ$, . . . ,$\overset{\circ}{\underset{\circ\circ}{\circ\circ}}$. Missing are the top sheets of signature sets B and F, parts of K and M, the bottom sheet of P, all of Q through KK, and whatever existed of UU through ZZ or after $\overset{\circ}{\underset{\circ\circ}{\circ\circ}}$.

Except for these missing parts, the Akademie der Wissenschaften der DDR has signature sets A through D and F through P; the Staatsbibliothek Preussischer Kulturbesitz has E; the Niedersächsische Staats- und Universitätsbibliothek Göttingen has LL through TT.

After the first signature set, page numbers appear only on the first page of each set (a continuous pagination in the Göttingen manuscript is a later addition). On the eighth set, properly designated H and 57, a previous J and 65 have been crossed out. What should be page 77 is marked 89 in the fragment of set K. As these particular sets show numerous corrections, it is reasonable to assume that these discrepancies are due to Gauss's reworking of that portion of the manuscript. Similarly, the fragmentary set M has substitute pages 90, and an inserted slip of paper.

The designation "Analysis Residuorum" appears next to the signatures on the Göttingen sets. Although it is apparently a later addition, I will follow tradition by using it to refer to the Göttingen portions of the manuscript, and also apply it to the newly found portions in Berlin.

CONTENTS

The *AR* is divided into eight chapters. The first is headed "Elements of the Doctrine of Residues." The following three are entitled

II. "Of Residues of a Function of the First Degree,"
III. "Of Residues of the Exponential Function,"
IV. "Of Residues of a Function of the Second Degree"

[translated from Akademie der Wissenschaften der DDR—Archiv-Signatur: Nachlass Dirichlet No. 55]. The title of the fifth chapter is not known. The last three are entitled

VI. "Solution of the congruence $x^m - 1 = 0$ and of the Equation $x^m - 1 = 0$; with Excursions in the Theory of Regular Polygons,"
VII. "Various Applications of the Preceding Investigations,"
VIII. "General Disquisition on Congruences"

[translated from Niedersächsische Staats- und Universitätsbibliothek Göttingen: Gauss. Manuskripte 50, 51] [4].

The first four chapters of the *AR* closely parallel the first four sections of the *DA*. The substance of the material is largely the same, as is much of the wording. The most notable omission in the *AR* is the fundamental theorem of §42 of the *DA*. Obvious differences are found in the numerical examples: many times the numbers have been changed; often no numeric example appears in the *DA* where one is given in the *AR*. There are some notable differences in terminology: "*quicunque*" has generally been replaced by "*indeterminata*" in modifying "*functio*"; "*potestas*" has taken the place of "*dignitas*"; use of the difference notation Δ (*AR*, §37) has been abandoned in the *DA*; Euler's famous ϕ-function, which is designated by ϕ in §38 of the *DA*, is still described in words, without the ϕ-symbol, in the corresponding §35 of the *AR*. The exposition in the *DA* has been tightened; occasional digressions in the *AR*, such as remarks concerning related work by others or differences in proof techniques, have been either shortened or omitted in the *DA*; proofs are often abbreviated. As may be expected, the *AR* contains errors and, especially after the third chapter, shows traces of the toil that certain proofs entailed.

The relationship of the last three chapters of the *AR* to Sections 6 and 7 of the *DA* and to the unpublished Section 8 was outlined by Dedekind in Volume 2 of Gauss's *Werke*, pp. 240–242. Although much of the material is the same, the approach to Chapters 6–8 of the *AR* differs from that to Sections 6–8 of the DA. The reason for this difference, which presents a striking contrast to the similarity of Chapters 1–4 and Sections 1–4, emerges if one regards the circumstances surrounding the creation of both versions.

CREATION OF THE *AR*

As is well known, Gauss discovered a substantial amount of number-theoretic results by induction in the years 1792–1795. Numerous relics of his occupation with prime numbers during these years of adolescence exist in the forms of tables and scattered notes; they confirm his later statements concerning these discoveries [5]. There is no evidence that he proved many of his results at that time.

It is not known to what extent his teachers in Braunschweig had first-hand knowledge of works of Euler, Lagrange, or Legendre [6]. We do know, however, that Gauss began his serious study of Euler the week he matriculated in Göttingen. On 19 October 1795 he wrote to his counselor Zimmermann that he had visited the library and taken home several volumes of the St. Petersburg *Commentarii*. He continued:

> I cannot deny that I am chagrined to find that I have made the largest part of my beautiful discoveries in indeterminate analysis only for the second time. What comforts me is this: all discoveries of Euler which I have found so far I have made also, and a few more. I have hit upon a general, and as I believe, natural point of view; I still see an immeasurable field before me and Euler made his discoveries over a period of many years, after having previous *tentaminibus*. [Translated from Niedersächsische Staats- und Universitätsbibliothek: Gauss-Briefe B (Zimmermann)][7]

From the record of his library use we know also that by January 1796 he was supplementing these studies with the reading of Lagrange's works in the *Memoirs* of the Berlin Academy [8].

On 29 March 1796 Gauss achieved the construction of the 17-gon and a week later he apparently obtained his first proof of the quadratic reciprocity law [9]. These two accomplishments, which occasioned his starting the journal that he kept until 1814, mark the beginning of a new stage of growth: from here on, the ingenious results derived from the table constructions and manipulations that had characterized his early endeavors as a boy were accompanied by the skillful proofs that marked the mathematician.

During the spring of 1796, Zimmermann apparently encouraged Gauss to publish the results of his studies. On 26 May, Gauss wrote to him that he was prepared to undertake such a project. He had been made confident by his recent reading of the 1785 memoir in which Legendre justified his statement of the quadratic reciprocity law by assuming that which Gauss had just succeeded in proving [Schlesinger 1933, 20–21]. He was also reassured by his reading the recently translated *Additions* by Lagrange to Euler's *Algebra*. He acknowledged that it would be suitable to write the book in Latin, but expressed hesitation and a desire first to work it out in German; the Latin would take him longer and make him subject to criticism "from another side." After speculating on the title he would like to give the book—favoring *Theory of Quadratic Residues with Allied Investigations*—he observed:

> Since I have an Euler and a Lagrange as predecessors I shall have to marshall great diligence for the composition itself; but I think I shall be able to complete it sufficiently soon if I am certain of a publisher. I am pleased with any conditions and if I am permitted to add some, it would be for a number of free copies that is not too small. [Translated from Niedersächsische Staats- und Universitätsbibliothek: Gauss. Briefe B (Zimmermann)]

Thus the text of the *AR* was created some time after May 1796. It is plausible that Gauss made a preliminary outline and started a partial draft in 1796, but did not write down the manuscript under consideration until 1797, completing it in late summer or fall. This is inferred largely from the following evidence:

1. In a footnote to §27 of the *AR*, the Kausler translation into German of Lagrange's *Additions* is referred to as having been received "in the preceding year" [10]. This was the translation mentioned in Gauss's letter of May 1796 quoted earlier. It has the publication date 1796 and was reviewed in the *Göttingische Anzeigen* of 7 July 1796, pp. 1,077–1,080. However, on a sheet of paper, which seems to be a discarded page from an earlier draft, the footnote reference to the *Additions* contains the phrase "of which a German version was recently [*nuper*] issued and others are expected" [translated from Niedersächsische Staats- und Universitätsbibliothek: Gauss. Varia 6 Bl. 43].
2. As noted above, the important theorem and proof of §42 of the *DA* is not present in the corresponding Chapter 3 of the *AR*. According to Gauss's journal, he established the theorem in August 1797 [11]. This suggests that the first three chapters of the *AR* had been written down by then.
3. The beginning of the *AR* displays unusually clear penmanship, typical of material that has been copied over. The handwriting deteriorates, however, and, as also noted earlier, by the time one reaches Chapter 4 the manuscript reflects the concentrated effort and toil that went into it.
4. In his journal, Gauss referred to the proofs of several theorems appearing in Chapters 6–8 of the *AR* as originating in July, August, and September of 1797 [12].
5. From a letter to Zimmermann of 22 November 1797 it is clear that Gauss had begun to prepare a final manuscript for the printer which constitutes a revision of a previous manuscript. References in this letter, together with the associated textual evidence, leave no doubt that the earlier manuscript is the *AR* version, the new manuscript that of the *DA*.

This letter also accounts for some of the stylistic differences between the two texts. Zimmermann was having the revised manuscript reviewed by the young philologist Meyerhoff, who subjected Gauss's Latin to considerable criticism and suggested improvements. Gauss rejected several of the suggestions, arguing that he was following the classical model of Huygens and that Meyerhoff was insufficiently acquainted with mathematics and mathematical terminology [13]. Nevertheless, Meyerhoff's advice left its mark.

CREATION OF THE *DA*

On the basis of several known letters and the 1801 publication we can outline some of the events surrounding the printing of the *DA*, and take note of certain material added to the revised manuscript which cannot have been included in the missing fifth chapter of the *AR* manuscript.

In March 1797, Gauss had sent to Zimmermann a detailed outline of the projected book. In the accompanying letter he projected the ultimate size of the work as being approximately 400 "Artikel" (§§), which he estimated would amount to roughly one quarto alphabet, perhaps less in octavo. In this letter he also suggested the title *Disquisitiones arithmeticae* [Gauss 1917, 19–21]. Having gained the support of Carl Wilhelm Ferdinand, the Duke of Brunswick, Zimmermann put Gauss in touch with the printer Kircher some time later in 1797. Kircher, originally from nearby Goslar, had been established in Braunschweig for a decade [Schmidt 1907; Haenselmann 1878]. We learn from Gauss's letter of 22 November that Kircher promised to process three folios per week. Gauss had just received Meyerhoff's comments on the first three signature sets of the revised manuscript. He suggested that, given the printer's estimate, it would be "time enough if he starts a quarter year before the Easter fair, in other words *about mid-January*, by which time . . . I should have more than half finished and be able to keep pace with him" [translated from Niedersächsische Staats- und Universitätsbibliothek: Gauss. Briefe B (Zimmermann)] [13].

Gauss followed through, sending Zimmermann signature sets K through P of the revised manuscript by Christmas, and promising another nine or ten sets within a fortnight.

If he had thought the book would be finished in time for the Easter book fair in Leipzig, he was disappointed. Printing only began in April 1798, and the printer proved to be "a very phlegmatic man with whom all remonstrations and pleas are of little help" [Schmidt-Staeckel 1899, 11]. One of Kircher's problems may have been that he was attempting to maintain several branches of his printing enterprise in nearby communities [Schmidt 1907]. At any rate, by September 1798, only five folios, comprising less than the first three sections, had been printed. By November, seven folios were completed and the eighth, taking him through the end of §128, was in press [Schmidt-Staeckel 1899, 11].

Gauss used the delay to rework the fifth Section. Writing to Bolyai on 29 November 1798, he noted:

> With this section, which is the most substantial in the entire work, I have had many *fata* already; the present version is the fourth; with each one I have succeeded to carry out the thing in a manner exceeding the boldest hopes entertained for the preceding one, and in a few days I shall have completed for the fourth time what, during the entire summer, I worked out for the third. [Translated from Schmidt-Staeckel 1899, 12]

With the new material of Section 5, the bulk of the work increased. This and the continuing delays caused by the printer pushed back the publication date even further. On 9 January 1799, writing to Bolyai, Gauss observed that it would be impossible to have 30 or more folios printed by Easter, since he was still awaiting proofs of the 11th. In other words, printing of Section 5 was barely beginning. Nevertheless, the publication was announced at the 1799 book fair [Schmidt-Staeckel 1899, 15].

Gauss continued to add important new results to the expanding Section 5. Some of these he singled out for mention in his journal notes and in dated marginal notations to an unbound copy of the *DA* [Gauss 1917, 539–556; Gauss 1863a, 476]. Thus we learn that the fundamental program of §266 emerged in February 1799 and that much of the material on the theory of ternary forms and class number determinations begun in 1798 was developed during 1799 and 1800.

He had ample time to include new material in the book. By April 1799 the printer had only reached §190; soon thereafter Kircher moved back to Goslar, as a result of which printing ceased over the summer. Only in December could Gauss resume reading proof [Schmidt-Staeckel 1899, 22, 35]. The process continued at a snail's pace. Typically, Gauss wrote to Professor Hellwig of Braunschweig on 28 July 1800 that he was expecting proof to which he must add a folio's worth of new material not contained in the first version of the work [Gauss 1929, 207]. In October 1799 Gauss had again estimated the book would come to between 30 and 40 folios. In the end his estimate was exceeded slightly, the final version reaching 43 folios without the introductory material.

CONCLUSION

The circumstances contributing to the growth of Section 5 account for the difference in size between the *AR* and the *DA*. There is a more fundamental difference between the two versions of the work, however. The manuscript called *Analysis residuorum* documents the hard work of a young genius; the printed text of the *Disquisitiones arithmeticae* presents the output of a mature mathematician. The *AR* was built around the concept of congruence and residue. Opening with the definition of congruence, it was to close with a chapter on the general theory of congruences. Major results, such as the quadratic reciprocity law and the solution of the cyclotomic equation, were to be placed in this framework, which provided a "general and . . . natural point of view" for Gauss's early discoveries. The deeper researches and profound discoveries that followed the completion of the *AR* and the printing of the first $3\frac{1}{2}$ chapters of the *DA* forced him to break out of this framework. This explains the seemingly pointless reshuffling of material in Sections 6 and 7, just as it accounts in part for the omission of the original conclusion to the work, which apparently remained incomplete. The *Analysis residuorum* was an eighteenth-century masterpiece; but only the *Disquisitiones arithmeticae* could become the seedcase for nineteenth- and twentieth-century number theory.

NOTES

1. A copy of this letter, which is dated 10 March 1897, along with Fuchs's letter, dated 8 March 1897, is in the Niedersächsische Staats- und Universitätsbibliothek: Gauss-Briefe A (Dirichlet).
2. Letter, Dedekind to Fuchs, of 28 April 1897: Akademie der Wissenschaften der DDR—Archiv-Signatur: Nachlass Dirichlet No. 76/10.
3. G. W. Dunnington knew of the correspondence and used Schering's account, however. See Dunnington 1955, 276.
4. Also Gauss 1863b, 199, 212, 240.
5. See, for instance, Bachmann 1922, 4–5 and Schlesinger 1933, 17.
6. Schlesinger 1933, 17.
7. This excerpt is reprinted by Schlesinger 1933, 19.
8. The list of books borrowed by Gauss was compiled by the editors of his *Werke* and is in the Niedersächsische Staats- und Universitätsbibliothek: Gauss-Archiv. It was published in Dunnington 1955, 398–404.
9. The proof of the construction was announced publicly in the spring of 1796. See Gauss 1917, 3–4, 125 (letter to Gerling of 6 January 1819), 488–489.
10. ". . . in supplem. quae M. LaGrange adversionem gallicam Algebrae Euleri adiecit, quarumque anno praec. versionem germanicam a Dm. *Kaussler* accepimus": Akademie der Wissenschaften der DDR—Archiv-Signatur: Nachlass Dirichlet No. 55.
11. Entry of 23 July and editors' notes in Gauss 1917, 520. Also see Dedekind 1931 and Note 14 following.
12. In addition to the 23 July entry, see, for example, those for 26, 30, 31 August and 9 September in Gauss 1917, 519–523. Also see Bachmann 1922.
13. Niedersächsische Staats- und Universitätsbibliothek: Gauss. Briefe B (Zimmermann). A translated excerpt appears in Dunnington 1955, 38.

REFERENCES

Bachmann, P. 1922. "Ueber Gauss' zahlentheoretische Arbeiten," in Gauss, C. F. 1922. *Werke*, *10*, pt. 2, no. 1.

Berlin. Akademie-Archiv. Akademie der Wissenschaften der DDR. Nachlass Dirichlet.

Berlin. Handschriftenabteilung. Staatsbibliothek Preussischer Kulturbesitz. Nachlass Dirichlet.

Dedekind, R. 1931. "Ueber einen arithmetischen Satz von Gauss," [1892]. In *Gesammelte Mathematische Werke*, *2*, 28–39.

Dunnington, G. W. 1955. *Carl Friedrich Gauss: Titan of Science* (New York: Exposition Press).

Gauss, C. F. 1863a. *Werke 1*.

Gauss, C. F. 1863b. *Werke 2*.

Gauss, C. F. 1917. *Werke 10*, pt. 1.

Gauss, C. F. 1927. *Werke 12*.

Göttingen. Handschriftenabteilung. Niedersächsische Staats- und Universitätsbibliothek. Gauss-Archiv.

Haenselmann, L. 1878. *Karl Friedrich Gauss. Zwoelf Kapitel aus seinem Leben* (Leipzig: Duncker & Humblot).

Schlesinger, L. 1933. "Ueber Gauss' funktionentheoretische Arbeiten." In Gauss, C. F. 1933. *Werke*, *10*, pt. 2, no. 2.

Schmidt, F. & P. Staeckel, ed. 1899. *Briefwechsel zwischen Carl Friedrich Gauss und Wolfgang Bolyai* (Leipzig: B. G. Teubner).

Schmidt, R. 1907. *Deutsche Buchhändler* (Eberswalde), *4*, 598–599.

The Smithsonian Institution
Washington, D.C.

Über die Anstöße zu Kummers Schöpfung der "Idealen Complexen Zahlen"*

OLAF NEUMANN

In the present article the question is considered, What motivated Kummer to introduce his "ideal numbers"? For this inquiry we make use of those primary sources which were quite recently discovered by Kurt-R. Biermann (GDR) [1967, 1973] and Harold M. Edwards (USA) [1975, 1977a] and of some new mathematical and chronological arguments. The calculus of Gaussian and Jacobian sums and the search for higher reciprocity laws for power residues of algebraic numbers allow us to explain in an entirely satisfactory manner Kummer's development of "ideal numbers" without any reference to Fermat's last theorem.

1. ZIELSTELLUNG

Ernst Eduard Kummer (1810–1893) ging in die Geschichte der Zahlentheorie und Algebra ein als der Schöpfer der "idealen complexen Zahlen" (heute: Divisoren) in Kreisteilungskörpern. Diese epochemachende Neuerung erlaubte ihm, die Kreisteilungs-Arithmetik (explizite Klassenzahlformeln, Struktur der Klassen- und Einheitengruppe) umfassend zu untersuchen und in zwei Problemkreisen einen Durchbruch zu erzielen: den Reziprozitätsgesetzen für höhere Potenzreste und der Fermatschen Vermutung [Kummer 1975].

* Herrn Professor Dr. Kurt-R. Biermann zum 60. Geburtstag am 5. Dezember 1979 gewidmet.

ISBN 0-12-204050-3

In der vorliegenden Note soll den Beweggründen, die Kummer zur Einführung der "idealen Zahlen" brachten, nachgespürt werden, und zwar an Hand der Briefe und der ersten zahlentheoretischen Arbeiten Kummers sowie derjenigen Primärquellen, die erst in neuerer Zeit von Kurt-R. Biermann (Berlin, DDR) [1967, 1973] und Harold M. Edwards (New York, N.Y., USA) [1975, 1977a,b] erschlossen wurden.

Der zweite und dritte Abschnitt dieses Aufsatzes behandeln das Verhältnis Kummers zur Fermatschen Vermutung und beleuchten kritisch die von verschiedenen Autoren dazu geäußerten Ansichten.

Mit Kummers Erfolgen in der Behandlung des Fermatschen Problems wurde spätestens um die Jahrhundertwende die heute weitverbreitete Ansicht verknüpft, dieser sei in erster Linie oder gar ausschließlich durch die Beschäftigung mit der Fermatschen Vermutung zur Einführung seiner "idealen complexen Zahlen" angeregt worden. Kein Geringerer als David Hilbert (1862–1943) sagte 1900 in seinem berühmten Vortrag über "Mathematische Probleme":

> Fermat hatte bekanntlich behauptet, daß die diophantische Gleichung—außer in gewissen selbstverständlichen Fällen—
>
> $$x^n + y^n = z^n$$
>
> in ganzen Zahlen x, y, z unlösbar sei [siehe Fermat 1932]; [. . .] durch die Fermatsche Aufgabe angeregt, gelangte Kummer zu der Einführung der idealen Zahlen und zur Entdeckung des Satzes von der eindeutigen Zerlegung der Zahlen eines Kreiskörpers in ideale Primfaktoren. [Hilbert 1900; 1935; 1971, 24–25]

Im gleichen Sinne wie Hilbert, aber wesentlich detaillierter äußerte sich 1910 Kurt Hensel (1861–1941) (siehe Abschnitt 2). Offensichtlich unter dem Einfluß von Hilbert und Hensel wird z. B. in den mathematikgeschichtlichen Darstellungen von Leonard Eugene Dickson (1874–1954) [1920, vol. II, p. 738], Felix Klein (1849–1925) [1926/27, I.S.321], Nicolas Bourbaki [1965; 1974] und I. G. Bašmakova et al. [1978, p. 93] die oben gekennzeichnete Meinung vertreten, der Anstoß zur Kummerschen Schöpfung der "idealen Zahlen" wäre im Fermatschen Problem zu suchen. Wir werden hier an Hand der heute zugänglichen Primärquellen zeigen, daß die zitierte Äußerung Hilberts nicht zutrifft. Es ist auch nicht bekannt, daß irgendeiner der Kummer nahestehenden zeitgenössischen Mathematiker wie Leopold Kronecker (1823–1891), Peter Gustav Lejeune Dirichlet (1805–1859) oder Carl Gustav Jacob Jacobi (1804–1851) die Einführung der "idealen Zahlen" auf den Einfluß des Fermatschen Problems zurückgeführt hätte. H. M. Edwards [1977b, 79] gelangte unlängst zu dem Schluß: "It is widely believed that Kummer was led to his 'ideal complex numbers' by his interest in Fermat's Last Theorem, but this belief is surely mistaken." Im vorliegenden Aufsatz werden die wichtigsten Schlußfolgerungen Edwards' durch neue mathematische und chronologische Argumente erhärtet, z.T. aber auch modifiziert. Im vierten Abschnitt zeigen wir, daß der Hauptanstoß zur Schöpfung der "idealen Zahlen" im Kalkül der Gaußschen und Jacobischen Summen und der Theorie der Potenzreste zu suchen ist. Man kann also Kummers Hinwendung zur Arithmetik der Kreisteilungskörper und seine Einführung der "idealen Zahlen" zwanglos und befriedigend *ohne* jeden Bezug auf die Fermatsche Vermutung erklären.

Im dritten und vierten Abschnitt stehen die zusammenhängende mathematische Interpretation und Wertung der Quellen und ihre Einbettung in Kummers Gesamtwerk und dessen "mathematische Umwelt" im Vordergrund. Hierbei werden einige algebraisch-zahlen-theoretische Zusammenhänge herangezogen, die offenbar in der einschlägigen Literatur bisher nicht beachtet worden sind.

An dieser Stelle möchte ich Herrn Professor Dr. Kurt-R. Biermann, der mich auf die Publikationen von H. M. Edwards aufmerksam machte und der mein Manuskript durchsah, herzlich danken. Für nützliche Hinweise zum Manuskript bin ich ebenfalls Herrn Dr. Herbert Pieper (Berlin, DDR) zu großem Dank verpflichtet.

Die hier verwendete Terminologie ist in der Regel die heute übliche, um die Darlegung nicht durch unwesentliche Details zu belasten. Die Symbole $\mathbb{Z}$ und $\mathbb{Q}$ werden für den Ring der ganzrationalen Zahlen und den Körper der rationalen Zahlen stehen. Für den Stand der Gleichungstheorie und der Algebra um 1845 sei der Leser auf einschlägige Darstellungen verwiesen [Bašmakova 1978, Dieudonné 1978, Kiernan 1972, Nový 1973, Purkert 1973, Wußing 1969]. Unter "complexen Zahlen" im Sprachgebrauch der Zahlentheoretiker um 1845 sind in der Regel algebraische Zahlen zu verstehen, unter "reellen (algebraischen) Zahlen" rationale Zahlen.

2. ÜBER EINIGE ÄUSSERUNGEN HENSELS

Kurt Hensel hat 1910 aus Anlaß des 100. Geburtstages von Ernst Eduard Kummer bei zwei Gelegenheiten [1910a,b] behauptet, Kummer hätte einst eine "druckfertige Abhandlung" [1910b, 13] mit einem Beweisversuch für die wohlbekannte Fermatsche Vermutung an P. G. Lejeune Dirichlet übergeben. "Nach einigen Tagen" [Hensel 1910a, 22] hätte Dirichlet das als fehlerhaft erkannte Manuskript an Kummer zurückgegeben. Im einzelnen sagte Hensel in seiner Berliner Rede, es wäre

> durch ganz einwandfreie Zeugnisse, unter anderen durch das des Herrn Gundelfinger belegt, der die Mitteilung dem Mathematiker Grassmann verdankte, daß Kummer in dieser Zeit des Vorwärtsstrebens wirklich einen vollständigen Beweis des großen

> Fermatschen Satzes gefunden zu haben glaubte und ihn im Manuskript Dirichlet vorgelegt hat . . . Nach einigen Tagen gab ihn Dirichlet mit dem Urteile zurück, der Beweis sei ganz ausgezeichnet, [aber nur dann] richtig, wenn es feststände, daß die Zahlen in α . . . in unzerlegbare Faktoren [zu zerlegen] . . . *nur auf eine Weise* möglich wäre. [. . .] leider schienen ihm die Zahlen in α jene Fundamentaleigenschaft wirklich nicht allgemein zu haben. [Hensel 1910a, 22]

Bei den hier genannten Zeugen handelte es sich um die Mathematiker Sigmund Gundelfinger (1846–1910), Professor am Polytechnikum Darmstadt, und Hermann Günther Grassmann (1809–1877), seit 1836 Lehrer, seit 1852 Gymnasialprofessor für Mathematik und Physik in Stettin. Mit dem Symbol α bezeichnete Hensel eine primitive λ-te Einheitswurzel, wobei unter λ eine ungerade Primzahl zu verstehen ist. α ist also eine komplexe Zahl mit der Eigenschaft $\alpha^\lambda = 1$. Die von Hensel angeführten "Zahlen in α" sind die ganzzahligen Linearkombinationen der Potenzen von α, d.h. die komplexen Zahlen

$$a_0 + a_1\alpha + \cdots + a_{\lambda-2}\alpha^{\lambda-2} \tag{2.1}$$

($a_0, a_1, \ldots, a_{\lambda-2}$ ganzrational). Diese Zahlen werden wir im folgenden auch "ganze Kreisteilungszahlen" nennen; sie bilden wegen der Relation

$$\frac{\alpha^\lambda - 1}{\alpha - 1} = \alpha^{\lambda-1} + \alpha^{\lambda-2} + \cdots + \alpha + 1 = 0 \tag{2.2}$$

einen Ring, für den wir das heute übliche Symbol $\mathbb{Z}[\alpha]$ gebrauchen. Die Theorie des Ringes $\mathbb{Z}[\alpha]$ werden wir im weiteren als "Kreisteilungs- Arithmetik" bezeichnen.

Hensel gab in seiner Berliner Gedenkrede nur die mündlichen Überlieferungen durch Graßmann und Gundelfinger als Quelle an, während in seinem Marburger Vortrag [1910b] keine wie auch immer gearteten Quellen genannt wurden. Aber die Marburger Rede enthielt einige wenige *Zeitangaben*, die im Berliner Vortrag fehlten und die hier diskutiert werden sollen.

Hensel sagte:

> Dieser Gedanke [des Arbeitens mit ganzen Kreisteilungszahlen] war es, den der Liegnitzer Gymnasiallehrer Ernst Eduard Kummer . . . erfaßte und in mehrjähriger angestrengter Arbeit zur Reife brachte. [1910b, 12]
>
> Nach mehrjährigen Versuchen gelingt es ihm [d.h., Kummer], den Beweis zu führen, daß wirklich auch in diesem erweiterten Bereiche [der ganzen Kreisteilungszahlen] jede Zahl in nicht weiter zerlegbare Elemente zerfällt, so daß er sich zu der Annahme berechtigt glaubt, daß nun auch alle Zahlgesetze der Gaußschen Arithmetik [also insbesondere die Eindeutigkeit der Primzerlegung] genau ebenso erfüllt sind, wie im Bereiche der natürlichen Zahlen. [1910b, 13]

> In seinem feurigen vorwärts strebenden Geiste hielt Kummer nun die Zeit für gekommen, mitinem Schlage das vielumworbene heißumstrittene Fermatsche Problem vollständig zu lösen. [1910b, 13]

Weiter erwähnte Hensel den Gedankenaustausch zwischen Kummer und Dirichlet über das Fermatsche Problem und Kummers Einführung der "idealen Zahlen." Er fuhr mit den Worten fort: "Nun konnte Kummer nach sechsjähriger weiterer Geistesarbeit als Meister in diesem Reiche der [ganzen Kreisteilungs-] Zahlen seinen damals versuchten Beweis des Fermatschen Satzes aufnehmen" [1910b, 16], und er war in der Lage, die Gültigkeit der Fermatschen Vermutung in vielen Fällen nachzuweisen. Offensichtlich dachte Hensel hier an Kummers bahnbrechende Arbeit aus dem Jahre 1847 [Kummer 1847]. Wenn man von 1847 um sechs Jahre zurückrechnet, gelangt man zu dem Schluß, daß Kummers Beweisversuch mit Hilfe der Kreisteilungs-Arithmetik und der Gedankenaustausch mit Dirichlet in das Jahr 1840 oder 1841 fielen. Damals war Kummer noch Gymnasiallehrer in Liegnitz [Kummer 1975, Vol. I, 15–30]. Die heute zugänglichen Primärquellen geben keine Antwort auf die Frage, ob sich Kummer schon während dieser Zeit mit der Kreisteilungs-Arithmetik beschäftigt hat. Der von Hensel herausgegebene Teil des Briefwechsels Kummers mit seinem Schüler und späteren Freund Leopold Kronecker [Kummer 1975, Vol. I, 76–132] setzt erst am 16.1.1842 ein. Im ersten uns bekannten Brief sprach Kummer von seinen ersten Studien über dritte Potenzreste und von der damals erkannten Notwendigkeit, sich mit den Zahlen, die aus den dritten Einheitswurzeln hervorgehen, mehr als bisher anzufreunden. (L. E. Dickson [1920, p. 738] hatte ohne Analyse behauptet, der fragliche Gedankenaustausch zwischen Kummer und Dirichlet hätte um 1843 stattgefunden.)

Es besteht heute keine Hoffnung mehr, die Originalbriefe Kummers an Kronecker wiederaufzufinden [Edwards 1978]. Übrigens setzte L. Kronecker in einer 1881 verfaßten Glückwunschadresse [Kronecker, 1881] Kummers erste Arbeit über kubische Reste [1842] an den Beginn der Periode der "grundlegenden und bahnbrechenden zahlentheoretischen Untersuchungen . . . , welche die folgenden zwei Jahrzehnte erfüllen".

Auf den ersten Blick wirken die oben angeführten Mitteilungen von Hensel recht überzeugend, nämlich durch ihre Detailangaben und die Rolle, die Dirichlet zugedacht war. In der Tat mußte Dirichlet besonders kompetent für die Fermatsche Vermutung erscheinen. Denn er hatte bekanntlich seine mathematische Laufbahn 1825 mit dem Beweis der Fermatschen Vermutung für den Exponenten 5 angetreten [Dirichlet 1825; Institut 1918, 239–241, 303] und später den Fall des Exponenten 14 erledigt [Dirichlet 1832]. Bei Hensels Angaben handelt es sich aber um eine Information aus zweiter Hand über ein Ereignis, das 1910 schon ein Menschenalter zurücklag. Nachforschungen jüngeren Datums von H. M. Edwards [1975, 1977a,b] in verschiedenen Archiven lieferten *keine* Belege für Hensels Angaben. Eine Beschäftigung Kummers mit der Fermatschen Vermutung zwischen 1837 und 1847 läßt sich an Hand der bisher bekannten Primärquellen nicht belegen. Diese Vermutung wurde von ihm in diesem Zeitraum nirgends erwähnt. Darüberhinaus gibt es gegen Hensels Version gewichtige mathematische Einwände, die wir in den nächsten Abschnitten diskutieren werden. Zusammenfassend muß man feststellen: Die entscheidenden Details von Hensels Äußerungen sind mit den Aussagen aller bisher zugänglichen Primärquellen nicht zu vereinbaren.

3. KUMMER UND DIE FERMATSCHE VERMUTUNG

Wir vergegenwärtigen uns jetzt die zeitliche Aufeinanderfolge der Kummerschen Entdeckungen, soweit sie das Fermatsche Problem berühren.

Es trifft zu, daß die erste zahlentheoretische Publikation von Kummer [1837] die Fermatsche Gleichung $x^{2\lambda} + y^{2\lambda} = z^{2\lambda}$ behandelte. Dies geschah jedoch durch Überlegungen, die nicht über den Körper der rationalen Zahlen hinausgingen. Kummer erwähnte einleitend, daß Leonhard Euler (1707–1783) [1747; 1770, §248], Adrien-Marie Legendre (1752–1833) [1827] und Peter Gustav Lejeune-Dirichlet (1805–1859) [1825, 1832] die Unlösbarkeit von $x^\lambda + y^\lambda = z^\lambda$ für die Werte $\lambda = 3, 4, 5, 14$ gezeigt hatten. Die Beweise dieser Autoren gingen hinsichtlich der Methoden nicht über die Zahlringe $\mathbb{Z}[\sqrt{-3}]$, $\mathbb{Z}[\sqrt{5}]$ und $\mathbb{Z}[\sqrt{-7}]$ hinaus. Diese Ringe sind nicht ganz-abgeschlossen und daher a priori keine Ringe mit eindeutiger Primzerlegung; sie besitzen den Führer 2. Sie sind jedoch Hauptidealringe "im eingeschränkten Sinn", d.h., jedes zu 2 teilerfremde Ideal ist Hauptideal. Diese Tatsache verbürgte aus heutiger Sicht den Erfolg der genannten Autoren. Am Schluß seiner Untersuchungen zog Kummer die Primzahlen der Form $\lambda\mu + 1$ (μ natürliche Zahl) heran, und zwar auf Grund der folgenden Tatsache, die auf Euler [1761, artt. 54–57] zurückgeht und auch in der früher vielgelesenen "Théorie des nombres" von Legendre [1830, artt. 158–165] ausführlich dargestellt ist.

> *Behauptung.* Es seien λ eine ungerade Primzahl, x, y ganze Zahlen mit $(x, y) = 1$. Jeder Primteiler p von $x^\lambda + y^\lambda$ ist dann ein Teiler von $x + y$ oder besitzt die Form $p = \lambda\mu + 1$. (3.1)

Der durchsichtige Beweisgedanke sei hier kurz mitgeteilt. Der Fall $p = 2$ ist klar. Wir nehmen deshalb an: $p \neq 2, p \nmid x, p \nmid (x + y)$, $p \mid (x^\lambda + y^\lambda)$. Daraus folgt: $p \nmid y$. Es gibt eine Restklasse a (mod p) mit $ya \equiv 1$ (mod p). Dann ist $(xa)^\lambda + 1 \equiv 0$ (mod p), $xa \not\equiv 1, -1$ (mod p), $(xa)^{2\lambda} \not\equiv 1$ (mod p).

Die Restklasse xa (mod p) hat also die Ordnung 2λ. Diese Ordnung ist ein Teiler von $p - 1$, was zu beweisen war.

Es ist interessant, daß wir den Primzahlen $\lambda\mu + 1$ bei Kummer [1844a,b] nochmals begegnen, und zwar *unabhängig* vom Fermatschen Problem, aber in wichtigem Zusammenhang mit der Kreisteilungs-Arithmetik (siehe 4. Abschnitt). Kummer veröffentlichte damals die ersten Beispiele für mehrdeutige Primzerlegungen der von Hensel erwähnten "Zahlen in α", insbesondere der bei $\lambda = 23$ auftretenden Zahlen. Dies legt außerordentlich nahe, daß der fragliche, von Hensel erwähnte Gedankenaustausch (innerhalb weniger Tage) zwischen Kummer und Dirichlet—unabhängig von Hensels Zeitangaben- spätestens ins Jahr 1844 fiel, als sich übrigens Dirichlet (seit Herbst 1843) in Italien aufhielt. (Dieser Italien-Aufenthalt von Dirichlet wurde von Kummer sogar in seiner Arbeit [1844b] erwähnt.)

Am 18.10.1845 gab Kummer in einem ausführlichen Brief an Kronecker die erste Darstellung seiner Theorie der "idealen complexen Zahlen" ([Kummer 1910, 64–68]). Erst anderthalb Jahre später (!), am 2. April 1847, kündigte er—wiederum in einem Brief an Kronecker—seine bahnbrechenden Anwendungen der Kreisteilungs-Arithmetik auf das Fermatsche Problem an [Kummer 1910, 75–80]. Fast am Schluß dieses Briefes lesen wir: "Der obige Beweis von $x^\lambda + y^\lambda = z^\lambda$ ist erst drei Tage alt, denn erst nach Beendigung der Recension [von [Jacobi 1846] im März 1847] fiel es mir ein wieder einmal diese alte Gleichung vorzunehmen, und ich kam diesmal bald auf den richtigen Weg" [Kummer 1910, 80]. Demnach war Kummer erst im März 1847 innerhalb weniger Tage oder Wochen zu seinen Ergebnissen gelangt. Darauf teilte er seine Resultate brieflich an Dirichlet mit, der sie noch im April 1847 in den Monatsberichten der Berliner Akademie publizieren ließ ([Kummer 1847c]). Wahrscheinlich ist unter anderem auch dadurch zu erklären, daß Dirichlets Name von Hensel [1910a,b] genannt wurde.

Beachtenswert ist, wie Kummer in seiner bahnbrechenden Mitteilung [1847c] das Fermatproblem für sich genommen bewertet hat: "Der Fermatsche Satz ist zwar mehr ein Curiosum als ein Hauptpunkt der Wissenschaft, dessen ungeachtet halte ich diese meine Beweisart für bemerkenswert . . ." Die dabei gewonnenen zahlentheoretischen Einsichten schienen ihm jedenfalls "viel wissenswerter zu sein, als der Fermatsche Satz selbst" [Kummer 1847c, 139]. Eigentümlicherweise fehlte in Hilberts "Zahlbericht" [1897] im Verzeichnis der Kummerschen Arbeiten gerade die eben erwähnte Mitteilung zur Fermatschen Vermutung.

Daß Kummer sich zwischen 1837 und 1847 erst *nach* der Einführung der "idealen complexen Zahlen" intensiv mit der Fermatschen Vermutung beschäftigt hatte, wird auch durch seinen Brief [1847d] vom 28.4.1847 an Joseph Liouville (1809–1882) nahegelegt: "Les applications de cette théorie à la démonstration du théorème de Fermat m'ont occupé depuis longtemps ..." Die Zeitspanne, auf die sich "depuis longtemps" bezieht, läßt sich leider nicht näher bestimmen.

Es gibt einen weiteren Umstand, der die Version Hensels außerordentlich unglaubwürdig macht. Wenn man voraussetzt, daß die Primelementzerlegung im Ring $\mathbb{Z}[\zeta_\lambda]$ (λ ungerade Primzahl, ζ_λ primitive λ-te Einheitswurzel), eindeutig sei, dann erfordert ein *vollständiger* Beweis der Fermatschen Vermutung für den Exponenten λ immer noch tiefgehende Aussagen über die Einheiten von $\mathbb{Z}[\zeta_\lambda]$ (oder ein Äquivalent solcher Aussagen). Es handelt sich hierbei um Tatsachen, deren Erkenntnis Kummer um 1843/44 noch nicht besaß und nicht besitzen konnte, wenn man nach der Abhandlung [1844b] urteilt. Vielmehr überblickte er den hier vorliegenden Sachverhalt (das sogenannte "Kummersche Lemma", vgl. [Borewicz-Šafarevič 1972]) unter der schwächeren Voraussetzung der "Regularität" von λ erst im September 1847 [1847c]. Man nehme nun für einen Augenblick sogar an, die Fehler in dem von Hensel erwähnten und angeblich an Dirichlet übergebenen Manuskript Kummers über die Fermatsche Vermutung seien sehr versteckt gewesen. Dann kommt man zu dem Schluß, daß Kummer nach Entdeckung dieser Fehler versucht hätte, "noch zu retten, was zu retten war". Es ist sehr unwahrscheinlich, daß solche Bemühungen oder die daraus gewonnenen Erkenntnisse—falls vorhanden gewesen—keine Spuren in Abhandlungen und Briefen dieser Periode (bis Anfang 1847) hinterlassen haben.

Offenbleiben muß die Frage, wann und wie Kummer zum ersten Male versucht hat, das Fermat-Problem mit Hilfe der Kreisteilungs-Arithmetik anzugreifen. Die Zerlegung des Polynoms $x^\lambda - y^\lambda$ über dem Körper $\mathbb{Q}(\zeta_\lambda)$ (ζ_λ primitive λ-te Einheitswurzel):

$$x^\lambda - y^\lambda = \prod_{i=0}^{\lambda-1} (x - \zeta_\lambda^i y)$$

war z.B. schon von Joseph-Louis Lagrange (1736–1813) [1769; 1868, 531–535] angegeben und mit Beweisideen für die Fermatsche Vermutung in Verbindung gebracht worden. Euler hatte de facto diese Zerlegung zum Beweis der Fermatschen Vermutung für $\lambda = 3$ benutzt (siehe [Bergmann 1966]). Auch Carl Friedrich Gauß (1777–1855) dachte an diesen Zusammenhang, als er 1816 an Wilhelm Olbers (1758–1840) schrieb:

> Ich gestehe zwar, daß das Fermatsche Theorem als isolierter Satz für mich wenig Interesse hat, denn es lassen sich eine Menge solcher Sätze leicht aufstellen, die man weder beweisen noch widerlegen kann. ... Allein ich bin überzeugt, wenn ... mir einige Hauptschritte in jener Theorie [der Arithmetik der Kreisteilungskörper] glücken, auch der Fermatsche Satz nur als eines der am wenigsten interessanten Corollarien dabei erscheinen wird. [Gauß 1917, 75–76; Olbers 1900, 629]

Kummer nannte in seinen Arbeiten zum Fermat-Problem seit 1847 weder Lagrange noch die Namen von Mathematikern, die einzelne Fälle der Fermatschen Vermutung erledigt hatten.

Man muß hier hervorheben, daß Kummer bereits Ende 1844 alle Mittel in der Hand hatte, die folgende interessante Teilaussage seiner späteren Ergebnisse zu beweisen:

> *Behauptung.* Wenn die ungerade Primzahl $\lambda \geqq 5$ so beschaffen ist, daß im Ring $\mathbb{Z}[\zeta_\lambda]$ jede Primzahl $p = \lambda\mu + 1$ $(\mu \in \mathbb{Z})$ in $(\lambda - 1)$ Faktoren zerfällt, dann kann für die Fermatsche Gleichung $x^\lambda + y^\lambda = z^\lambda$ mit paarweise teilerfremden ganzen Zahlen x, y, z nur der sogenannte zweite Fall eintreten, d.h., genau eine der Zahlen x, y, z muß durch λ teilbar sein. (3.2)

Eine Beweisskizze im Stile von Kummers damaligen Arbeiten werden wir im vierten Abschnitt geben. Bei Kummer selbst finden wir jedoch 1844/45 keinen Hinweis auf ähnliche Überlegungen. Auch hier ist es äußerst unwahrscheinlich, daß der publikations- und hypothesenfreudige, gegenüber Kronecker recht mitteilsame Kummer jegliche Andeutung interessanter Ergebnisse zur Fermatschen Vermutung unterdrückte, falls er solche vor Anfang 1847 besessen hätte.

Wir können uns deshalb nicht der Meinung von H. M. Edwards anschließen, daß "in all probability [Kummer] was aware all along [since 1844] that his factorization theory would have implications for Fermat's Last Theorem" [1977b, 79].

Es drängt sich der Schluß auf, daß für Kummer bei der Einführung der "idealen Zahlen" das Fermat-Problem weder als wichtiges zu bewältigendes Problem noch als "Katalysator" eine nachweisbare Rolle gespielt hat. Nicht übersehen und nicht unterschätzt werden darf jedoch die stimulierende Rolle, die die Fermatsche Vermutung bei der Ausgestaltung der Kreisteilungs-Arithmetik (Klassenzahlformeln, Einheitentheorie) seit 1847 gespielt hat. Wie schon erwähnt, begann Dirichlet seine mathematische Laufbahn mit einer Arbeit [1825] über die Fermat-Vermutung für die Exponenten 5. Diese Tatsache nahm Kummer [1860] zum Anlaß, die Wirkung des "Fermatschen Satzes" in der Geschichte der Mathematik mit folgenden Worten zu würdigen:

> Dieser Satz . . . kann zwar, als eine aus ihrem wissenschaftlichen Zusammenhange herausgenommene Einzelheit, keinen besonderen Werth beanspruchen, aber er hat dadurch eine ungewöhnlich hohe Bedeutung gewonnen, daß er . . . [um 1825] . . . auf die Richtung, welche die Zahlentheorie in ihrer geschichtlichen Entwicklung genommen hat, von dem entschiedensten Einfluß gewesen ist. [Kummer 1860, 8]

Für weitere Details zur Geschichte der Fermatschen Vermutung sei der Leser auf die umfangreiche Literatur verwiesen [Bachmann 1919; Hasse 1970; Mordell 1972]. Natürlich ist dabei nicht zu vergessen, daß in einem Teil dieser Literatur die eingangs zitierte, aber nun unglaubwürdig gewordene Version von Hensel ungeprüft wiederholt wird.

4. KUMMERS WEG ZU DEN "IDEALEN COMPLEXEN ZAHLEN"

Was drängte nun Kummer wirklich zur Einführung der "idealen Zahlen", wenn es nicht die Fermatsche Vermutung war? Natürlich wird man zuerst versuchen, eine direkte Antwort auf diese Frage in Kummers eigenen Veröffentlichungen zu finden. Nach der Einführung der "idealen Zahlen" sprach Kummer in seiner ersten Mitteilung [1846b] von der nun erreichten Analogie der Teilbarkeitsgesetze für die ganzen Kreisteilungszahlen mit denen der Gaußschen ganzen Zahlen $a + b\sqrt{-1}$ (a, b ganzrational) [Gauß 1832]. Damit war nach Kummers Worten ein befriedigender Ersatz gefunden für das in der Breslauer Schrift [1844b] zum ersten Male konstatierte und bedauerte Fehlen der eindeutigen Primzahlzerlegung im Bereich der ganzen Kreisteilungszahlen. Mit ähnlichen Worten würdigte Kronecker [1881] die Einführung der "idealen Zahlen". Damit wird man auf Kummers Abhandlung [1844b] zurückverwiesen, in der er die Kreisteilungstheorie und die Theorie der höheren Potenzreste ausdrücklich als Zielgebiete weiterer Untersuchungen nannte. Indirekt bezeichnete er diese Gebiete auch als Ausgangsstationen seiner Untersuchungen, indem er auf Jacobis Arbeiten [1837, 1839] hinwies. Diese Tatsachen stehen im Einklang mit Äußerungen von Dirichlet und Emil Lampe (1840–1918), Professor der Mathematik an der Technischen Hochschule Berlin-Charlottenburg. Dirichlet sprach 1855, als er Kummer als seinen Nachfolger auf dem Berliner Lehrstuhl an erster Stelle vorschlug, bei dieser Gelegenheit mit höchster Wertschätzung von "Kummers Arbeiten auf diesem wichtigen Gebiete [nämlich der Kreisteilungs-Arithmetik], welches der mathematischen Spekulation vor einem halben Jahrhundert von Gauß erschlossen und später von Jacobi, Cauchy und anderen Mathematikern durch Einführung neuer Principien wesentlich erweitert worden ist . . ." [Biermann 1973, 174].

Von der Fermatschen Vermutung war in diesem Gutachten nirgends die Rede; vielmehr unterstrich Dirichlet die "innere" Weiterentwicklung der Kreisteilungstheorie. E. Lampe orientierte sich 1893 an der Glückwunschadresse der Akademie [Kronecker 1881], als er seinen Nachruf auf den ihm befreundet gewesenen Kummer verfaßte, und begnügte sich bezüglich der ersten zahlentheoretischen Arbeiten Kummers mit der Äußerung, daß zu diesen Arbeiten "außer den Disquisitiones arithmeticae besonders die Abhandlungen über die biquadratischen Reste von Gauß die ersten Gesichtspunkte geliefert haben" [Lampe 1893, 16].

Erst in neuerer Zeit wurde unsere Quellenkenntnis entscheidend dadurch bereichert, daß Kurt-R. Biermann 1967 ein bislang unbekanntes Kummer-Manuskript auffand [Biermann 1967]. Deshalb lohnt es sich, hier Kummers zahlentheoretische Untersuchungen seit 1841/42 auf dem Hintergrund seiner "mathematischen Umwelt" zu verfolgen.

Wir werden im einzelnen zeigen, daß der Hauptanstoß zur Schöpfung der "idealen Zahlen" in der Theorie der höheren Potenzreste und im Kalkül der Gaußschen und Jacobischen Summen zu suchen ist. Dabei spielten bei Kummer die Primzahlen der Form $\lambda\mu + 1$ (λ ungerade Primzahl, $\mu \in \mathbb{Z}$) von Anfang an eine besondere Rolle. Der erwähnte Kalkül, ein Bestanteil der Kreisteilungstheorie, wurde von Gauß (um 1808, siehe [1876b]), Carl Gustav Jacob Jacobi (1804–1851) [Brief an Gauß v.8.2.1827 = 1891b, 393–400; 1837], Augustin-Louis Cauchy (1789–1857) [1829] und Gotthold Eisenstein (1823–1852) [1844] unabhängig voneinander begründet und spielte in der Zahlentheorie des 19. Jahrhunderts eine große Rolle (Satz von Kronecker–Weber, Relationen in Klassengruppen, Reziprozitätsgesetze) [Smith 1859–1865, Weil 1974, Neumann 1980]. Seine Verallgemeinerungen nahmen ihren Anfang in den Arbeiten von Kummer und Eisenstein und sind auch heute noch wichtig.

1841/42 hatte Kummer "bei etwas [ihm] ganz neuem [angefangen], nämlich bei den Cubischen Resten der Primzahlen $6n + 1$" [Kummer 1910, 76]. Als Ziel seiner neuen Untersuchungen betrachtete er ein "Reziprozitätsgesetz" und ein Analogon zur Vorzeichenbestimmung der "quadratischen" Gaußschen Summen, das ihm, dem damaligen Analytiker, besonders nahelag. Er mußte aber Anfang 1842 bald feststellen, daß seine ersten Überlegungen über kubische Reste zum größten Teil schon in den Arbeiten von Jacobi [1827, 1837] enthalten waren. Gauß (1831/32, siehe [1832; 1876a, 93–148, 169–178]) und (fast unabhängig von ihm) Jacobi [1827, 1837, 1839], aber auch Cauchy [1829] hatten als erste in ihren Publikationen die Theorie der λ-ten Potenzreste mit den Ringen der λ-ten Einheitswurzeln in Verbindung gebracht. Unter dem Ring der λ-ten Einheitswurzeln, abgekürzt: $\mathbb{Z}[\zeta_\lambda]$, verstehen wir den Ring aller ganzzahligen Linearkombinationen der λ-ten Einheitswurzeln. Durch den Zusammenhang mit den Potenzresten war das eingehende Studium dieser Ringe in starkem Maße motiviert.

Ein gleichermaßen wichtiger und folgenreicher Anstoß zur Untersuchung der Ringe $\mathbb{Z}[\zeta_\lambda]$ kam aus der von Gauß [1801] entdeckten Auflösbarkeit der Kreisteilungsgleichungen durch Radikale. Es seien: p eine Primzahl, ζ eine primitive p-te Einheitswurzel, g eine Primitivwurzel modulo p, λ ein

Teiler von $(p-1)$, α eine primitive λ-te Einheitswurzel. Die Grundidee von Gauß bestand darin, die allgemeine "Gaußsche Summe"

$$(\alpha,\zeta) := \zeta + \alpha\zeta^{g} + \cdots + \alpha^{p-2}\zeta^{g^{p-2}} \tag{4.1}$$

zu bilden. Die λ-te Potenz dieser Summe liegt schon im Ring $\mathbb{Z}[\alpha]$:

$$\omega(\alpha) := (\alpha,\zeta)^{\lambda} \in \mathbb{Z}[\alpha]. \tag{4.2}$$

$\omega(\alpha)$ ist von der Wahl von ζ unabhängig. ζ selbst ist gleich dem "Mittelwert" der Summen (α,ζ):

$$\zeta = \frac{1}{p-1}\sum_{\alpha}(\alpha,\zeta) \tag{4.3}$$

wobei α alle $(p-1)$-ten Einheitswurzeln (also λ alle möglichen Teiler von $(p-1)$) durchläuft. Kummer hielt es für sehr wichtig, die Zahlen (α,ζ) und $\omega(\alpha)$ möglichst genau zu beschreiben. Er erörterte dieses Thema in einer Reihe seiner Arbeiten und Briefe und stieß schon im Falle $\lambda = 3$ auf eine Frage, die erst seit 1978 geklärt ist [1842] [Heath-Brown 1979].

Die Hauptlinie der weiteren Enwicklung knüpfte jedoch an eine allgemeinere Frage an. Gauß (um 1808, siehe [1876b]), Jacobi (1827, siehe [1837]), und Cauchy [1829] studierten die Zahlen (α,ζ) in ihrer Abhängigkeit von α und bemerkten, daß für jedes Exponentenpaar ($s.$ mod λ, t mod λ) der Quotient

$$(\alpha^{s},\zeta)(\alpha^{t},\zeta)(\alpha^{s+t},\zeta)^{-1}$$

in $\mathbb{Z}[\alpha]$ liegt. Diese Quotienten von Gaußschen Summen nennen wir Jacobische Summen, weil sie spezielle Darstellungen als Summen von Potenzen von α besitzen (vgl. [Hasse 1964]). Setzt man insbesondere

$$\psi_{s}(\alpha) := (\alpha^{s},\zeta)(\alpha,\zeta)(\alpha^{s+1},\zeta)^{-1} \qquad (s \not\equiv 0, -1 \mod \lambda) \tag{4.4}$$

so erhält man gewisse von der Wahl der Wurzel ζ unabhängige Zahlen $\psi_1(\alpha), \ldots, \psi_{\lambda-2}(\alpha)$ aus $\mathbb{Z}[\alpha]$. Nach Gauß, Jacobi und Cauchy bestehen nun die Relationen

$$\omega(\alpha) = (\alpha,\zeta)^{\lambda} = (-1)^{\mu}\cdot p\cdot\psi_1(\alpha)\cdots\psi_{\lambda-2}(\alpha) \tag{4.5}$$

$$\psi_s(\alpha)\cdot\psi_s(\alpha^{-1}) = p \qquad (s = 1, \ldots, \lambda-2), \tag{4.6}$$

wobei $p = \lambda\mu + 1$ gesetzt ist. Insbesondere handelt es sich hier um $(\lambda-2)$ explizite Faktorenzerlegungen einer beliebigen ganzrationalen Primzahl p mit $p \equiv 1 \pmod{\lambda}$ im Ring der λ-ten Einheitswurzeln. Die Frage nach dem gegenseitigen Zusammenhang dieser Zerlegungen gab einen überaus wichtigen Anstoß zur Einführung der "idealen complexen Zahlen" durch Kummer, was noch zu belegen sein wird.

Jacobi hatte bereits in seiner ersten Publikation über den Kalkül der von ihm eingeführten Summen auf die Konsequenzen für die Theorie der höheren Potenzreste hingewiesen [1837]. Er konnte nicht nur das quadratische Reziprozitätsgesetz, sondern auch die Reziprozitätsgesetze der kubischen und biquadratischen Reste aus seinem Kalkül elegant ableiten (vgl. [Smith 1859–1865; Weil 1974]). Dies war Kummer 1844 sicherlich auch aus den Nachschriften der Zahlentheorie-Vorlesungen Jacobis (1836/37) im Detail bekannt (vgl. [Jacobi 1837, Abdruck 1846, S.172, Fußnote]). Kummer [1847b, 808–809] rühmte Jacobis Arbeit [1837] als den wichtigsten Fortschritt in der Kreisteilungstheorie seit Gauß' "Disquisitiones arithmeticae".

Es wurde für Kummer besonders bedeutsam, daß Jacobi für die λ-Werte, die den Kreisteilungskörpern vom Grade 4 entsprechen, also für $\lambda = 5$, 8 und 12, die oben aufgeworfene Frage nach dem Zusammenhang der Zerlegungen (4.6) erledigen konnte [Jacobi 1839]. Er zeigte nämlich, daß für diese λ-Werte jede Primzahl, die $\equiv 1 \bmod \lambda$ ist, im Ring $\mathbb{Z}[\zeta_\lambda] = \mathbb{Z}[\alpha]$ in genau vier Primfaktoren zerfällt und daß die verschiedenen Zerlegungen (4.6) durch verschiedene Kombinationen dieser Primfaktoren zustandekommen. Außerdem war Jacobi in der Lage, daraus das Reziprozitätsgesetz der λ-ten Potenzreste zwischen zwei Zahlen aus $\mathbb{Z}[\zeta_\lambda]$ abzuleiten, falls eine dieser Zahlen ganzrational war. Im Zentralen Archiv der Akademie der Wissenschaften der DDR (Berlin, DDR) befindet sich ein unveröffentlichtes umfangreiches Manuskript Jacobis mit dem Titel "Zerlegung der Primzahlen von der Form $8n + 1$ in 4 complexe Factoren". Für den Hinweis auf dieses Manuskript bin ich Herrn Dr. H. Pieper zu Dank verpflichtet. Es sei weiter erwähnt, daß zu ähnlichen Untersuchungen auch der junge Charles Hermite (1822–1901) durch Jacobi angeregt wurde. In einem seiner Briefe an Jacobi berichtete er ausführlich, wie er die Zerlegbarkeit aller Primzahlen der Form $5n + 1$ bzw. $7n + 1$ in vier bzw. sechs Faktoren im Ring der fünften bzw. siebenten Einheitswurzeln bewiesen hatte [Hermite 1850, 261–278].

Eine Briefstelle, auf die H. M. Edwards [1975, 234] offenbar zuerst aufmerksam gemacht hat, gibt erste Auskunft darüber, in welche Richtung Kummer die Ergebnisse Jacobis weiterzutreiben versuchte. Am 10.4.1844 schrieb Kummer [1910, 53] an Kronecker von einem "Beweise, daß jede Primzahl p sich in $\lambda - 1$ complexe Factoren zerlegen läßt . . . ."

Dabei war mit λ eine Primzahl und mit p eine Primzahl der Form $\lambda\mu + 1$ gemeint. Kummer bewies ausführlich und einwandfrei, daß die Restklassen nach einer Zahl π aus $\mathbb{Z}[\zeta_\lambda]$, deren Norm eine Primzahl p mit $p \equiv 1 \pmod{\lambda}$ ist, durch ganzrationale Zahlen repräsentiert werden. Mit heutigen Worten: Der Restklassenring $\mathbb{Z}[\zeta_\lambda]/(\pi)$ ist zum endlichen Körper $\mathbb{Z}/(p)$ isomorph. Von Kronecker angeregt, leitete Kummer daraus ab, daß die Zerlegung von p in $(\lambda - 1)$ Faktoren im wesentlichen eindeutig ist. Aus seinen Überlegungen folgt sofort, daß jede ganze Kreisteilungszahl (d.h., jede Zahl aus $\mathbb{Z}[\zeta_\lambda]$)

der Primzahl-Norm p, $p \equiv 1 \pmod{\lambda}$, genau dann Teiler eines Produkts ist, wenn sie Teiler eines Faktors ist. Kummer hätte also folgende naheliegende Schlußfolgerung ziehen können:

> *Behauptung.* Jedes Potenzprodukt von Zahlen aus $\mathbb{Z}[\zeta_\lambda]$, deren Norm jeweils eine Primzahl p mit $p \equiv 1 \pmod{\lambda}$ ist, besitzt eine eindeutige Primzerlegung. (4.7)

Diese oder ähnliche Aussagen und ihre Folgerungen für die Fermatsche Vermutung (s.unten) tauchten aber bei Kummer 1844 weder in den uns bekannten Briefen noch anderswo auf. Dies spricht dafür, daß bei ihm damals das Interesse an der Fermatschen Vermutung nicht im Vordergrund stand.

Wir stellen zusammenfassend fest: Kummer glaubte Anfang 1844 bewiesen zu haben, daß im Ring der λ-ten Einheitswurzeln (λ ungerade Primzahl) jede Primzahl $p = \lambda\mu + 1$ eindeutig in $(\lambda - 1)$ Primfaktoren der Norm p zerfällt. In heutigen Termini bedeutete dies: Jedes Primideal 1. Grades wäre dann ein Hauptideal. Kummer glaubte ein Ergebnis zu besitzen, das in der Tat inhaltlich gleichbedeutend mit der eindeutigen Primzerlegung in den λ-ten Kreisteilungskörpern war. Denn jedes Ideal ist bekanntlich äquivalent zu einem Produkt von Primidealen 1. Grades, wie Kummer selbst allerdings erst zwei Jahre später [1847a, §10] ausgesprochen und korrekt bewiesen und zu weitgehenden Aussagen über die Klassengruppe benutzt hat. (Jedes Ideal ist sogar äquivalent zu einem einzelnen Primideal 1. Grades, wie aus den Eigenschaften der Dedekindschen Zetafunktion folgt.)

Wenig später als den oben zitierten Brief an Kronecker sandte Kummer im April 1844 ein Manuskript [1844a] an die Berliner Akademie. Es wurde von Kummer zur Veröffentlichung in den Abhandlungen oder Monatsberichten der Akademie eingereicht, dann jedoch von ihm zurückgezogen [Biermann 1967, Anm. [10]]. In diesem Manuskript wurde expressis verbis an Jacobis Untersuchungen angeknüpft; die in der oben erwähnten Briefstelle ausgesprochene Behauptung wurde "bewiesen" und als wichtiges Ergebnis in den Mittelpunkt gestellt. Wir haben hier nichts weniger als die erste—allerdings nach Edwards [1977a] mit einem ernsten Fehler behaftete—Arbeit Kummers zur Arithmetik in den Körpern beliebiger λ-ter Einheitswurzeln (λ ungerade Primzahl) vor uns. Jacobi war am 17. Juni 1844 aus Rom kommend in Berlin eingetroffen [Ahrens 1907, 117] und verhinderte den Druck von Kummers Arbeit, wie aus einem Brief Jacobis an Dirichlet vom Januar 1845 hervorgeht [Edwards 1977a, 393–394]. Demnach hatte Kummer im Sommer 1844 anscheinend auf Jacobis Initiative sein Manuskript von der Veröffentlichung zurückgezogen. Leider können wir nicht mit Sicherheit entscheiden, ob Jacobi nur den Fehler in Kummers allgemeinen Überlegungen entdeckt hatte—was ja die Möglichkeit einer Korrektur offengelassen hätte—oder ob er schon numerische Gegenbeispiele zur eindeutigen Primzerlegung angeben konnte.

Wie H. M. Edwards bemerkt [1975, 233], meinte zweifellos auch G. Eisenstein das Manuskript von Kummer, als er annähernd im Juli 1844 an M. A. Stern (1807–1894) schrieb:

> "Prof. Kummer hat zum Glück seine schöne Theorie der complexen Zahlen noch bei Zeiten durch Encke an der Akademie zurücknehmen lassen; . . . man kann durch dieselbe beweisen, daß zu jeder Diskriminante nur *eine* quadratische Form gehört und dgl. Unsinn mehr. Kummer hofft die Theorie leicht zu ergänzen . . . ." [Eisenstein 1895, 173] (Johann Franz Encke (1791–1865), Astronom, war Mitglied und Sekretär der Berliner Akademie).

Sicherlich hatte Eisenstein folgenden in die heutigen Begriffe übersetzten Schluß vollzogen: Wenn sich jede Primzahl $p = \lambda\mu + 1$ im Ring $\mathbb{Z}[\zeta_\lambda]$ in $(\lambda - 1)$ Faktoren zerlegen läßt, dann besitzt der quadratische Teilkörper $\mathbb{Q}(\sqrt{\lambda^*})$ von $\mathbb{Q}(\zeta_\lambda)$ die Klassenzahl Eins ($\lambda^* = \pm\lambda \equiv 1 \pmod{4}$). Diese Schlußfolgerung ist richtig; wir wissen jedoch nicht, wie Eisenstein im einzelnen zu ihr gelangt ist. Er hatte jedenfalls die klare Einsicht, daß

> "bei den höheren complexen Zahlen . . . eigentlich gar keine complexen *Primzahlen* mehr [existieren]. Gibt man den Satz zu, daß das Product zweier complexer Zahlen nicht anders durch eine Primzahl teilbar sein kann, als wenn mindestens ein Factor durch die Primzahl teilbar ist, was ganz evident erscheint [!], so hat man die ganze Theorie auf einen Schlag; aber dieser Satz ist *total falsch* . . . ." [Eisenstein 1895, 173]

Eisenstein und Jacobi sahen den Ausweg aus dieser Situation im Ausbau der Theorie der Formen.

In der schon angeführten Breslauer Abhandlung [1844b] fand sich dann bei Kummer die Erkenntnis, daß die aus seinem Brief vom 10.4.1844 und seinem Manuskript [1844a] oben zitierte Behauptung über die ausnahmslose Zerlegbarkeit der Primzahlen $\lambda\mu + 1$ falsch war. Kummer nutzte eine ähnliche (aber etwas abgeschwächte) Schlußweise wie Eisenstein, indem er z.B. die Tatsache verwendete, daß die Klassenzahl in $\mathbb{Q}(\sqrt{-23})$ größer als Eins ist. Ob dies auf einen direkten Einfluß Eisensteins zurückging, läßt sich nicht entscheiden.

Kummer rechnete [1844b] aus, daß in den Ringen $\mathbb{Z}[\zeta_\lambda]$ für $\lambda = 5, 7, 11, 13, 17, 19$ jede Primzahl $\lambda\mu + 1$ unter 1000 in $(\lambda - 1)$ Faktoren zerfällt. Für $\lambda = 23$ fand er, daß $p = 47, 139, 277, 461, 967$ jeweils nur in 11 Faktoren zerfallen und daher keine eindeutige Primelementzerlegung besitzen können. Jacobi zitierte in seinem Brief [Edwards 1977a] eine Reihe dieser Gegenbeispiele, ohne selbst Anspruch auf die Erkenntnis zu erheben, daß in bestimmten Kreisteilungskörpern die Primelementzerlegung mehrdeutig sein kann. Wie weit Jacobis und Dirichlets Einsichten in die hier vorliegende Situation schon vor Kummers Arbeiten gingen, läßt sich an Hand der bisher zugänglichen Quellen nicht klären. Jedenfalls kann man Jacobis Brief entnehmen, daß die Abhandlung Kummers [1844b] tatsächlich eine korrigierte und erweiterte Fassung des 1844 an die Berliner Akademie eingesandten Manuskripts [1844a] darstellt. Anscheinend war diese Abhandlung spätestens im September 1844 abgeschlossen, weil Kummer [1910, 57] in einem Brief vom 2.10.1844 an Kronecker schon von seiner nächsten Abhandlung sprach und außerdem in einem bestimmten Zusammenhang den "§9 meiner Dissertation" anführte. Die zitierte Stelle stimmte inhaltlich mit der endgültigen Fassung überein.

Wie schon erwähnt, hatte Kummer nun alle Mittel in der Hand, die Behauptung (3.2) zu beweisen. In der Tat kann man wie folgt verfahren: Im sogenannten ersten Fall der Fermatschen Gleichung:

$$x^\lambda + y^\lambda = z^\lambda, \qquad x, y, z \text{ nicht durch } \lambda \text{ teilbar,}$$

müssen die Zahlen $x + y$, $(x^\lambda + y^\lambda)/(x + y)$ zueinander teilerfremd, also für sich genommen λ-te Potenzen sein:

$$x + y = u^\lambda, \qquad \frac{x^\lambda + y^\lambda}{x + y} = v^\lambda, \qquad (u, v) = 1.$$

Dies war Kummer spätestens seit 1835 gut bekannt. Nach Behauptung (3.1) enthält die Zahl v nur Primteiler der Form $\lambda\mu + 1$. Mit Hilfe von Behauptung (4.7) und seiner damaligen Erkenntnisse über Kreisteilungs-Einheiten hätte Kummer den ersten Fall zu einem Widerspruch führen können, und zwar ebenso, wie er es dann (unter der schwächeren Voraussetzung, daß λ "regulär" ist) in seinem Brief [1910, 76] vom 2.4.1847 an Kronecker getan hat.

Diesen Schritt hat Kummer 1844 jedoch nicht vollzogen. Er bemühte sich aber 1844/45, wenigstens für $\lambda = 5, 7, 11, 13, 17$, und 19 die Zerlegbarkeit *aller* Primzahlen der Form $\lambda\mu + 1$ in $(\lambda - 1)$ Faktoren zu beweisen [1910, 58–62]. Im Februar 1845 beschrieb Kummer in einem Brief an Dirichlet, der sich damals in Florenz aufhielt, diesem offenbar zum erstenmal (so nach [Edwards 1975, 236]) seine Untersuchungen von 1844/45 zur Kreisteilungs-Arithmetik.

Am 16.10.1844 schrieb Kummer an Kronecker, er hätte "alle Hauptresultate [seines] Programms [d.h., der Abhandlung [1844b]] jetzt auch für diejenigen complexen Zahlen bewiesen, welche nicht aus den Wurzeln der Gleichung $\alpha^\lambda = 1$, sondern aus den Perioden dieser Wurzeln gebildet sind." [Kummer 1910, 62] Leider sind uns diese Verallgemeinerungen der "Hauptresulate" in den Einzelheiten nicht bekannt. Die nächste Abhandlung, die im Druck erschien [1846a], enthielt schon einige Schlußweisen, die dann bei der Begründung der Theorie der "idealen Zahlen" wiederkehrten. Insbesondere benutzte L. Kronecker in seiner Dissertation [1845, §2] expressis verbis eine Überlegung von Kummer [1846a], die wenig später für die Begründung der Theorie der "idealen Zahlen" unerläßlich wurde und zu deren strengem Beweis Kronecker damals auf eine Arbeit von Th. Schönemann (1812–1868) [1839, 306] verwies. Deshalb darf man im Gegensatz zu Edwards [1975, 231] annehmen, daß auch der mit Kroneckers Dissertation vertraute Kummer diese Arbeit Schönemanns gekannt hat, obwohl er in seinen Arbeiten nirgends auf sie verwies. Man vergleiche dazu Kummers Brief vom 7.12.1846 an Kronecker, wo vom "19. Band von Crelles Journal" die Rede ist [Kummer 1910, 73]).

Schließlich gab Kummer am 18.10.1845, wie schon erwähnt, die erste Darstellung seiner Theorie der "idealen Zahlen" [Kummer 1910, 64–67], der bald die Mitteilung [1846b] folgte. Aus Kummers Argumentation in dieser Veröffentlichung kann man schließen, daß er mit großer Wahrscheinlichkeit die Zerlegbarkeit in "ideale Primfaktoren" zuerst an den "reellen" Primzahlen der Form $\lambda\mu + 1$ bemerkt hat.

Übrigens geht aus dem vierten der im Jahre 1850 publizierten Briefe Hermites an Jacobi hervor, daß Jacobi seinen Briefpartner damals (1846?) auf Kummers "ideale Zahlen" hingewiesen hatte (Hermite [1850, 308–315]). Es wäre interessant, den Wortlaut dieses Jacobischen Briefs kennenzulernen.

Den Ausgangspunkt von Kummers Weg zu den "idealen Zahlen" bildeten, wie hier gezeigt wurde, die Theorie der höheren Potenzreste und der Kalkül der Gaußschen und Jacobischen Summen- zwei Problemkreise, die Kummer stets im Auge behielt. In seinen beiden ersten Mitteilungen [1846b; 1847a] über seine neue Theorie erledigte er nun auch die Frage nach der Faktorenzerlegung der Gaußschen und Jacobischen Summen. Wie Kummer selbst sagte, hätte er damit die "innere Natur" dieser Summen "durch ihre Zerlegung in die wahren complexen Primfaktoren aufgeschlossen" [1847b, 808–809]. Es wurde bereits darauf hingewiesen, daß die ersten Anwendungen der "idealen Zahlen" auf die Fermatsche Vermutung erst auf das Jahr 1847 fielen. 1848 erriet Kummer dann an Hand von umfangreichen numerischen Rechnungen das allgemeine Reziprozitätsgesetz der λ-ten Potenzreste in λ-ten Kreisteilungskörpern [1850]. Die so gewonnene Vermutung konnte er allerdings erst 1858 (nur) für "reguläre" Primzahlen beweisen [1858]. Inzwischen hatte Eisenstein [1850] mit seinem Reziprozitätsgesetz die erste herausragende und völlig streng begründete Anwendung der neuen Theorie der "idealen Zahlen" auf die alte Problematik der höheren Potenzreste gegeben.

ZITIERTE LITERATUR

Ahrens, W. ed. 1907. *Briefwechsel zwischen C. G. J. Jacobi und M. H. Jacobi* (Leipzig: B. G. Teubner).

Bachmann, P. 1919. *Das Fermatproblem in seiner bisherigen Entwicklung* (Berlin–Leipzig: W. de Gruyter & Co.; reprint, New York–Heidelberg–Berlin (West): Springer–Verlag, 1976).

Bašmakova, I. G. i A. N. Rudakov pri učastii A. N. Paršina i E. I. Slavutina. 1978. "Algebra i algebraičeskaya teoriya čisel." In: *Matematika XIX veka*. Pod redakciej A. N. Kolmogorova i A. P. Yuškeviča. (Moskva: Izd-vo "Nauka"). Glava vtoraya, 39–122.

Bergmann, G. 1966. "Über Eulers Beweis des großen Fermatschen Satzes für den Exponenten 3," *Mathematische Annalen*, *164*, 159–175.

Biermann, Kurt-R. 1967. "Zur Geschichte mathematischer Einsendungen an die Berliner Akademie," *Monatsbericht der Deutschen Akademie der Wissenschaften zu Berlin*, *9* (3), 216–222.

——— 1973. *Die Mathematik und ihre Dozenten an der Berliner Universität. 1810–1920* (Berlin: Akademie-Verlag).

Borewicz, Z. I., I. R. Šafarevič. 1972. *Teoriya čisel* (Moskva: Izdanie toroe. Izd-vo "Nauka").

Bourbaki, N. 1965. *Algèbre commutative. Note historique* (Paris: Hermann).

——— 1974. *Eléments d'histoire des mathématiques* (Nouv. éd., Paris: Hermann).

Cauchy, A. L. 1829. "Mémoire sur la théorie des nombres. Lu à l' Ac. roy. sci, le 21 sept. 1829," *Bulletin des sci. math. etc.*, *12* (Sept. 1829), Nr. 125, 205–221.

——— 1840. "Mémoire sur la théorie des nombres," [Datiert v. 31.5.1830], *Mém. Ac. sci.*, vol. *17* (1840), 249–768 = *Oeuvres* (I), vol. *3* (Paris, 1911), 5–450.

Dickson, L. E. 1920. *History of the Theory of Numbers. Vol. II: Diophantine Analysis* (Washington: The Carnegie Institution).

Dieudonné, J. 1978. *Abrége d'histoire des mathématiques 1700–1900, Tome I* (Paris: Hermann).

Dirichlet, P. G. L. 1825. "Mémoire sur l'impossibilité de quelques équations indéterminées du cinquiéme degré. Lu à l'Acad. R. Sc. [Paris] 1825," In: Dirichlet 1897, 1–20.

——— 1832. "Démonstration du théorème de Fermat pour le cas des 14ièmes puissances," *Journal für die reine und angewandte Mathematik*, *9*, 390–393. In: Dirichlet 1897, 191–194.

——— 1897. *Werke* (Band 1, Berlin: Reimer).

Edwards, Harold M. 1975. "The Background of Kummer's Proof of Fermat's Last Theorem for Regular Primes," *Archive for History of Exact Sciences*, *14* (3), 219–236.

——— 1977a. "Postscript to 'The Background of Kummer's Proof . . . ,'" *Archive for History of Exact Sciences*, *17* (14), 381–394.

——— 1977b. *Fermat's Last Theorem. A Genetic Introduction to Algebraic Number Theory* (Graduate Texts in Mathematics 50, New York–Heidelberg–West-Berlin: Springer-Verlag).

——— 1978. "On the Kronecker Nachlass," *Historia Mathematica*, *5*, 419–426.

Eisenstein, G. 1844. "Beiträge zur Kreistheilung," *Journal für die reine und angewandte Mathematik*, *27*, 269–278 = *Math. Werke*, *1* (1975), 45–54.

——— 1850. "Beweis der allgemeinsten Reciprocitätsgesetze zwischen reellen und complexen Zahlen," *Monatsbericht der Deutsche Akademie der Wissenschaften zu Berlin*, 189–198 = *Math. Werke*, *2* (1975), 712–721.

——— 1895. "Briefe von G. Eisenstein an M. A. Stern," Hrsg. v. A. Hurwitz und F. Rudio, *Zeitschr. f. Math. u. Physik*, *40* (Supplem. Zugleich Abhandl. zur Gesch. d. Math. Wiss., Heft 7. Leipzig), 169–203 = *Math. Werke 2* (1975), 791–823.

Ellison, W. J. et F. 1978. "Théorie des nombres." In: J. Dieudonné 1978, Chap. V, 165–334.

Euler, L. 1747. "Theorematum quorundam arithmeticorum demonstrationes," *Comment. ac. sc. Petropol*, *10* (1738) 1747, 125–146. In: Euler 1915, 38–58.

——— 1761. "Theoremata circa residua ex divisione potestatum relicta," *Novi Comment. ac. sc. Petropol*, *7* (1758/59) 1761, 49–82. In: Euler 1915, 493–518.

——— 1770. *Vollständige Anleitung zur Algebra* (St. Petersburg; Neue Ausgabe, Leipzig: Verlag Ph. Reclam).

——— 1915. *Opera omnia*. Ser. I, vol. *2*. (Berlin–Leipzig: B. G. Teubner).

Fermat, P. de 1891–1912. *Oeuvres*. Ed. Ch. Henry et P. Tannery. 4 vol. (Paris).

——— 1932. *Bemerkungen zu Diophant* Übers. u. hrsg. v. Max Miller. *Ostwald's Klassiker*, Band *234* (Leipzig: Akad. Verlagsgesellschaft).

Gauβ, C. F. 1801. *Disquisitiones arithmeticae* (Lipsiae) = *Werke*, Bd. *1* (2. Aufl. Göttingen-Berlin 1870). Deutsche Übersetzung: *Arithmetische Untersuchungen*, in: Gauß 1889, 1–453.

——— 1832. "Theoria residuorum biquadraticorum. Comment. secunda," [Datiert v. 15.4.1831] *Comment. soc. reg. scient. Gotting. recent.*, vol. *1* (Göttingen) = Gauß 1876a, 93–148. Deutsche Übersetzung: "Theorie der biquadratischen Reste. Zweite Abhandlung." in: Gauß 1889, 534–586

Gauß, C. F. 1876a. *Werke* (Band *2*, 2. Aufl. Göttingen–Berlin).

——— 1876b. "Disquisitionum circa aequationes puras ulterior evolutio." In: Gauß 1876a, 243–265. Deutsche Übersetzung: "Weitere Entwicklung der Untersuchungen über die reinen Gleichungen." In: Gauß 1889, 632–652.

——— 1889. *Untersuchungen über höhere Arithmetik von C. F. Gauß* (Deutsch hrsg. v. H. Maser, Berlin: Springer–Verlag; reprint, New York: Chelsea Publishing Company).

——— 1917. *Werke* (Band *10*(1), 1. Nachlaß und Briefwechsel zur reinen Mathematik, Leipzig: B. G. Teubner).

Hasse, H. 1964. *Vorlesungen über Zahlentheorie* (2., erweit. Auflage (West-)Berlin–Heidelberg–New York: Springer–Verlag).

——— 1970. *Bericht über neuere Untersuchungen und Probleme aus der Theorie der algebraischen Zahlkörper. Teil II: Reziprozitätsgesetz.* (3. Aufl., Würzburg–Wien: Physica–Verlag).

Heath-Brown, D. R., and Patterson, S. J. 1979 "The distribution of Kummer sums at prime arguments." *Journal für die reine und angewandte Mathematik*, 111–130.

Hensel K. 1910a. "Kummer und sein Lebenswerk. Gedächtnisrede auf Ernst Eduard Kummer." In: Kummer 1910, 1–37 = Kummer 1975, *1*, 33–69.

——— 1910b. *Ernst Eduard Kummer und der große Fermatsche Satz. Marburger akademische Reden* (Nr. 23, Marburg: N. G. Elwert'sche Verlagsbuchhandlung, 22S.)

Hermite, C. 1850. "Extraits de lettres de M Ch. Hermite à M. Jacobi sur différents objets de la théorie des numbres," *Journal für die reine und angewandte Mathematik*, *40*, 261–315 = C. Hermite: *Oeuvres*, *1*, (Paris, 1905), 100–163 = C. G. Jacobi: *Opuscula math.*, *2*, (Berlin, 1851), 221–275.

Hilbert, D. 1897. "Die Theorie der algebraischen Zahlkörper," *Jahresbericht der Deutschen Mathematiker Vereinigung*, *18*, 175–546. In: Hilbert 1932, 63–363.

——— 1900. Vortrag über Mathematische Probleme," *Intern. Mathematiker—Kongress Paris 1900*. In: Hilbert 1935, 290–329; und Hilbert 1971.

——— 1932. *Gesammelte Abhandlungen* (Band *1* Berlin: Springer–Verlag).

——— 1935. *Gesammelte Abhandlungen* (Band *3*, Berlin: Springer–Verlag).

——— 1971. *Die Hilbertschen Probleme*. Ed. H. Wußing. *Ostwalds Klassiker*, Band *252* (Leipzig: Akad. Verlagsgesellschaft).

Institut. 1918. *Institut de France. Académie des sciences. Procèsverbaux des séances de l'Académie*, Tome 8. (Année 1824–1827, Bassess-Pyrénées: Hendaye).

Jacobi, C. G. J. 1827. "De residuis cubicis commentatio numerosa," *Journal fur die reine und angewandte Mathematik*, *2*, 66–69. In: Jacobi 1891a, 233–237.

——— 1837. "Über die Kreistheilung und ihre Anwendung auf die Zahlentheorie," *Monatsbericht der academie der Wissenschaften zu Berlin*, 127–136. Abdruck in: Journal für die reine und angewandte mathematik, 30 (1846), 166–182; Jacobi 1846; Jacobi 1891a, 254–274.

——— 1839. "Über die complexen Primzahlen, welche in der Theorie der Reste der 5-ten, 8-ten und 12-ten Potenzen zu betrachten sind," *Monatsbericht der Academie der Wissenschaften zu Berlin*, 86–91. Abdruck in: Journal für die reine *und angewandte Mathematik*, *19* (1839), 314–318, und in: Jacobi 1891a, 275–280.

——— 1846. *Mathematische Werke von C. G. Jacobi. Erster Band.* Auch unter dem Titel: *Opuscula mathematica* (Vol. *1* Berlin: Reimer).
——— 1891a. *Gesammlte Werke* (Band 6., Berlin: Reimer).
——— 1891b. *Gesammelte Werke* (Band 7., Berlin: Reimer).
Kiernan, B. M. 1972. "The Development of Galois Theory from Lagrange to Artin," Archive for History of Exact Sciences, *8* (1971/72), 40–154.
Klein, F. 1926/27. *Vorlesungen über die Entwicklung der Mathematik im 19. Jahrhundert. I, II* (Berlin: Springer-Verlag. Reprint, New York: Chelsea, 1950).
Kronecker, L. 1845. *De unitatibus complexis* (Dissert. Univ. Berlin). Abdruck in Journal für die reine und angewandte Mathematik, 93 (1882), 1–52, und in *Werke* (Band *1*, Leipzig: B. G. Teubner, 1895), 5–73.
——— 1881. "Glückwunschadresse der Kgl. Pr. Akad. d. Wiss. zum fünfzigjährigen Doctor-Jubiläum von E. E. Kummer," zuerst anonym publiziert in *Monatsbericht Kgl. Pr. Akad. d. Wiss. zu Berlin*, 895–898. Wiederabdruck in Kronecker, L. 1930. *Werke*, Bd. *5* (Leipzig–Berlin) 461–464.
Kummer, E. E. 1837. "De aequatione $x^{2\lambda} + y^{2\lambda} = z^{2\lambda}$ per numeros integros resolvenda," [Datiert v. Okt. 1835], *Journal für die reine und angewandte Mathematik*, *17*, 203–209. In Kummer 1975, *1*, 135–141.
——— 1842. *De residuis cubicis disquisitiones nonnullae analyticae* (Univ. Breslau, Abdruck in *Journal für die reine und angewandte Mathematik*, *32* (1846), 341–359, und in Kummer 1975, *1*, 145–163.
——— 1844a. "Über die complexen Primfactoren der Zahlen, und deren Anwendung in der Kreisteilung," [Datiert v. 20.4.1844], *Archiv Akad. Wiss. DDR, Berlin* (*DDR*). Aufgefunden von Biermann 1967; publiziert in Edwards 1977a, 388–393.
——— 1844b. "De numeris complexis, qui radicibus unitatis et numeris integris realibus constant." In *Gratulationsschrift d. Univ. Breslau zur Jubelfeier d. Univ. Königsberg* (Breslau. Abdruck in *J. math. pures et appl.*, *12* (1847), 185–212, und in Kummer 1975, *1*, 165–192).
——— 1846a. "Über die Divisoren gewisser Formen der Zahlen, welche aus der Theorie der Kreistheilung entstehen," *Journal für die reine und angewandte Mathematik*, *30*, 107–116. In Kummer 1975, *1*, 193–202.
——— 1846b. "Zur Theorie der complexen Zahlen," *Monatsbericht der Akademie der Wissenschaften zu Berlin*, 87–96. Abdruck in *Journal für die reine und angewandte Mathematik*, *35* (1847), 319–326, und in Kummer 1975, *1*, 203–210.
——— 1847a. "Über die Zerlegung der aus Wurzeln der Einheit gebildeten complexen Zahlen in ihre Primfactoren," [Datiert v. Sept. 1846], *Journal für die reine und angewandte Mathematik*, *35*, 327–367. In Kummer 1975, *1*, 211–251.
——— 1847b. "Anzeige des ersten Bandes der Mathematischen Werke von C. G. Jacobi," *Neue Jenaische allgemeine Literatur-Zeitung 6*, 801–812. Abdruck in Kummer 1975, *2*, 695–705.
——— 1847c. "Beweis des Fermatschen Satzes der Unmöglichkeit von $x^\lambda + y^\lambda = z^\lambda$ für eine unendliche Anzahl von Primzahlen λ," *Monatsbericht der Akademie der Wissenschaften zu Berlin*, 132–141, 305–319. In Kummer 1975, *1*, 274–297.
——— 1847d. "Extrait d'une lettre de M. Kummer à M. Liouville," *J. Math. pures et appl*, *12*, 136. In Kummer 1975, *1*, 298.
——— 1850. "Allgemeine Reciprocitätsgesetze für beliebig hohe Potenzreste," *Monatsbericht der Akademie der Wissenschaften zu Berlin*, 154–165. In Kummer 1975, *1*, 345–357.
——— 1858. "Über die allgemeinen Recipocitätsgesetze der Potenzreste," *Monatsbericht der Akademie der Wissenschaften zu Berlin*, 158–171. In Kummer 1975, *1*, 673–687.
——— 1860. "Gedächtnisrede auf Gustav Peter Lejeune Dirichlet," *Abhandl. Akad. Wiss. Berlin*, 1–36. In Kummer 1975, *2*, 721–756.

——— 1910. *Festschrift zur Feier des 100. Geburtstages Eduard Kummers mit Briefen an seine Mutter und an Leopold Kronecker* (Leipzig–Berlin: B. G. Teubner). In Kummer 1975, *1*, 31–133.

——— 1975. *Collected Papers*. Ed. A. Weil. Vol. *1*. *Contributions to Number Theory*. Vol. *2*. *Function Theory, Geometry, and Miscellaneous*. (Berlin–Heidelberg–New York: Springer–Verlag).

Lagrange, J.-L. 1769. "Sur la solution des problèmes indéterminés du second degré," *Mém. Ac. Berlin*, t. *23*. In Lagrange 1868, 377–535.

——— 1868. *Oeuvres*, t. *2* (Paris: Gauthier–Villars).

Lampe, E. 1893. "Nachruf für Ernst Eduard Kummer," *Jahresbericht der Deutschen Mathematiker Vereinigung*, *3*, 13–28. In Kummer 1975, *1*, 76–82.

Legendre, A. M. 1827. "Sur quelques objets d'analyse indéterminée et particulièrement sur le théorème de Fermat," *Mém. Ac. Roy. Sc. de l'Institut de France*, *6*, (Paris).

——— 1830. *Théorie des nombres* (Paris).

Mordell, L. J. 1921. *Three lectures on Fermat's Last Theorem* (Cambridge, England: Cambridge University Press. Reprint in Mordell 1972, 11–56).

——— 1972. *Two papers on number theory* (Berlin: VEB Deutscher Verlag der Wissenschaften).

Neumann, O. 1980. "Zur Genesis der algebraischen Zahlentheorie. Bemerkungen aus heutiger Sicht über Gauß' Beiträge zu Zahlentheorie, Algebra und Funktionentheorie (Fortsezung)," *NTM-Schriftenr. f. Gesch. d. Naturw., Technik u. Medizin* (Leipzig: Akad. Verlagsgesellschaft), 17(1), 32–48, 17(2), 38–58.

Nový, L. 1973. *Origins of Modern Algebra* (Prague: Academia).

Olbers, W. 1900. *Sein Leben und seine Werke*, Band *2* (1) (hrsg.v. C. Schilling, Berlin).

Purkert, W. 1973. "Zur Genesis des abstrakten Körperbegriffs. Teil 1, *NTM*, *10* (1), (1973), 23–37; Teil 2, *NTM*, *10* (2), (1973), 8–20.

Schönemann, T. 1839. "Theorie der symmetrischen Functionen der Wurzeln einer Gleichung. Allgemeine Sätze über Congruenzen nebst einigen Anwendungen derselben," *Journal für die reine und angewandte Mathematik*, *19*, 231–243, 289–308.

Smith, H. J. S. 1859–65. *Report on the Theory of Numbers*, Part I–VI London: British Association for the Advancement of Science) = *Collected Math. Papers* vol. *1*, (Oxford, 1894), 38–364. Reprint. New York: Chelsea, 1965.

Weil, A. 1974. "La cyclotomie jadis et naguère," *Sém. Bourbaki*, 26e année, 1973/74, n° 452. (21 S., Paris). = *L'Enseignement mathém.* (2) *20* (1974), 247–263.

Wußing, H. 1969. *Die Genesis des abstrakten Gruppenbegriffes* (Berlin: VEB Deutscher Verlag der Wissenschaften).

Sektion Mathematik
Friedrich-Schiller-Universität
Jena, Deutsche Demokratische Republik

Leibniz on the Probable*

IVO SCHNEIDER

The major sources of Leibniz's contributions to a mathematical–philosophical theory of probability are to be found in his unpublished notes, manuscripts and correspondence. This paper draws upon such materials, including publications by Kurt-R. Biermann and Ian Hacking, to evaluate the various contributions of Leibniz to the development of the calculus of probabilities and the philosophy of probability. Among topics considered are Leibniz's interests in jurisprudence, as well as his attempts to mathematize the probable and to estimate uncertainty. His influence on the theory of probability in the 18th century is also discussed, including the work of figures like the Bernoulli's, de Moivre, and Laplace.

1. INTRODUCTION

It is not easy to determine the role of Leibniz in the development of those areas later known as inductive logic, probability theory, and statistics. Because Leibniz scholars often consider him to be the last universal genius, it does not surprise one that they find in his voluminous papers ideas concerning probability. More interesting is the fact that some historians of

* For K.-R. Biermann on his 60th birthday. I am very much indebted to M. Norton Wise for his assistance in translating my paper into English.

ISBN 0-12-204050-3

mathematics and philosophy who are less committed to Leibniz have given him considerable credit for the development of the calculus of probabilities and the philosophy of probability. The publications of Leibniz cannot alone justify such attributions. What then is of interest in Leibniz's contribution to the discussion of probability? The literature offers more than one answer. Disregarding those who deny any merit to Leibniz, we have, on the one hand, those who emphasize his influence on the pioneers of probability theory and, on the other, those who treat Leibniz as a witness for the existence of definite concepts of probability and as an advocate of a program for their application. Kurt-R. Biermann, on the basis of his attempts to document Leibniz's role in the early growth of the calculus of probabilities,[1] is the most important representative of the first group; while Ian Hacking in the second group parades Leibniz as the most important of those who prepared and made possible the quantification of the probable.[2]

However one regards Leibniz's contribution to a mathematical–philosophical theory of probability, the main source for it is to be found in his unpublished notes and manuscripts as well as in his abundant correspondence. In addition we must not exclude the possibility that Leibniz influenced some of his contemporaries through oral communication, especially since he repeatedly mentioned in his letters his long cherished but never realized program of a *doctrina de gradibus probabilitatis*. The potential for such influence is indicated by the attitude of expectation evident in his correspondence with Montmort and Niklaus Bernoulli, both significant contributors to probability theory in the eighteenth century. In what follows I shall attempt to evaluate on the basis of the available sources the various contributions of Leibniz to the discussion of the probable.

2. LEIBNIZ'S INTERESTS IN JURISPRUDENCE AS HIS STARTING POINT

Leibniz concerned himself with problems of probability from a variety of perspectives, which qualified him as an informed discussant with his contemporaries and in his later writings resulted in a more or less unified concept. According to Leibniz's own testimony, his early interest in probability derived from jurisprudence, which he took up as a student. Already in his baccalaureate dissertation *De conditionibus*, written at age nineteen, the connection between probability and law appears.[3] He dealt here with the subject of conditional rights (*jus conditionale*). He placed the value of a conditional right between 0 and 1, where 0 represents no claim (*jus nullum*) and 1 represents absolute right (*jus purum*). Hacking offers a formal interpretation:[4]

In general, consider propositions of the form 'if q then r' where q is a disjunction of mutually exclusive alternatives. Now consider a set of conditions each of which is sufficient for r. Three cases may arise. Every disjunct of q may preclude each of these conditions. In that case Leibniz calls the condition for r impossible, and we have *jus nullum*, no right at all. If every element of q entails some condition sufficient for r, then the condition is called necessary, and we have *jus purum*. However if some disjuncts entail a condition for r, while the rest entail a condition for not-r, then we have only a conditional right, and the condition is called uncertain (*incerta*, in the 1665 version) or contingent (*contingens*, in the 1672 version).

Although a numerical value for the *jus conditionale* could be given, presupposing, for example, equal weight for every condition favourable to r, in no concrete case in the known juridical papers of Leibniz is a conditional right evaluated numerically. The second part of the tract, *De conditionibus* contains a fictive example for a relative ordering of claims. Presupposing that a person receives 100 Taler under condition A and 200 Taler under condition B, then the conditional right depending on B is assigned a higher value than that depending on A, if the occurrence of A and B is equally uncertain.[5] Also in the *Specimina juris* of 1669 Leibniz remarked only generally that the value of a conditional right increases with the number of possibilities for the existence of the relevant condition.[6]

Finally one finds in a 1671 draft of the *Elementa juris naturalis* an example of a decision based on a numerical evaluation. It concerns a choice between two actions A and B, taking into account the probability of the occurrence of an effect stemming from the action and the weight of this effect. His interpretation of effect as advantage motivated Leibniz's principle of decision: to choose that action for which the product of probability and effect is a maximum. It is worth noting in Leibniz's example not only that the probabilities are not normalized, but also that the numbers involved are purely fictive, corresponding to no concrete case. This, together with his inability to respond to James Bernoulli's request for examples of concrete law cases which could be regarded as applications of probability calculations, shows that Leibniz had no method available for evaluating conditional right. Leibniz confirms this himself in a letter of 1698 to Gabriel Wagner, in which he discussed in general terms the determination of *gradus probabilitatis*:[7]

wie man die anzeigungen, so keinen vollkommenen Beweiß machen und gegen einander laufen (*indicantia et contra-indicantia*, wie die *Medici* reden) abwegen und schäzen solle, umb den außschlag zu geben. Denn man insgemein gar wohl sagt, *rationes non esse numerandas sed ponderandas*, man müße die anzeigungen nicht zehlen, sondern wägen, aber niemand hat noch dazu die Wage gezeiget, wiewohl Keine dem werck näher gekommen und mehr hülffe an hand gegeben als die *Juristen*, daher ich auch der *materi* nicht wenig nachgedacht, und dermahleins den mangel in etwas zu ersezen hoffe.

Naturally one must distinguish Leibniz at the turn of the eighteenth century from the young law student of the 1660s. Still, the idea appearing in the letter to Gabriel Wagner of a comprehensive probability theory occurs very early in Leibniz's work and apparently grew out of his involvement with questions of law. This is indicated, for example, in a letter of December 18 (28), 1670 from Friedrich Nitzsche to Leibniz encouraging him to carry out his announced project for a *doctrina de gradibus probabilitatis.*[8]

Later letters show, however, that Leibniz, to the end of his life, was not able to realize such a *doctrina*—or "new kind of logic," as he called it[9]—with degrees of probability as its object. At the same time one sees from these letters that in his later years Leibniz hoped more and more for a mathematician who would be able to give a comprehensive representation of games. In this vein he wrote in his *Nouveaux essais*, which are a commentary on John Locke's *Essay Concerning Human Understanding*:[10]

> J'ay dit plus d'une fois qu'il faudroit une *nouvelle espece de Logique*, qui traiteroit des degrés de probabilité, puisqu'Aristote dans ses Topiques n'a rien moins fait que cela, et s'est contenté de mettre en quelque ordre certaines regles populaires, distribuées selon les lieux communs, qui peuvent servir dans quelque occasion, où il s'agit d'amplifier le discours et de luy donner apparence, sans se mettre en peine de nous donner une balance necessaire pour peser les apparences et pour former là dessus un jugement solide. Il seroit bon que celuy qui voudroit traiter cette matiere, poursuivit l'examen des *jeux de hazard*; et generalement je souhaiterois qu'un habile Mathematicien voulût faire un ample ouvrage bien circonstancié et bien raisonné sur toute sorte de jeux.

From the *Nouveaux essais* it is clear that Leibniz's understanding of probability, though motivated by his juridical work, reaches directly back to antiquity. We see, namely, that the sources for Leibniz's conception of probability, or better, his distinction of degrees of probability, are to be found in the *Topics* of Aristotle, which he cites several times, and also in the skeptical writings of the Middle Academy. The allusions in the *Nouveaux essais* to the Skeptics[11] suggests a close connection between the degrees of probability distinguished by Leibniz and the ideas of the Skeptics concerning the probable, as available in Sextus Empiricus. As background for the existence of a skeptical attitude among seventeenth-century natural philosophers and mathematicians, most prominently in Robert Boyle, I would point to the authority problem which arose in the realm of religious controversies. In the *Nouveaux essais* it appears also that his distinction of different degrees of probability allowed Leibniz only to give qualitative differences, but not quantitatively assessible gradations. Here Leibniz proceeds according to the distinction between speculative propositions and propositions of fact, the latter depending on human testimony:[12]

Mais lorsque les temoignages se trouvent contraires au cours ordinaire de la nature, ou entre, eux, les degrés de probabilité se peuvent diversifier à l'infini, d'où viennent ces degrès que nous appellons *croyance*, *conjecture*, *doute*, *incertitude*, *defiance*; et c'est là où il faut de l'exactitude pour former un jugement droit et proportionner nostre assentiment aux degrés de probabilité.

Les Jurisconsultes en traitant des preuves, presomtions, conjectures et indices, ont dit quantité de bonnes choses sur ce sujet, et sont allés à quelque detail considerables.

Here, as in the examples from civil law which follow in his text, Leibniz is not thinking of a numerical gradation of the probable, but of a qualitative ordering, corresponding to the view of the Skeptics. They too had entertained the notion of a continuum of possibilities of modality.[13] This holds also for the medical situation, on which Leibniz comments in the same connection:[14] "Les Medecins encor ont quantité de degrés et de differences de leur *signes* et *indications*, qu'on peut voir chex eux. Les Mathematiciens de nostre temps ont commencé à estimer les hazards à l'occasion de jeux."

The final sentence refers to the hoped-for mathematician who, in contrast to the usual practice in law and medicine of distinguishing degrees of probability only qualitatively, will be able to attain quantitative evaluations, starting from the odds in games of chance. Leibniz himself had occasionally attempted calculations on games of chance and mortality. His investigations qualified him for discussions with, especially, James Bernoulli, who, independently of Leibniz became the originator of the calculus of probabilities. James Bernoulli succeeded in finding a measure for probability, conceived as degree of certainty, and founded on this measure his calculus of probabilities, which the aging Leibniz regarded as the best approach to his youthful dream of a *doctrina de gradibus probabilitatis*.[15]

Before offering an evaluation of Leibniz's efforts to formulate a theory of games of chance, reference should be made to the fact that his merits as a probability theorist appear somewhat more significant from a philosophical perspective. Ian Hacking has designated him as the "first philosopher of probability," as the originator of a new form of logic, which already contained the essential elements of an inductive logic newly created in our own century. This Leibnizian logic meant for its creator a "natural jurisprudence" in which probability was conceived as an objective relation between hypotheses and evidence. According to Hacking, Leibniz had combined this concept of probability with an understanding of probability as a matter of physical propensities, dating from his experience with a calculus of chances.[16]

Nearly nothing is known about the extent to which Leibniz's form of inductive logic influenced a logic of the probable in eighteenth-century Germany.[17] Though his philosophical investigations of the probable are interesting, more influential in the development of a mathematical probability theory is Leibniz's work in combinatorial analysis, which is part of pure mathematics.[18] To this belongs first of all his *Dissertatio de arte combinatoria*,[19] a work of his youth. It appeared increasingly worthless to him as he grew older, but it represents one of the few realized components in his great project for an *ars inveniendi* based on a *characteristica universalis*. The discrepancy between this ambitious project and its poor realization may well explain Leibniz's later dissatisfaction with this dissertation as well as his plans for a considerably extended version. Thus in 1680 Leibniz formulated his ideas for a new work, *De arte combinatoria* in which he identified probability theory and the theory of games of chance as important areas of application.[20]

Prerequisites for the inclusion of these areas of application, not mentioned in the dissertation of 1666, were Leibniz's studies of problems concerning games of chance and mortality during and after his stay in Paris from 1672 to 1675. Independent of Leibniz's own ideas about the possible applications of his *ars combinatoria*, James Bernoulli later recognized him as a pioneer of the combinatorial methods necessary for probability theory.[21]

3. LEIBNIZ'S ATTEMPT TO MATHEMATIZE THE PROBABLE, TO ESTIMATE THE UNCERTAIN

Leibniz's interest in problems of games of chance was first aroused during his Paris period by the Duke of Roannez. At the beginning of this work around the end of 1675, Leibniz already knew Huygens's tract *De ratiociniis in ludo aleae*[22], as well as Pascal's *Triangle arithmétique*[23] and, because of his access to Pascal's papers, perhaps even the correspondence with Fermat. Leibniz studied the division problem at this time, as well as special problems of dicing, which he tried to solve with combinatorial methods. Again stimulated by the Duke of Roannez, Leibniz concerned himself with different possibilities for an ordering of mortality[24] based on the data published by John Graunt.[25] More important is the manuscript *De incerti aestimatione*, composed in Hannover in September 1678.[26] In it he took a decisive step forward using the relation between favourable and possible cases as a means to determine expectation, "*spes*," understood as *probabilitas habendi*. At the beginning of this work Leibniz discusses, in the rambling style typical of his drafts, the principle of a fair game already treated by Huygens:[27]

> Justus ludus est si spei et metus utrinque eadem ratio sit. In justo ludo tanti est spes quanti emta est, quia justum est rem tanti emi quanti est, et tantus est metus quantum spei pretium est.
>
> Axioma: Si ludentes similia agunt ita ut nullum discrimen inter ipsos assignari possit, nisi quod in solo eventu consistat, eadem spei metusque ratio est.

> Potest demonstrari ex Metaphysicis; nam ubi quae apparent eadem sunt, idem de iis judicium formari possunt, id est eadem est ratio opinandi de futuro eventu; opinio autem de futuro eventu spes metusve est. Si sors communi ludentium contributione aequali formetur, et unusquisque eodem modo ludat, et eidem eventui idem praemium eademve poena statuatur, justus ludus est.

This passage has been interpreted as a justification of expectation that goes beyond Huygens work.[28] In any case, it shows that Leibniz assigned the same central role to the concept of expectation that Huygens had. Leibniz determines expectation in general as follows:[29] "Si diversos eventus utiles disjunctim habere possit negotium, spei aestimatio erit summa utilitatum possibilium ex omnibus eventibus collectarum, divisa per numerum eventuum."

His notation of fear, "*metus*," for the expectation in cases where all the possible outcomes of the game bring losses, seems a bit awkward, especially since he does not cover the possibility of a lottery with negative and positive "gains."

After this determination of the expectation value Leibniz points to another way to justify expectation. He starts this part with four statements similar to definitions:

> Probabilitas est gradus possibilitatis.
> Spes est probabilitas habendi.
> Metus est probabilitas amittendi.
> Aestimatio rei tanta est, quantum est jus cujusque in rem.

The last sentence makes the estimation of a thing equal to the right or claim for the same thing. It leads to the second deduction of the expectation value on the basis of Leibniz's early ideas about conditional right. In the juridical formulation, expectation is determined in the following way: "Potestas habendi rem in aliquem eventum est ad potestatem habendi rem in omnem eventum, ut possibilitas eventus unius ad possibilitatem eventuum omnium."

The right or claim of a thing depending on the happening of a certain event is to the right or claim for a thing in any case as the possibility of this event to the possibility of all events. A numerical evaluation of expectation requires further assumptions. Formulas can be given, for example, if the events are *aeque faciles* as in Huygens's calculus of chances. In order to apply this to his juridical formulation, Leibniz needs to explain the relation of a set of events described as *aeque faciles* and the possibility of an event. Leibniz's solution is simple: he states the equivalence of *aeque facilis* and *aeque possibilis*:[30] "*Si eventus sint aeque faciles seu aeque possibiles* . . . ."

With the help of the other equivalence introduced previously—probability is degree of possibility—Leibniz now is able to formulate his finding in a probability terminology. As a result he brings up two theorems, the second of which implies that the relation between favorable and possible cases measures probability:

> Si plures sunt eventus aeque faciles, et aliquot eventibus rem habiturus sum, aliquot aliis re cariturus, spei aestimatio erit portio rei quae ita sit ad rem totam, ut numerus eventuum qui favere possunt ad numerum omnium eventuum. Nempe S/R aequ. F/n seu S aequ. $(F/n)R$.

The remaining part of Leibniz's theory of expectation contains as the main result a deduction of Huygens's third proposition in a generalized form: If there are $n = a + b + c$ events described as *aeque faciles*, out of which a results in winning A, b in winning B, and c in winning C, then the expectation, *spes* $(=S)$, is

$$S = \frac{aA + bB + cC}{n}.$$

This formula for the expectation S is not changed if there are d events more, the occurrence of which brings no gain. In this case $n = a + b + c + d$ and

$$S = \frac{aA + bB + cC + d0}{n} = \frac{aA + bB + cC}{n}.$$

The second part of the *De aestimatione incerti* deals with the division problem. It shows that, as in an earlier attempt of 1676, Leibniz was not able to go beyond his predecessors from the mathematical point of view.[31] However, the juridical background of Leibniz's approach to this problem deserves attention. Presumably because Leibniz could not transcend the foundation of expectation, the *De aestimatione incerti* was buried among a mass of other papers inaccessible to his contemporaries. Still, it reveals the decisive step to a concept of probability understood as degree of possibility, which was applied to the realm of games of chance. His familiarity with this concept since at least 1678 enabled Leibniz to discuss critically James Bernoulli's main theorem and the consequences connected with it. Leibniz's impact on the theory of probability in the eighteenth century is mainly based on this discussion with James Bernoulli, who, independently of Leibniz, had found his way to an understanding of *probabilitas* as degree of certainty.

4. LEIBNIZ'S INFLUENCE ON THE THEORY OF PROBABILITY IN THE EARLY EIGHTEENTH CENTURY

It is true that the secondary literature shows no unanimity concerning the importance of James Bernoulli's *Ars conjectandi* for the calculus of probabilities in the eighteenth century. Contrary to common opinion, which takes Montmort[32] as a witness, that the *Ars conjectandi* was published too late to exert influence on the generation of mathematicians following James Bernoulli, neither the development of the De Moivre–Laplace limit theorem nor de Moivre's changed conceptions manifest in the different approaches of his *Mensura sortis* of 1711 and his *Doctrine of Chances* of 1718 can be explained without the *Ars conjectandi.*[33] Among the four books of this work, the unfinished last one was always considered the most interesting. James Bernoulli worked on it until his death. James Hermann confirms this in his letter to Leibniz of October 1705.[34] According to Hermann, a few months more would have sufficed to complete the work, which had occupied James Bernoulli again and again in his last years. Hermann was presumably the first mathematician who had access to the papers of his former teacher James Bernoulli and the most important man for the diffusion of James Bernoulli's ideas and results[35] before the publication of the *Ars conjectandi.* In an earlier letter to Leibniz he had emphasized the obligation James Bernoulli felt to Leibniz.[36] Though this obligation is expressed only in general terms, it includes Leibniz's share in the *Ars conjectandi,* since he had discussed questions concerning the last part of the work with James Bernoulli on several occasions between 1703 and 1705.

In what follows, the attempt is made to account for Leibniz's possible influence on James Bernoulli's conceptions of probability and especially on the fourth book of the *Ars conjectandi.*[37] Leibniz's exchange of ideas with James Bernoulli concerning mathematization of probabilities begins with a letter from Leibniz of April 1703.[38] In his postscript Leibniz remarked that he had heard of Bernoulli's involvement with a *doctrina de aestimandis probabilitatibus.* In this well-known prelude to a longer discussion of the topic, it is interesting to find Leibniz's hope briefly expressed, corresponding to his early program, that someone like Bernoulli would treat mathematically the different kinds of games, in which beautiful examples of such a *doctrina* could be found. That means that at least by this time it had become natural for Leibniz to regard the mathematical treatment of games of chance as part of a theory of probability estimation. The secondary literature has justifiably emphasized that James Bernoulli, who had worked for many years on questions of games of chance and probability[39] calculations, took the postscript as an invitation to communicate his most cherished ideas. This is all the more understandable as James Bernoulli had long sought in vain for a suitable correspondent in this area of study and Leibniz appeared to be able to fulfill that role. This explains, too, why Bernoulli already presented in his answer the central problem of his research, as well as his most important result, his main theorem. Bernoulli was concerned with the determination of such probabilities as that of a young man of age twenty surviving a man of sixty. Bernoulli called these probabilities *a posteriori*, because they can be determined only in retrospect on the basis of numerous observations of the occurrence of a relevant event. Bernoulli's main theorem was supposed to establish that with an increasing number of observations the estimated value of the probability approaches the true value, at least with probability. That James Bernoulli, who had expressed his astonishment about Leibniz's information concerning his research at the beginning of his letter of October 3, 1703,[40] nevertheless knew of Leibniz's activities in the field is demonstrated by his request that Leibniz should send to him juridical material to which one could, in Leibniz's judgment, apply *a posteriori* determination of probabilities. At the same time Bernoulli was interested in obtaining the assessment[41] of Jan de Witt, Raadspensionaris of Holland, in which the advantage of buying and selling life annuities was determined on the basis of hypotheses about life expectancies at different ages.

Leibniz answered this detailed report of James Bernoulli in a letter of December 3, 1703.[42] He emphasized at first the extraordinary utility of the *aestimandae probabilitates*, only to add immediately the qualification that in the area of jurisprudence and politics, which was so important for Bernoulli's program, no such extended calculations were usually required, since an enumeration of the relevant conditions would suffice. Considering the request of Bernoulli, this implies that Leibniz was not in the position to offer concrete juridical material to which the methods of Bernoulli's probability theory could be applied. It is relatively certain that if Leibniz had found nontrivial evaluations in the realm of conditional right, he did not remember them in 1703. After Leibniz's statement that James's brother John was not his informant for James Bernoulli's investigations, but an unnamed third party,[43] Leibniz tried to shake James Bernoulli's self-confidence, which was founded above all on his discovery of the main theorem. Against the possibility of attaining a better approximation to a sought after probability with an increasing number of observations, Leibniz suggested that contingent events—here identified with dependence on infinitely many conditions—could not be determined by a finite number of experiments. As a foundation Leibniz added that, to be sure, nature has her conventions, which follow from the permanent repetition of causes. That this holds only as a rule which permits exemptions, is expressed by the classical greek term *hos epi to poly*. In this sense Bernoulli's presupposition of the absolute determinability of a probability *a posteriori* seemed already questionable, because it implied the invariance of such a probability with time. For Leibniz the appearance of new diseases could change the probability of survival of a twenty-year-old relative to a sixty-year-old. Leibniz attempted to lend greater weight to his objection through the example of determining the orbits of comets. These were always found under the assumption that the orbit was a conic section. But if this presupposition is dismissed, then there would be infinitely many different curves that fit the observations. Bernoulli was understandably not particularly pleased with Leibniz's objections. In a letter of April 20, 1704,[44] he emphasized that the mere enumeration of conditions in law did not suffice; rather, calculations were required just as for games of chance. Bernoulli referred to problems of insurance, life annuities, marriage contracts, *praesumptiones*, and others. He put off until later supplying Leibniz with an illustration of such calculations, perhaps because Leibniz had disappointed him regarding the requested juridical material. He attempted to clarify his main theorem anew, using the example of an urn containing white and black stones in the ratio of 2 : 1. In this case Bernoulli claimed to be able to determine exactly the number of draws (with replacement) for which it would be ten times, a hundred times, a thousand times, etc., more probable that the ratio of white to black stones found by drawing would fall inside rather than outside a given interval about the true value, for example, (199/100, 201/100). Although Bernoulli could only prove this assertion for probabilities from *a priori* considerations, as in the urn model, he was convinced that he had also shown, with his main theorem, the solubility of the reverse problem, namely, the determination of unknown, *a posteriori* probabilities. This false conclusion becomes understandable through Bernoulli's implication that it would make no difference for the behavior of the observed ratio whether the person drawing the stones knew the true ratio or not. The possibility that two urns containing different ratios of white to black stones would yield the same ratio for an equal number of draws appeared conceivable to Bernoulli only for a small number of draws, while for a large number such a result would be excluded by the "moral certainty" *secured* through the main theorem.

In his way of thinking Bernoulli saw no problem with applying the urn model to human mortality, with the stones corresponding to diseases with which a person can be taken ill.[45] However, he was prepared to concede that with the data then available, the life expectancy of the antediluvians could not be found. For him it was only important to be able to determine the validity of the approximation, since in any concrete case the data would be only finite, and he proceeded on the assumption that the probability to be determined would remain stable over a sufficiently long time. Leibniz's objections did not hit on the nonapplicability of the main theorem to the reverse problem; they were concerned with the applicability of the urn model to areas like human mortality. James Bernoulli's research program was not affected by these objections. This research program stood firm after Bernoulli's discovery of his main theorem in 1689.

It consisted mainly in integrating into the calculus of probabilities all areas operating with the concept of *probabilitas*. James Bernoulli, who at the beginning of his discussion with Leibniz had only two more years to live, was not prepared to change essential parts of his program. Rather, he expected from Leibniz the confirmation of the validity of his ideas and complementary material suitable for supporting his results for a realization of his program. In this sense he requested from Leibniz a copy of de Witt's assessment and in his letter from August 2, 1704,[46] additional material dealing with conditional right. In order to get more concrete descriptions from Leibniz, Bernoulli asked him for examples in which Leibniz should give his view of the evaluation of conditional right and of the problem of joint annuities. For this Bernoulli apparently expected from Leibniz no calculations, but only elucidation from the juridical side, since he himself had never been engaged professionally in questions of law. Leibniz's answer of November 28, 1704,[47] hardly met Bernoulli's hopes. He not only could not find de Witt's *Waerdye*, but he had no copy remaining of his twice-issued investigation on conditional right.[48]

After all, Leibniz was not convinced that an increase in the number of observations would in all cases improve the certainty of the attained result. Leibniz acknowledged that for pure mental games and games of chance one could calculate the chance of winning, even though with some difficulty, while in most cases, on the basis of reflection alone, one could determine only who has the better position There were, certainly, inventive players who without calculating made their decisions, as in military matters and in medicine, on the basis of a multitude of judgments. Leibniz appreciated this way of thinking as an *ars*.

To what degree Leibniz stimulated Bernoulli's treatment of the qualitative evaluation of probabilities, as presented in the fourth part of the *Ars conjectandi*, can no longer be determined.[49] That correspondence, brought to an untimely end by James Bernoulli's death, shows that Leibniz could fulfill Bernoulli's desire for a congenial correspondent only in part. After pressing Leibniz once more to send the papers he had repeatedly requested, in his letter of February 28, 1705,[50] he again described briefly the basic problem the solution of which he thought would also meet Leibniz's approval, once it became accessible in print. This indicates that Bernoulli in late winter of 1705 had finished the last chapter of the *Ars conjectandi*, which contains the main theorem. In this chapter Bernoulli reacted to Leibniz's objections.

Leibniz had written James Bernoulli for the last time in March 1705.[51] There he sought to disqualify as inappropriate or not new the works of de Witt and himself that Bernoulli desired, and put off sending them until later. Even if one takes into account that Leibniz was working in completely different areas at that time and that his main interest in mathematics was absorbed by the infinitesimal calculus he had created, it is difficult to understand why he did not inform James Bernoulli, who urgently needed such material, about Halley's mortality table.[52] Halley's work on annuities was based on the data, known to Leibniz, of the city of Breslau and went far beyond de Witt.

After Bernoulli's death on August 16, 1705, Leibniz needed no longer to question his authority as arbiter in discussions of probability. He could now accept respect and devotion from the distance of the grand old man and enjoy the role of judge, instigator, and promoter. This period of his life is interesting insofar as Leibniz repeatedly related his own earlier works and unrealized views to the activities of James Bernoulli's successors.

The meagre references by James Hermann to the content of the *Ars conjectandi* in a letter of October 28, 1705,[53] told Leibniz nothing new. They indicated, nevertheless, that the entire Bernoulli clan, among whom Hermann is to be counted, was so impressed with this unfinished manuscript of James Bernoulli that further critical remarks from Leibniz would have little significance for them. Only after some years did Leibniz, in his correspondence with John Bernoulli, return once more to the realm of games of chance and to the calculus of probabilities. The opportunity arose through their regular exchange of information concerning scientific projects and new published works, as well as the personal situation of their authors.

A few years after James Bernoulli's death, a new generation appeared, which was ready to follow in his footsteps. In this generation one can find James Bernoulli's nephew Niklaus Bernoulli, the Frenchman Pierre Rémond de Montmort, and Abraham de Moivre, a Huguenot living in England. Leibniz's first opportunity to play the role of instigator was offered in a letter of June 27, 1708, to Bernoulli. In response to John Bernoulli's reference to Montmort's intention of writing a book on the calculus of games of chance, Leibniz expressed his wish, with the tone of the great organizer of science, that the subject should be worked out carefully ("*vellem hoc argumentum bene tractaretur*"). At the same time Leibniz urged publication of James Bernoulli's *Ars conjectandi*, at least a selection from it. With the cautious suggestion that the manuscript of the *Ars conjectandi* might possibly contain something noteworthy, Leibniz left all possibilities open relative to John Bernoulli's views. As John Bernoulli, however, repeatedly referred to the

Ars conjectandi very favorably [54] in comparison with Montmort's book,[55] which had appeared in the meantime, Leibniz's opinion became firm. In the postscript to his letter of September 6, 1709, Leibniz asserted without qualification that in his view the *Ars conjectandi* deserved to be published, and that he had several times pressed that view on James Bernoulli while he still lived. Leibniz's transformation into a promoter of, and almost a participant in, James Bernoulli's work was by now complete. Thus he wrote in his postscript: "Again and again I reminded him that we are lacking that part of logic which deals with degrees of probability (*de gradibus verisimilitudinis*)." Leibniz added that in his view these degrees had to be estimated from the degrees of possibility equivalent to the number of equal possibilities.

In order to document his active role in this development he referred to a political paper, written at the command of his sovereign, in which he had indicated that certain of these estimations of degrees of probabilities should be made by addition and others by multiplication.[56] One cannot exclude the possibility that Leibniz, who certainly had no knowledge of James Bernoulli's *Ars conjectandi* at the time he wrote this political treatise,[57] had formulated in a preliminary way the basic rules of a probability algebra: the sum rule for disjunctive events and the product rule for the simultaneous occurrence of several events. It is surprising in retrospect that Leibniz did not mention a single word about this political paper and the results contained in it in his letters to James Bernoulli. However, Leibniz's willingness to communicate had been stimulated again by John Bernoulli's announcement that his nephew's dissertation, *De usu artis conjectandi in jure*, would be published soon.[58]

In this connection John Bernoulli called attention to the problems of taking missing persons for dead and of life annuities. He regarded the dissertation of his nephew as a work paralleling his own application of mathematics to medicine. John Bernoulli's additional remark that Niklaus's dissertation would offer something new and unusual to the jurists, who had normally regarded such considerations as superfluous, may have induced Leibniz in his response to refer to his own repeated emphasis on the significance of a probability theory for jurisprudence and to his earlier work on such a theory.[59]

Leibniz remembered his law studies, during which he dealt with conjectures, indices, expectancies, and degrees of confirmation depending on partial or complete evidence. Leibniz conceded that the jurists had treated the subject better than anybody else, but nevertheless without sufficient basis and methods.

In later letters to John and Niklaus Bernoulli, such reminiscences no longer appear. Leibniz limited himself to distributing praise and encouragement. He was not prepared to discuss details of the works available by 1713 in the calculus of probabilities, which included, next to Niklaus Bernoulli's dissertation, de Moivre's *Mensura sortis*,[60] the second edition of Montmort's *Essay*,[61] and, finally, the posthumous edition of the *Ars conjectandi*. Leibniz's self-adopted role as instigator culminated in his contention in 1714 that James Bernoulli had worked out his *Ars conjectandi* according to Leibniz's exhortations using the contributions of Pascal, Fermat, Huygens, de Witt, and Hudde as a basis.[62]

5. CONCLUSION

Laplace does not mention Leibniz in the final chapter, dealing with the historical development of probability theory, of his *Essai philosophique sur les probabilités*.[63] Biermann is certainly right when he justifies[64] this omission by the relative insignificance of Leibniz's only two publications in this realm, two papers on the mathematical treatment of games.[65]

On the other hand Laplace points to Leibniz more than once in his *Essai*. Laplace knew perhaps better than anybody else that the most important tools of an "analytical" theory of probability had been shaped by the Leibnizian form of the infinitesimal calculus. In this sense he had honored Leibniz as a mathematician similar to James Bernoulli, who in his *Ars conjectandi* explicitly mentioned Leibniz's contribution to combinatorics. Leibniz is even the first whom Laplace introduced in the initial chapter "On Probability" of his *Essai*; Laplace does this in connection with the so-called principle of sufficient reason. Here we can see Leibniz's influence as a philosopher. Where and how far this influence could be extended outside a mathematical theory of probability, we do not know. The philosopher Leibniz could color the calculus of probability proper only by his correspondence with James Bernoulli.

Whether Leibniz's short allusion to equipossibility in connection with de Witt's tract[66] could have illuminated James Bernoulli considerably[67] is doubtful. Following Huygens, James Bernoulli had already used in the 1680s, as can be seen from his *Meditationes*, the terms *aeque facilis* and *aeque in proclivi* and later (in the *Ars conjectandi*) *aeque facilis* synonymously with *aeque possibilis*.

From the *Meditationes*, James Bernoulli's independence of Leibniz concerning the concept of probability, understood as degree of certainty, becomes clear, too.[68] Leibniz could only have confirmed Bernoulli in that respect.

More important are Leibniz's objections against the *a posteriori* determination of probabilities. Not only did they induce James Bernoulli to reactions in the *Ars conjectandi*, but they can also be considered as the starting point of de Moivre's considerations about the theorem named after him. De Moivre was convinced that he had demonstrated with his theorem the stability, questioned by Leibniz, of probabilities determined *a posteriori*; at the same time de Moivre's interpretation marks his partisanship for Clarke in the controversy with Leibniz.[69]

ABBREVIATIONS

AE *Acta Eruditorum*, Leipzig (from 1682).
JBW *Die Werke von Jakob Bernoulli* (ed. Naturforschende Gesellschaft in Basel, Basel from 1969).
LMG G. W. Leibniz, *Mathematische Schriften* (ed. C. I. Gerhardt) vols. I–VII (Berlin/Halle, 1849–63).
LNE G. W. Leibniz, *Nouveaux Essais sur l'entendement humain* (ed. W. v. Engelhardt and H. H. Holz, two vols., Frankfurt/Main, 1961).
LPG G. W. Leibniz, *Die philosophischen Schriften* (ed. C. I. Gerhardt) vols. I–VII (Berlin, 1875–90).
LSSB G. W. Leibniz, *Sämtliche Schriften und Briefe* (Darmstadt/Leipzig/Berlin, from 1923).
PT *Philosophical Transactions of the Royal Society in London* (from 1665).

NOTES

1. The numerous publications of Kurt-R. Biermann will be referred to separately in the individual sections below.
2. See especially Hacking, Ian. 1975. *The Emergence of Probability* (Cambridge).
3. "Probability and the Law" is the title of chapter 10 in Hacking 1975, *The Emergence of Probability*, which is devoted to Leibniz.
4. See Hacking 1975, p. 88.
5. *LSSB*, ser. VI, vol. I, p. 140.
6. *LSSB*, ser. VI, vol. I., p. 426.
7. *LPG* vol. VII, p. 521.
8. *LSSB*, ser. II, vol. I, p. 72. The letter is still dated in the old style, which differed from the new by ten days at that time.
9. See the letter to Duke Johann Friedrich written in the autumn of 1679; *LSSB*, ser. II, vol. I, p. 489.
10. *LNE*, vol. II, p. 514f.
11. See *LNE*, vol. II, pp. 268–273, where Leibniz alludes to his controversy with the Skeptic Foucher.
12. *LNE*, vol. II, pp. 508–511.
13. See Schneider, Ivo. 1977. "The contributions of the sceptic philosophers Arcesilas and Carneades to the development of an inductive logic," *Indian Journal of History of Science*, *12*, 173–180.
14. *LNE*, vol. II, p. 512f.
15. For my view of Leibniz's achievement in the calculus of probabilities, especially compared with James Bernoulli, which contradicts in part that of Hacking, see Ivo Schneider "Why do we find the origin of a calculus of probabilities in the seventeenth century," in *Prob abilistic Thinking, Thermodynamics, and the Interpretation of the History and Philosophy of Science* (J. Hintikka *et al.*, eds.), pp. 3–24. (Boston: Dordrecht, 1981).

16. See chapter 15 "Inductive Logic" of Hacking 1975, where Hacking attempts an integration of Leibniz's philosophical probability theory into his metaphysics. A counterposition to Hacking is offered by Wilson, Margaret. 1971. "Possibility, propensity and chance: some doubts about the Hacking thesis," *Journal of Philosophy*, *68*, 610–617.
17. See for example Kahle, L. M. 1735. *Elementa logicae probabilium methodo mathematica, in usu scientiarum et vitae adornata* (Halle).
18. See Knobloch, Eberhard. 1973. *Die mathematischen Studien von G. W. Leibniz zur Kombinatorik* (Wiesbaden).
19. Leipzig, 1666, and nonauthorized reprint Frankfurt/Main 1690.
20. Knobloch 1973, 57f.
21. See Bernoulli, Jakob. 1713. *Ars conjectandi* (Basel), 73.
22. Published as an appendix to van Schooten, Frans. 1657. *Exercitationes mathematicae* (Leiden).
23. Pascal, Blaise. 1665. *Traité du triangle arithmétique* (Paris).
24. For the relevant manuscripts of Leibniz see the following publications of Kurt-R. Biermann:
    1. "Über die Untersuchungen einer speziellen Frage der Kombinatorik durch G. W. Leibniz," *Forschungen und Fortschritte*, *28* (1954), pp. 357–361.
    2. "Über eine Studie von G. W. Leibniz zu Fragen der Wahrscheinlichkeitsrechnung", *Forschungen und Fortschritte*, *29* (1955), pp. 110–113.
    3. "Eine Untersuchung von G. W. Leibniz über die jährliche Sterblichkeitsrate," *Ibid.*, pp. 205–208.
25. Graunt, John. 1662. *Natural and political observations mentioned in a following index and made upon the bills of mortality*, (London, and later editions).
26. See Biermann, Kurt-R. and Margot Faak. 1957. "G. W. Leibniz, 'De incerti aestimatione'," *Forschungen und Fortschritte*, *31*, 45–50.
27. Biermann–Faak 1957, 47.
28. Hacking 1975, 126f. Hacking finds in Leibniz two justifications of expectation, one based on insufficient reason, and one based on physical equipossibility. This interpretation ascribes to Leibniz considerable insight in modern concepts of probability.
29. Biermann–Faak 1957, 47.
30. Hacking sees this statement, on the contrary, as the origin of the understanding of equipossibility, which Leibniz transmitted to James Bernoulli; see Hacking 1975, 127.
31. See Biermann, Kurt-R. 1955. "Über eine Studie von G. W. Leibniz zu Fragen der Wahrscheinlichkeitsrechnung," *Forschungen und Fortschritte*, *29*, 110–113.
32. Letter Montmort to Niklaus Bernoulli of December 12, 1713; *JBW*, vol. III, p. 400f.
33. See Schneider, Ivo. 1968. "Der Mathematiker Abraham de Moivre (1667–1754)," *Archive for history of exact sciences*, *5*, 177–317, and Schneider, Ivo. 1972. *Die Entwicklung des Wahrscheinlichkeitsbegriffs in der Mathematik von Pascal bis Laplace* (*Habilitationsschrift*, München), 57–64.
34. Letter James Hermann to Leibniz of October 28, 1705; *LMG*, vol. IV, p. 285.
35. See K. Kohli, "Zur Publikationsgeschichte der Ars Conjectandi," *JBW*, vol. III, pp. 391–401.
36. Letter of August 19, 1705; *LMG*, vol. IV, p. 283.
37. Leibniz' impact has been emphasized especially by Gini, Corrado. 1946. "Gedanken zum Theorem von Bernoulli," *Schweizerische Zeitschrift für Volkswirtschaft und Statistik*, *82*, 401–413, and by Ian Hacking in Hacking 1975.
38. *LMG*, vol. III/1, pp. 62–71, especially p. 71.
39. See the relevant sections of the *Meditationes* published in *JBW*, vol. III, pp. 21–89.
40. *LMG*, vol. III/1, p. 77f.

41. de Witt, Jan. 1671. *Waerdye van Lyf-Renten naer Proportie van Los-Renten* ('s Graven-Hage; reprint in *JBW*, vol. III, pp. 327–350). A detailed description of the role de Witt's tract played for Leibniz, partly represented in the correspondence with James Bernoulli, can be found in Biermann, Kurt-R. and Margot Faak. 1959. "G. W. Leibniz und die Berechnung der Sterbewahrscheinlichkeit bei I. de Witt," *Forschungen und Fortschritte*, *33*, 168–173. For an evaluation of the correspondence between Leibniz and James Bernoulli see K. Kohli, "Aus dem Briefwechsel zwischen Leibniz und Jakob Bernoulli", *JBW*, vol. III, pp. 509–513.
42. *LMG*, vol. III/1, pp. 79–86, especially p. 83f.
43. It is possible that Leibniz, because of the strained relations between the Bernoulli brothers, prevaricated slightly, for John Bernoulli had already informed Leibniz of James' activities concerning an *ars conjectandi* in his letter of February 16, 1697, *LMG*, vol. III/1, p. 367. It is not clear in this case, however, why Leibniz began the discussion with James Bernoulli concerning probabilities only in 1703.
44. *LMG*, vol. III/1, pp. 86–89, esp. pp. 87–89.
45. See *Ars Conjectandi*, 226.
46. *LMG*, vol. III/1, pp. 90–92, esp. p. 91.
47. *LMG*, vol. III/1, pp. 92–95, esp. p. 93f.
48. See Leibniz' dissertation, *De conditionibus*, published in two parts, Leipzig 1665 and with extensions, Nürnberg, 1666.
49. For this area of James Bernoulli's interests, especially the occurrence of non-additive probabilities, see Shafer, Glenn. 1978. "Non-Additive Probabilities in the Work of Bernoulli and Lambert," *Archive for history of exact sciences*, *19*, 309–370.
50. *LMG*, vol. III/1, p. 97.
51. *Ars Conjectandi*, 227f.
52. Halley, Edmond. 1693. "An estimate of the degrees of the mortality of mankind, drawn from curious tables of the births and funerals at the city of Breslaw, with an attempt to ascertain the price of annuities upon lives," *PT*, *17*, 596–610, 654–656.
53. *LMG*, vol. IV, p. 185f.
54. See letters of John Bernoulli to Leibniz of September 1, 1708, October 1, 1709, and April 26, 1710.
55. *Essay d' Analyse sur les Jeux de Hazard*, anon. (Paris, 1708).
56. *LMG*, vol. III/2, p. 845.
57. My attempts to find this political tract, including a search by the Leibniz-Archiv in Hannover, have so far been unsuccessful.
58. See the letter of April 15, 1709, *LMG*, vol. III/2, p. 842.
59. See the letter of John Bernoulli of June 6, 1710; *LMG*, vol. III/2, p. 850.
60. De Moivre, Abraham. 1711. "De Mensura Sortis, Seu, de Probabilitate Eventuum in Ludis a Casu Fortuito Pendentibus," *PT*, *27*, 213–264.
61. Paris, 1713.
62. See the letter to Bourguet of March 22, 1714, *LPG*, vol. III, p. 570.
63. Paris 1814 and later.
64. See Biermann, Kurt-R. 1967. "Überblick über die Studien von G. W. Leibniz zur Wahrscheinlichkeitsrechnung," *Sudhoffs Archiv*, *51*, 79–85.
65. 1. "Ad ea, quae vir clarissimus I. B. mense majo nupero in his actis publicavit, responsio," *AE*, *7*, 1690, pp. 338–360.
    2. "Annotatio de quibusdam ludis, imprimis de ludo quodam Sinico, differentiaque Scachici et Latrunculorum, et novo genere ludi navalis", *Miscellanea Borolinensia*, *1* (1710), pp. 22–26.

66. See the letter of December 3, 1703 (*LMG*, vol. III/1. p. 84), where he speaks of the "*possibilitas aequalis casuum aequalium*."
67. Hacking holds that James Bernoulli took over the concept of equipossibility from Leibniz and also, indirectly at least, his understanding of probability as degree of certainty; see Hacking 1975, 125 and 145.
68. See article 77 of James Bernoulli's *Meditationes* written in the winter of 1685/86; *JBW*, vol. III, p. 47.
69. See Schneider, Ivo. 1979. "Die Mathematisierung der Vorhersage künftiger Ereignisse in der Wahrscheinlichkeitstheorie vom 17. bis zum 19. Jahrhundert," *Berichte zur Wissenschftsgeschichte*, *2*, 101–112, esp. 108.

Institut für Geschichte der Naturwissenschaften
der Universität München
München, Bundesrepublik Deutschland

Von Pascals Dreieck zu Eulers Gamma-Funktion. Zur Entwicklung der Methodik der Interpolation*

CHRISTOPH J. SCRIBA

The problem of interpolating number sequences may be solved in a variety of ways, depending upon the mathematical information that serves as background. This paper considers historical examples which demonstrate how mathematicians of the 17th and 18th centuries approached the problem to produce broader definitions through interpolation of given series of whole numbers. Cases considered include John Wallis's quadrature of the circle and his related developments of the representation of $\pi/4$ as an infinite product and the interpolation of binomial coefficients. The binomial series discovered by Isaac Newton as a possibility for the interpola-tion of certain integrals of non-integer powers is also considered, as is Euler's interpolation producing an analytic expression for the Γ-function.

Das Problem der Interpolation von Zahlenreihen kann sich in verschiedener Weise stellen, je nachdem, welches Ausmaß an mathematischer Information als Hintergrund gegeben ist. Es kann vorkommen, daß lediglich eine Anzahl empirisch bestimmter Beobachtungswerte (astronomische Daten etwa) vorliegt, für die bestimmte Zwischenwerte rechnerisch ermittelt werden sollen. Dabei mag die mathematische Gestalt der Funktion, durch die diese Zahlenwerte beschrieben werden können, unbekannt und auch gar nicht gefragt sein. Ein anderer Sachverhalt liegt vor, wenn prinzipiell die Funktion

* Herrn Prof. Dr. Kurt-R. Biermann zum 60. Geburtstag gewidmet.

ISBN 0-12-204050-3

bekannt ist, die mit Hilfe von Interpolationsverfahren tabelliert werden soll (z.B. die trigonometrischen Funktionen). Dann könnte man im Prinzip jeden gewünschten Wert gemäß der Funktionsvorschrift genau berechnen, was allerdings sehr aufwendig sein kann. Deshalb begnügt man sich in der Praxis oft damit, bestimmte Eckwerte mit Hilfe der Funktionsvorschrift zu bestimmen, die Zwischenwerte aber durch Interpolation zu berechnen.

Von diesen beiden Aufgabenstellungen verschieden ist das Problem, das hier betrachtet werden soll. Es geht um die Interpolation ganzzahliger Zahlenfolgen, die durch solche Vorschriften gegeben sind, daß eine Ausweitung auf Zwischenwerte nicht unmittelbar möglich ist, weil dort die Definition versagt. Ein typisches Beispiel ist die Funktion "n-Fakultät", $n! = 1 \cdot 2 \cdot 3 \cdots n$. Sie ist durch die Vorschrift "Produkt der ersten n natürlichen Zahlen" so definiert, daß sie auf nicht-ganze Zahlen nicht unmittelbar verallgemeinert werden kann. Ein anderes, etwas verschieden geartetes Beispiel liefern die Binomialkoeffizienten $\binom{n}{k}$. Ein drittes, noch weiter gehendes, hat schon Leibniz formuliert: den Prozeß des Differenzierens einer Funktion so zu verallgemeinern, daß der Begriff der n-ten Ableitung auch für unganze n erklärt ist.

Im folgenden soll an historischen Beispielen dargestellt werden, wie Mathematiker des 17. und 18. Jahrhunderts das Problem anpackten, durch Interpolation von nur ganzzahlig gegebenen Zahlenfolgen zu erweiterten Definitionen zu kommen. Behandelt werden die Kreisquadratur von John Wallis (sie ergab das nach ihm benannte unendliche Produkt für $\pi/4$ und lief methodisch auf die Interpolation der Binomialkoeffizienten hinaus), die im Anschluß daran von Isaac Newton gefundene Binomialreihe als Möglichkeit der Interpolation gewisser Integrale von nicht-ganzen Potenzen und Leonhard Eulers Interpolation der Fakultät, d.h. die Gewinnung eines analytischen Ausdrucks für die Γ-Funktion.

I

Der Oxforder Mathematiker John Wallis (1616–1703) stellte sich in seinem "Arithmetica infinitorum" (1656) betitelten Buch[1] die Aufgabe, die Quadratur des Kreises in neuer Weise durchzuführen. Sein eigentliches Ziel war die Berechnung des Integrals

$$\int_0^1 (1 - x^2)^{1/2}\, dx = \frac{\pi}{4} \tag{1}$$

(selbstverständlich stand ihm diese, die Leibnizsche Schreibweise noch nicht zur Verfügung). Sein Weg dahin führte über das Studium einer Schar verwandter Funktionen, für die in bestimmten Fällen das Integral ausgerechnet werden kann:

$$f(p, n) = 1 : \int_0^1 (1 - x^{1/p})^n\, dx. \tag{2}$$

Darunter befindet sich der eigentlich gesuchte Wert für $p = \frac{1}{2}$, $n = \frac{1}{2}$:

$$f(\tfrac{1}{2}, \tfrac{1}{2}) = 1 : \int_0^1 (1 - x^2)^{1/2}\, dx = 4 : \pi = \square. \tag{3}$$

Den reziproken Wert führte Wallis ein, weil sich auf diese Weise gerade das Pascalsche Dreieck der Binomialkoeffizienten ergibt:

$p \backslash n$	0	1	2	3	4
0	1	1	1	1	1
1	1	2	3	4	5
2	1	3	6	10	15
3	1	4	10	20	35
4	1	5	15	35	70

(4)

Er fand also, daß für ganzzahlige p und n für die gemäß (2) definierte Funktionenschar gilt:

$$f(p, n) = \binom{n + p}{p}. \tag{5}$$

Da er an dem für $p = n = \frac{1}{2}$ gegebenen Funktionswert interessiert war, schob er in die Tabelle (4) Zeilen und Spalten für halbzahlige Argumente ein, wobei er das oben eingeführte Symbol $\square$ für $4 : \pi$ verwendete:

$p \backslash n$	0	$\frac{1}{2}$	1	$\frac{3}{2}$	2	$\frac{5}{2}$	3
0	1		1		1		1
$\frac{1}{2}$		$\square$					
1	1		2		3		4
$\frac{3}{2}$							
2	1		3		6		10
$\frac{5}{2}$							
3	1		4		10		20

(4)

Hierin ist auf augenfällige Weise zum Ausdruck gebracht, worum es geht: um die Interpolation der Binomialkoeffizienten $f(p, n) = \binom{n+p}{p}$ für halbzahlige Werte von p und n. Wie bewältigte der Oxforder Mathematiker dieses Problem?

Unter Rückgriff auf die Darstellung

$$f(p, n-p) = \binom{n}{p} = \frac{n(n-1)(n-2)\cdots(n-p+1)}{1\cdot 2\cdot 3\cdots p} \tag{6}$$

konnte er in diesem Schema zunächst in den Zeilen für $p = 0, 1, 2, \ldots$ auch die Funktionswerte für $n = \frac{1}{2}, \frac{3}{2}, \frac{5}{2}, \ldots$ eintragen und dann dank der Spiegelsymmetrie der Tabelle auch jeden zweiten Wert in den Zeilen für halbzahliges n ergänzen. Das führte auf

$p \backslash n$	$-\frac{1}{2}$	0	$\frac{1}{2}$	1	$\frac{3}{2}$	2	$\frac{5}{2}$	3		n
$-\frac{1}{2}$		1		$\frac{1}{2}$		$\frac{3}{8}$		$\frac{15}{48}$	...	
0	1	1	1	1	1	1	1	1	...	1
$\frac{1}{2}$		1	□	$\frac{3}{2}$		$\frac{15}{8}$		$\frac{105}{48}$	...	
1	$\frac{1}{2}$	1	$\frac{3}{2}$	2	$\frac{5}{2}$	3	$\frac{7}{2}$	4	...	$n+1$
$\frac{3}{2}$		1		$\frac{5}{2}$		$\frac{35}{8}$		$\frac{315}{48}$	...	
2	$\frac{3}{8}$	1	$\frac{15}{8}$	3	$\frac{35}{8}$	6	$\frac{63}{8}$	10	...	$\frac{(n+1)(n+2)}{1\cdot 2}$
$\frac{5}{2}$		1		$\frac{7}{2}$		$\frac{63}{8}$		$\frac{693}{48}$	...	
3	$\frac{15}{48}$	1	$\frac{105}{48}$	4	$\frac{315}{48}$	10	$\frac{693}{48}$	20		$\frac{(n+1)(n+2)(n+3)}{1\cdot 2\cdot 3}$

(7)

Die Tatsache, daß in den Zeilen für halbzahlige p bisher nur jeder zweite Wert erscheint, führte Wallis schließlich dazu, Bildungsgesetze zu suchen, die von einem Wert einer Zeile zum übernächsten Wert derselben Zeile führen. Dann müssen—wenn diese Vermutung richtig ist—pro Zeile zwei Bildungsgesetze angebbar sein: eines, das mit dem Anfangswert für $n = -\frac{1}{2}$ einsetzt (der allgemein mit A bezeichnet sei), und eines, das mit dem Anfangswert für $n = 0$ beginnt, der in allen Zeilen übereinstimmend gleich 1 ist.

Betrachten wir z.B. die Zeile für $p = 1$. Der erste Anfangswert ist $A = \frac{1}{2}$, der übernächste Wert ist $\frac{3}{2}$; er geht aus A durch Multiplikation mit $\frac{3}{1}$ hervor. Eine weitere Multiplikation mit $\frac{5}{3}$ führt auf den fünften Wert ($\frac{5}{2}$) dieser Zeile, und allgemein wird also jeder zweite Wert aus A durch eine schrittweise Multiplikation erzeugt nach der Regel

$$A \cdot \tfrac{3}{1} \cdot \tfrac{5}{3} \cdot \tfrac{7}{5} \cdot \tfrac{9}{7} \cdots,$$

wobei mit jeder Stelle ein weiterer dieser Faktoren hinzukommt. Die dazwischenliegenden Zahlen 1, 2, 3, ... könnte man sich entsprechend gemäß der Regel

$$1 \cdot \tfrac{2}{1} \cdot \tfrac{3}{2} \cdot \tfrac{4}{3} \cdots$$

aus 1 gebildet denken, doch durch Vergleich mit den entsprechenden Gesetzen der übrigen Zeilen stellte Wallis fest, die Vorschrift füge sich besser in das allgemeine Muster ein, wenn man sie in der Form

$$1 \cdot \tfrac{4}{2} \cdot \tfrac{6}{4} \cdot \tfrac{8}{6} \cdots$$

schreibt. Sinngemäß ausgedehnt auch auf die Zeile für $p = 0$ und schließlich sogar auf die Zeile für $p = -\frac{1}{2}$, erhielt er so die beiden Regelfolgen

$p \backslash n$	$-\frac{1}{2}$	0		1. Regel	2. Regel
$-\frac{1}{2}$	$A = ?$	1	...	$A \cdot \frac{0}{1} \cdot \frac{2}{3} \cdot \frac{4}{5} \cdots$	$1 \cdot \frac{1}{2} \cdot \frac{3}{4} \cdot \frac{5}{6} \cdots$
0	$A = 1$	1	...	$A \cdot \frac{1}{1} \cdot \frac{3}{3} \cdot \frac{5}{5} \cdots$	$1 \cdot \frac{2}{2} \cdot \frac{4}{4} \cdot \frac{6}{6} \cdots$
$\frac{1}{2}$	$A = ?$	1	...	$A \cdot \frac{2}{1} \cdot \frac{4}{3} \cdot \frac{6}{5} \cdots$	$1 \cdot \frac{3}{2} \cdot \frac{5}{4} \cdot \frac{7}{6} \cdots$
1	$A = \frac{1}{2}$	1	...	$A \cdot \frac{3}{1} \cdot \frac{5}{3} \cdot \frac{7}{3} \cdots$	$1 \cdot \frac{4}{2} \cdot \frac{6}{4} \cdot \frac{8}{6} \cdots$
...	...	...	...	...	...

(8)

Unter der Voraussetzung (welche Wallis aus "Kontinuitätsgründen" als erfüllt ansah), daß diese ganz gesetzmäßig gebildeten Regeln allgemeingültig sind, konnte er nun leicht die Zeile für $p = \frac{1}{2}$ vollständig ausfüllen, wobei stets an den noch offenen Plätzen das Zeichen $\square$ eingehen muß:

$$\tfrac{1}{2}\square, \quad 1, \quad \square, \quad \tfrac{3}{2}, \quad \tfrac{4}{3}\square, \quad \tfrac{15}{8}, \quad \tfrac{8}{5}\square, \quad \tfrac{105}{48}, \ldots. \tag{9}$$

Da die Spiegelsymmetrie die gleichen Werte in der Spalte für $n = \frac{1}{2}$ erfordert und damit in jeder Zeile ein Wert zur Verfügung steht, ist in Verbindung mit der ersten Regel das Voranschreiten in jeder Zeile möglich, kann also das gesamte Schema ausgefüllt werden. Für die Herleitung des unendlichen Produktes

$$\square = \frac{4}{\pi} = \frac{3}{2} \cdot \frac{3}{4} \cdot \frac{5}{4} \cdot \frac{5}{6} \cdot \frac{7}{6} \cdots \tag{10}$$

benötigte Wallis sogar nur die eben angegebene Zeile (9) für $p = \frac{1}{2}$. Die endgültige Ausfüllung des Schemas setzte aber dessen Kenntnis voraus, da ja der Faktor $\square$ bei jedem zweiten Wert auftritt.

Die grundlegende Beobachtung, welche Wallis jetzt die Herleitung des unendlichen Produktes ermöglichte, war die Tatsache, daß für je drei aufeinander folgende Glieder a, b, c einer Zeile für ganzzahliges p nicht nur $a < b < c$ gilt sondern genauer $b^2 > a \cdot c$. Wiederum aus Kontinuitätsgründen nahm er an, diese Beziehung müsse dann auch für die Zeilen mit halbzahligem p gültig sein:

a	b	c			
$\frac{1}{2}\square$	1	$\square$	$1^2 > \frac{1}{2}\square^2$	oder	$\square < \sqrt{2}$
1	$\square$	$\frac{3}{2}$	$\square^2 > 1 \cdot \frac{3}{2}$	oder	$\square > \sqrt{\frac{3}{2}}$
$\square$	$\frac{3}{2}$	$\frac{4}{3}\square$	$\frac{3}{2} \cdot \frac{3}{2} > \frac{4}{3}\square^2$	oder	$\square < \frac{3}{2}\sqrt{\frac{3}{4}} = \frac{3}{2} \cdot \frac{3}{4}\sqrt{\frac{4}{3}}$
$\frac{3}{2}$	$\frac{4}{3}\square$	$\frac{15}{8}$	$\cdots$		$\square > \frac{3}{2} \cdot \frac{3}{4}\sqrt{\frac{5}{4}}$

(11)

Wallis fuhr fort, bis er bei

$$\frac{3}{2} \cdot \frac{3}{4} \cdot \frac{5}{4} \cdot \frac{5}{6} \cdot \cdots \cdot \frac{13}{12} \cdot \frac{13}{14} \cdot \sqrt{\frac{15}{14}} < \square < \frac{3}{2} \cdot \frac{3}{4} \cdot \cdots \cdot \frac{13}{14} \cdot \sqrt{\frac{14}{13}}$$

angelangt war. Damit ist sowohl klargemacht, wie es weitergehen muß, wie auch gezeigt, daß man die Grenzen der einschließenden Intervalle beliebig dicht zusammenrücken lassen kann.

Ziel wie Ergebnis dieser genialen Überlegungen war es. einen bestimmten, nicht unmittelbar gegebenen Funktionswert einer zweiparametrigen Schar von Werten zu bestimmen. Methodisch gesehen, hat Wallis' Vorgehen Züge, die dem Entziffern chiffrierter Briefe ähnlich sind, worin er sich bereits als Kaplan in London hervorgetan hatte. Sind es hier der logische Sinn und semantische Bedeutungszusammenhang, welche den Weg zum Ergänzen von Lücken weisen, so dienen bei der mathematischen Interpolation der Glaube an Kontinuität und stetige funktionale Zusammenhänge und die Hoffnung, explizit formulierbare Gesetzmäßigkeiten angeben zu können, als Wegweiser.

II

Als Student in Cambridge war Isaac Newton (1643–1727) im Winter 1664/65 mit der Lektüre von Wallis' "Arithmetica infinitorum" beschäftigt. Newton las mit der Feder in der Hand, und da neben seinen späteren Berichten auch die damals aufgeschriebenen Überlegungen [2] noch in der Universitätsbibliothek in Cambridge erhalten sind, läßt sich sein Vorgehen im einzelnen verfolgen.

Die von Wallis erfolgreich behandelte Fragestellung erweiternd, interessierte sich Newton für die Kreisfläche wie für die Hyperbelfläche (letztere in zwei verschiedenen Darstellungen); diese Flächen sind in den Funktionenscharen

$$\int_0^x (a^2 - t^2)^n\,dt, \qquad \int_0^x (a^2 + t^2)^n\,dt, \qquad \int_0^x a^2(b + t)^n\,dt \quad \text{(12a,b,c)}$$

enthalten für $n = \frac{1}{2}$ (in den beiden ersten Fällen) bzw. $n = -1$. Im Gegensatz zu Wallis hielt Newton die obere Grenze der Integration offen.

In der zweiten Niederschrift tabellierte Newton aus methodischen Gründen zunächst den Integranden $a^2(b + t)^n$ des dritten Integrals (12c) für ganzzahlige n:

n	-1	0	1	2	3
	$\frac{a^2}{b+t}$	$1 \cdot a^2$	$1 \cdot a^2b$	$1 \cdot a^2b^2$	$1 \cdot a^2b^3$
			$1 \cdot a^2t$	$2 \cdot a^2bt$	$3 \cdot a^2b^2t$
				$1 \cdot a^2t^2$	$3 \cdot a^2bt^2$
					$1 \cdot a^2t^3$

(13)

Der Kern dieser Darstellung ist wieder das sog. Pascalsche Dreieck. Newton setzte es mittels des bekannten additiven Bildungsgesetzes (jeder Koeffizient ist die Summe des darüberstehenden und des links neben ihm stehenden) nach links fort

$$\begin{array}{ccccccccccccccc}
\ldots & 1 & 1 & 1 & 1 & 1 & 1 & 1 & 1 & 1 & 1 & 1 & 1 & \ldots \\
\ldots & -4 & -3 & -2 & -1 & 0 & 1 & 2 & 3 & 4 & 5 & 6 & 7 & \ldots \\
\ldots & 10 & 6 & 3 & 1 & 0 & 0 & 1 & 3 & 6 & 10 & 15 & 21 & \ldots \\
\ldots & -20 & -10 & -4 & -1 & 0 & 0 & 0 & 1 & 4 & 10 & 20 & 35 & \ldots \\
\ldots & 35 & 15 & 5 & 1 & 0 & 0 & 0 & 0 & 1 & 5 & 15 & 35 & \ldots \\
\ldots & -56 & -21 & -6 & -1 & 0 & 0 & 0 & 0 & 0 & 1 & 6 & 21 & \ldots \\
\ldots & 84 & 28 & 7 & 1 & 0 & 0 & 0 & 0 & 0 & 0 & 1 & 7 & \ldots
\end{array} \quad (14)$$

Da außerdem im Schema (13) beim Übergang von einer Spalte zur nächsten ein Faktor b hinzuzufügen ist, konnte er aus der Tabelle (14) ablesen, daß der Ausdruck $a^2(b + t)^{-1}$ folgendermaßen als unendliche Reihe darstellbar ist:

$$\frac{a^2}{b+t} = \frac{a^2}{b} - \frac{a^2t}{b^2} + \frac{a^2t^2}{b^3} - \frac{a^2t^3}{b^4} + \cdots,$$

das heißt, er hatte die Entwicklung

$$\frac{a^2}{b}\left(1 + \frac{t}{b}\right)^{-1} = \frac{a^2}{b}\left(1 - \frac{t}{b} + \frac{t^2}{b^2} - \frac{t^3}{b^3} + \cdots\right) \quad (15)$$

durch Interpolation gewonnen. Newton betonte später wiederholt, er habe das übliche algebraische Divisionsverfahren nur nachträglich zur Kontrolle herangezogen.

Während aber das Integral (12c) lediglich die Ausdehnung des Schemas der Binomialkoeffizienten nach rückwärts erfordert hatte (mit anschließender Integration der Reihenglieder), verlangten die beiden anderen Integrale (12a und 12b), die ja ebenfalls nur für natürliche Zahlen n auswertbar waren, eine Interpolation desselben. Denn da der Integrand wiederum ein Binom ist, treten bei der Ausmultiplikation von $(a^2 + t^2)^n$ für ganzzahlige n wieder die Binomialkoeffizienten auf. Unter sofortiger Einbeziehung der unbestimmten Integration lautet also das zu interpolierende Schema der Koeffizienten der integrierten Potenzen von x (bis auf Potenzen von a):

n	0	$\frac{1}{2}$	1	$\frac{3}{2}$	2	$\frac{5}{2}$	3	$\frac{7}{2}$	4	$\frac{9}{2}$
1. Glied: $+x$	1		1		1		1		1	
2. Glied: $-\frac{1}{3}x^3$	0		1		2		3		4	
3. Glied: $+\frac{1}{5}x^5$	0		0		1		3		6	
4. Glied: $-\frac{1}{7}x^7$	0		0		0		1		4	
5. Glied: $+\frac{1}{9}x^9$	0		0		0		0		1	
...	⋮		⋮		⋮		⋮		0	

(16)

Hatte Wallis nach einer multiplikativ zu formulierenden Gesetzmäßigkeit gefragt, welche auf die gesuchten Zwischenwerte führte—seine Regeln (8)—, so achtete Newton auf allgemeine Gesetzmäßigkeiten im additiven Aufbau. Er fand diese dahingehend, daß beim Übergang von einer Spalte zur nächsten zum jeweils vorhandenen Wert neue Anteile zu addieren sind in der durch folgendes Schema zum Ausdruck gebrachten Weise:

$$\begin{array}{llllll}
a & a & a & a & a & \ldots \\
b & b+c & b+2c & b+3c & b+4c & \ldots \\
d & d+e & d+2e+f & d+3e+3f & d+4e+6f & \ldots \\
g & g+h & g+2h+i & g+3h+3i+k & g+4h+6i+4k & \ldots \\
l & l+m & l+2m+n & l+3m+3n+p & l+4m+6n+4p+q & \ldots \\
r & r+s & r+2s+t & r+3s+3t+v & r+4s+6t+4v+w & \ldots
\end{array} \tag{17}$$

Ein Beispiel mag das erläutern. Es sei die erste Spalte in (17) die zu $n = 3$ gehörende in (16). Dann ist $a = 1, b = 3, d = 3, g = 1, l = 0$, und $r = 0$. Durch Vergleich der zweiten Spalte in (17) mit derjenigen für $n = 4$ in (16) ergibt sich $c = 1, e = 3, h = 3, m = 1$, und $s = 0$. Durch Vergleich der einander entsprechenden folgenden Spalten folgt $f = 1, i = 3, n = 3$, und $t = 1$, entsprechend bei einem weiteren Schritt $k = 1, p = 3, v = 3$, und schließlich $q = 1, w = 3$.

Die gleiche Gesetzmäßigkeit hat aber auch statt, wenn man im Pascalschen Dreieck (16) den Übergang jeweils von einer Spalte zur übernächsten vollzieht. Beginnt man z.B. wie eben (Spalte für $n = 3$) mit $a = 1, b = 3$,

$d = 3, g = 1, l = 0, r = 0$, so wird $c = 2, e = 7, h = 9, m = 5$, und $s = 1$; denn dies führt auf die Spalte 1, 5, 10, 10, 5, 1. Mit $f = 4, i = 16, n = 25$, und $t = 19$ gelangt man zur Spalte 1, 7, 21, 35, 35, 21; $\cdots$; die weiteren Summanden werden $k = 8, p = 36, v = 66, q = 16$, und $w = 80$.

Diese Einsicht befähigte Newton, die durch (17) zum Ausdruck gebrachte Gesetzmäßigkeit auch beim Fortschreiten um Halbschritte oder um irgendeine andere rationale Schrittspanne vorauszusetzen. Mit Newton sei das an der dritten Zeile des erweiterten Pascalschen Dreiecks (14) in Verbindung mit der entsprechenden, nach links erweiterten dritten Zeile von (17) für Schritte im Abstand von $\frac{1}{2}$ gezeigt (wobei aus drucktechnischen Gründen hier die beiden Zeilen in Spaltenform gesetzt sind):

Elemente der 3. Zeile aus (14)	Formen der erweiterten 3. Zeile aus (17)
...	...
3	$d - 4e + 10f$
?	$d - 3e + 6f$
1	$d - 2e + 3f$
?	$d - e + f$
0	d
?	$d + e$
0	$d + 2e + f$
?	$d + 3e + 3f$
1	$d + 4e + 6f$
?	$d + 5e + 10f$
3	$d + 6e + 15f$
...	...

(18)

Aus irgend drei Gleichungen, für die links ein Wert angegeben ist, kann man die drei Unbekannten bestimmen zu $d = 0, e = -\frac{1}{8}, f = \frac{1}{4}$ und daraus die zuvor unbekannten, zu interpolierenden Größen berechnen gemäß der rechts stehenden Formen. In diesem Beispiel erhält man die Folge

$$\cdots, 3, \ \tfrac{15}{8}, \ 1, \ \tfrac{3}{8}, \ 0, \ -\tfrac{1}{8}, \ 0, \ \tfrac{3}{8}, \ 1, \ \tfrac{15}{8}, \ 3, \cdots.$$

Allgemein gab Newton das nach links extrapolierte und im Innern durch Halbschritte interpolierte Pascal-Dreieck an:

$$\begin{array}{cccccccccccccc}
\cdots & 1 & 1 & 1 & 1 & 1 & 1 & 1 & 1 & 1 & 1 & 1 & \cdots \\
\cdots & -2 & -\frac{3}{2} & -1 & -\frac{1}{2} & 0 & \frac{1}{2} & 1 & \frac{3}{2} & 2 & \frac{5}{2} & 3 & \cdots \\
\cdots & 3 & \frac{15}{8} & 1 & \frac{3}{8} & 0 & -\frac{1}{8} & 0 & \frac{3}{8} & 1 & \frac{15}{8} & 3 & \cdots \\
\cdots & -4 & -\frac{35}{16} & -1 & -\frac{5}{16} & 0 & \frac{1}{16} & 0 & -\frac{1}{16} & 0 & \frac{5}{16} & 1 & \cdots \\
\cdots & 5 & \frac{315}{128} & 1 & \frac{35}{128} & 0 & -\frac{5}{128} & 0 & \frac{3}{128} & 0 & -\frac{5}{128} & 0 & \cdots \\
\cdots & -6 & -\frac{693}{256} & -1 & -\frac{63}{256} & 0 & \frac{7}{256} & 0 & -\frac{3}{256} & 0 & \frac{3}{256} & 0 & \cdots
\end{array} \qquad (19)$$

Speziell enthält die 6. Spalte dieser Tabelle die Aussage, daß

$$(a^2 - t^2)^{1/2} = a \cdot \left(1 - \frac{t^2}{a^2}\right)^{1/2}$$
$$= a \cdot \left(1 - \frac{1}{2} \cdot \frac{t^2}{a^2} - \frac{1}{8} \cdot \frac{t^4}{a^4} - \frac{1}{16} \cdot \frac{t^6}{a^6} - \frac{5}{128} \cdot \frac{t^8}{a^8} - \cdots\right). \quad (20)$$

Durch gliedweise Integration von 0 bis x ergibt sich daraus die Reihenentwicklung für (12a), und speziell für $x = a$ die Newtonsche Kreisreihe

$$\int_0^a (a^2 - t^2)^{1/2}\, dt = \frac{\pi}{4} \cdot a^2 = a^2 \cdot \left(1 - \frac{1}{6} - \frac{1}{40} - \frac{1}{112} - \frac{5}{1152} - \cdots\right). \quad (21)$$

Erst nachdem er auf diese Weise eine auf additiver Grundlage aufgebaute Interpolationsmethode entwickelt hatte, vollzog Newton den Übergang zur unmittelbaren Bildung der Binomialkoeffizienten durch Ausdehnung der Bildungsregel (6), indem er $\binom{n}{p}$ für unganze, aber rationale Werte $n = r/s$ nach der Vorschrift

$$\binom{r/s}{p} = \frac{r \cdot (r - s)(r - 2s) \cdots (r - (p - 1)s)}{s \cdot 2s \cdot 3s \cdot \cdots \cdot ps}$$

direkt berechnete.

Wo Wallis einen einzelnen Funktionswert bestimmt hatte, stieß Newton also zu einem allgemeinen Verfahren der Interpolation beliebiger Zwischenwerte vor. Da er zudem die obere Grenze der Integration variabel hielt, wo sie Wallis numerisch festgelegt hatte, ergaben sich unendliche Reihen anstelle von konkreten Zahlenwerten—ein "Nebeneffekt", der Newtons Überlegungen weit größere Allgemeinbedeutung verlieh als denjenigen seines Vorgängers.

III

Ebenfalls angeregt durch Wallis' "Arithmetica infinitorum", beschäftigte sich 1729 der junge Leonhard Euler (1707–1783) mit dem Problem der Interpolation einer ganzzahlig definierten Folge, nämlich derjenigen der Fakultät: 1, 2, 6, 24, . . . , $n!$. Hier sei noch skizziert, wie er in der Abhandlung mit dem Titel "De progressionibus transcendentibus seu quarum termini generales algebraice dari nequeunt" [3] vorging. Euler beginnt darin unmittelbar mit dem Ausdruck

$$n! = \frac{1 \cdot 2^n}{1 + n} \cdot \frac{2^{1-n} \cdot 3^n}{2 + n} \cdot \frac{3^{1-n} \cdot 4^n}{3 + n} \cdot \frac{4^{1-n} \cdot 5^n}{4 + n} \cdots, \quad (22)$$

den er auf dem Weg

$$n! = 1 \cdot 2 \cdot 3 \cdots n = \frac{1 \cdot 2 \cdot 3 \cdots}{(n+1)(n+2)(n+3) \cdots} = \frac{2}{n+1} \cdot \frac{3}{n+2} \cdot \frac{4}{n+3} \cdots$$
$$= \frac{2^n}{n+1} \cdot \frac{2^{1-n} \cdot 3^n}{n+2} \cdot \frac{3^{1-n} \cdot 4^n}{n+3} \cdots$$

gefunden haben mag. (Das Zeichen $n!$ wird hier nur der Kürze wegen verwendet, ebenso wie zuvor bei der Behandlung von Wallis und Newton das Integralzeichen!).

Euler fragt als erstes nach dem Wert für $n = \frac{1}{2}$, der zwischen $0! = 1$ und $1! = 1$ zu interpolieren ist, und findet

$$\left(\frac{1}{2}\right)! = \sqrt{\frac{2 \cdot 4}{3 \cdot 3} \cdot \frac{4 \cdot 6}{5 \cdot 5} \cdot \frac{6 \cdot 8}{7 \cdot 7} \cdots} = \sqrt{\frac{\pi}{4}}; \tag{23}$$

dabei bezieht er sich ausdrücklich auf die Herleitung des unendlichen Produktes für $\pi/4$ in Wallis' Werk.

Genügt der Ausdruck (22) auch, um für rationale Werte von n die Funktion $n!$ zu berechnen, so ist sie doch noch nicht ohne Einschränkung berechenbar. Darum suchte Euler nach einer Integraldarstellung. Dazu ging er von der Beobachtung aus, daß das Integral

$$J(e, n) = \int_0^1 x^e (1 - x)^n \, dx \tag{24}$$

(das sog. Eulersche Integral 1. Art) für ganzzahlige $n > 0$ und $e + 1 > 0$ die Werte

$$J(e, 0) = \frac{1}{e+1}, \qquad J(e, n) = \frac{1 \cdot 2 \cdot 3 \cdots n}{(e+1)(e+2) \cdots (e+n+1)} \tag{25}$$

annimmt, wie sich durch gliedweise Integration des in eine binomische Reihe entwickelten Integranden ergibt. Eulers Ziel ist nun, die Zähler in (25) zu isolieren.

Als erstes setzte er e rational an, $e = f/g$; so ergab sich

$$J(f/g, n) = \int_0^1 x^{f/g} (1 - x)^n \, dx = \frac{g^{n+1}}{f + (n+1)g} \cdot \frac{1 \cdot 2 \cdot 3 \cdots n}{(f+g)(f+2g) \cdots (f+ng)}$$

oder

$$\frac{1 \cdot 2 \cdot 3 \cdots n}{(f+g)(f+2g) \cdots (f+ng)} = \frac{f + (n+1)g}{g^{n+1}} \int_0^1 x^{f/g} (1 - x)^n \, dx. \tag{26}$$

Für $f = 1$, $g = 0$ würde das formal einen Ausdruck für $n!$ liefern; Euler schreibt, es wäre also $n!$ gleich dem (zwischen den Grenzen 0 und 1 genommenen) Integral

$$\int \frac{x^{1/0}(1-x)^n\,dx}{0^{n+1}}.$$

Wir würden das heute aufschreiben als

$$n! = \lim_{g\to 0} \int_0^1 \frac{x^{1/g}(1-x)^n\,dx}{g^{n+1}}. \tag{27}$$

Zur Auswertung des Integrals führte Euler in (26) die Transformation

$$x = t^{g/(f+g)}, \qquad dx = \frac{g}{f+g}\cdot x^{-f/(f+g)}\,dx \tag{28}$$

durch und erhielt

$$\frac{1\cdot 2\cdots n}{(f+g)(f+2g)\cdots(f+ng)} = \frac{f+(n+1)g}{(f+g)^{n+1}} \int_0^1 \left(\frac{1-t^{g/(f+g)}}{g/(f+g)}\right)^n dt.$$

Setzt man jetzt wiederum mit Euler $f = 1$, $g = 0$, so läuft diese Gleichung hinaus auf die Bestimmung des Grenzwertes

$$n! = \lim_{h\to 0} \int_0^1 \left(\frac{1-t^h}{h}\right)^n dt. \tag{29}$$

Unter Anwendung der l'Hospitalschen Regel auf den Integranden fand Euler schließlich das sog. Eulersche Integral 2. Art

$$n! = \int_0^1 (-\ln t)^n\,dt. \tag{30}$$

Insbesondere ist darin der Fall

$$\int_0^1 \sqrt{-\ln t}\,dt = \sqrt{\frac{\pi}{4}} \tag{31}$$

enthalten.

Aus verständlichen Gründen mit der Darstellung (30) noch nicht zufrieden, da der natürliche Logarithmus im Integranden stört, greift Euler nochmals auf (26) zurück. Indem er dort $f = n$, $g = 1$ setzt, erhält er

$$\frac{1\cdot 2\cdot 3\cdots n}{(n+1)(n+2)\cdots 2n} = (2n+1)\int_0^1 (x-x^2)^n\,dx. \tag{32a}$$

Für $n = 1, 2, 3, \cdots$ liefert also die rechte Seite die Ausdrücke

$$\frac{1}{2}, \quad \frac{1\cdot 2}{3\cdot 4}, \quad \frac{1\cdot 2\cdot 3}{4\cdot 5\cdot 6}, \cdots \text{ oder } \quad \frac{1\cdot 1}{1\cdot 2}, \quad \frac{1\cdot 2\cdot 1\cdot 2}{1\cdot 2\cdot 3\cdot 4}, \quad \frac{1\cdot 2\cdot 3\cdot 1\cdot 2\cdot 3}{1\cdot 2\cdot 3\cdot 4\cdot 5\cdot 6}, \cdots . \tag{32b}$$

Diese Ausdrücke, in der zweiten Form geschrieben, erlauben die Interpolation für halbzahlige n; denn die Zähler sind $(n!)^2$, während die Nenner jeweils einen Faktor überspringen, der bei halbzahliger Interpolation zu ergänzen ist. Es müsse also, schließt Euler, insbesondere, wenn A den Wert $(\frac{1}{2})!$ bedeutet, in der obigen Darstellung (32b) dem der Ausdruck $A \cdot A/1$ entsprechen, d.h.

$$\frac{A\cdot A}{1} = 2\int_0^1 \sqrt{x - x^2}\, dx$$

sein. Da das Integral die halbe Fläche des Kreises mit Durchmesser 1 repräsentiert, also den Wert $\pi/8$ besitzt, wird

$$A = \left(\frac{1}{2}\right)! = \sqrt{\frac{\pi}{4}}$$

in Übereinstimmung mit (23).

Setzt man entsprechend für das zu $n = \frac{3}{2}$ gehörende Glied in der Folge $1, 2, 6, 24, \ldots$ den Buchstaben B, so wird $B \cdot B/(1 \cdot 2 \cdot 3)$ das zugehörige Zwischenglied in (32b), also

$$B^2 = 1\cdot 2\cdot 3\cdot 4 \int_0^1 (x - x^2)^{3/2}\, dx,$$

und allgemein erhält man das zu $p/2$ gehörende Glied als

$$\left(1\cdot 2\cdot 3 \cdots p\cdot (p+1)\int_0^1 (x - x^2)^{p/2}\, dx\right)^{1/2}. \tag{33}$$

Will man jeweils zwei Zwischenwerte zwischen benachbarten Einträgen finden, d.h. mit der Schrittweite $\frac{1}{3}$ interpolieren, so hat man—auch das zeigt Euler ausführlich—in (26) $f = 2n$, $g = 1$ zu setzen. Das ergibt

$$(3n+1)\int_0^1 (x^2 - x^3)^n\, dx = \frac{1\cdot 2\cdots n}{(2n+1)(2n+2)\cdots 3n}, \tag{34a}$$

liefert somit die Folge

$$\frac{1}{3}, \quad \frac{1\cdot 2}{5\cdot 6}, \quad \frac{1\cdot 2\cdot 3}{7\cdot 8\cdot 9}, \quad \frac{1\cdot 2\cdot 3\cdot 4}{9\cdot 10\cdot 11\cdot 12}, \cdots . \tag{34b}$$

Multipliziere man nämlich jetzt das Integral (34a) mit dem Integral (32a), so erhalte man

$$(2n+1)(3n+1)\int_0^1 (x-x^2)^n\,dx \cdot \int_0^1 (x^2-x^3)^n\,dx; \tag{35a}$$

dieses Produkt erzeugt für $n = 1, 2, 3, \ldots$ die Ausdrücke

$$\frac{1\cdot 1\cdot 1}{1\cdot 2\cdot 3}, \quad \frac{1\cdot 2\cdot 1\cdot 2\cdot 1\cdot 2}{1\cdot 2\cdot 3\cdot 4\cdot 5\cdot 6}, \quad \frac{1\cdot 2\cdot 3\cdot 1\cdot 2\cdot 3\cdot 1\cdot 2\cdot 3}{1\cdot 2\cdot 3\cdot 4\cdot 5\cdot 6\cdot 7\cdot 8\cdot 9}, \ldots \tag{35b}$$

Den gleichen Gedanken wie oben für die Schrittweite $\frac{1}{3}$ verwendend, zieht Euler hieraus den Schluß, das zu $n = \frac{1}{3}$ gehörende Glied A lasse sich jetzt berechnen aus

$$\frac{A^3}{1} = \left(\frac{2}{3}+1\right)(1+1)\int_0^1 (x-x^2)^{1/3}\,dx \cdot \int_0^1 (x^2-x^3)^{1/3}\,dx,$$

und allgemein ergebe sich der zu $n = p/3$ gehörende Wert aus

$$\left(1\cdot 2\cdots p\cdot \frac{2p+3}{3}\cdot (p+1)\int_0^1 (x-x^2)^{p/3}\,dx \cdot \int_0^1 (x^2-x^3)^{p/3}\,dx\right)^{1/3} \tag{36}$$

Nach einem weiteren Zwischenschritt gibt er endlich die allgemeine Formel zur Berechnung desjenigen Gliedes an, das zu $n = p/q$ gehört:

$$\left(1\cdot 2\cdots p\cdot\left(\frac{2p}{q}+1\right)\left(\frac{3p}{q}+1\right)\left(\frac{4p}{q}+1\right)\cdots(p+1)\right)$$
$$\cdot \int_0^1 (x-x^2)^{p/q}\,dx \cdot \int_0^1 (x^2-x^3)^{p/q}\,dx \cdots \int_0^1 (x^{q-1}-x^q)^{p/q}\,dx. \tag{37}$$

Dies stellt das in der genannten Abhandlung erzielte Endergebnis dar, wenngleich Euler abschließend noch kurz auf die eingangs von mir erwähnte, von Leibniz aufgeworfene Frage eingeht, wie man für das Differential $d^{1/2}t$ eine Deutung geben könne.

Bei Eulers Behandlung des Interpolationsproblems der Gamma-Funktion —darum handelt es sich ja bei der Fakultät, obgleich ich diese Bezeichnungsweise hier bewußt vermieden habe—fällt vor allen Dingen die große Gewandtheit im Umgang mit den Integralen und den darunterstechenden Funktionen auf. Euler erweist sich als ein Meister im Umformen der auftretenden Ausdrücke: das gilt sowohl für die Schritte, die notwendig sind, um zur Integraldarstellung (30) zu gelangen, wie für die folgenden Überlegungen, die in der allgemeinen Formel (37) gipfeln, die freilich—weil sie die transzendente Funktion ln x vermeidet, viel weniger kompakt ist.

Wenn man in Eulers Darstellung etwas vermißt, so ist es der Übergang von der Darstellung (30) zur sog. Legendreschen Form dieses Integrals:

$$n! = \int_0^\infty e^{-t} t^n \, dt; \tag{38}$$

sie ist leicht mittels der Substitution $\ln x = -t$, d.h. $x = e^{-t}$ zu finden. In anderem Zusammenhang hat Euler auch diese Integraldarstellung verwendet. —Wie man die Wallisschen Überlegungen unter Verwendung der Beta- und der Gamma-Funktion modern interpretieren kann, hat A. Prag in seiner Studie [4] sehr schön gezeigt, während für Newton ausdrücklich auf die Edition [2] und die Arbeit von D. T. Whiteside [6] verwiesen sei.

LITERATUR

[1] Wallis, John. 1656. *Arithmetica infinitorum* (Oxford, (1655) 1656). Wiederabdruck in Wallis, John. 1695. *Opera mathematica*, Bd. *2*, (Oxford).

[2] *The mathematical papers of Isaac Newton*, ed. by D. T. Whiteside (Bd. 1, Cambridge, 1967), 96–142: "Annotations on Wallis," insbes. 104–111.

[3] Euler, Leonhard. 1738. "De progressionibus transcendentibus seu quarum termini generales algebraice dari nequeunt," *Comm. acad. sci. Petropol.*, *5* (1730/31) 1738, 36–57. Wiederabdruck in *Leonhardi Euleri Opera omnia* (1) *14* (Leipzig u. Berlin, 1925), 1–57.

[4] Prag, Adolf. 1930. "John Wallis, 1616–1703. Zur Ideengeschichte der Mathematik im 17. Jahrhundert," *Quell. Stud. Gesch. Math.*, *1*, 381–412.

[5] Scott, John Frederick. 1938. *The mathematical work of John Wallis, D.D., F.R.S.* (*1616–1703*) (London). Insbes. Kap. 4.

[6] Whiteside, Derek Thomas. 1961. "Newton's discovery of the general binomial theorem," *Math. Gazette*, *45*, 175–180.

Institut für Geschichte der Naturwissenschaften,
Mathematik und Technik
Universität Hamburg
Hamburg, Bundesrepublik Deutschland

Neue geometrische Texte aus Byzanz*

KURT VOGEL

This article discusses new Byzantine geometrical manuscripts. The contribution of Byzantium to mathematics consists largely in its preservation of classical works, passed on to the West by means of Sicily and Italy, and then in translations through the Muslims. If their own contributions were limited, one can still identify periods of considerable activity. As early as 425 the school of Theodosius II must be considered, as the transformation from Hellenistic to Byzantine science was being accomplished, largely through the efforts of figures like Eutocius, Athenaios of Tralleis, and Isidoros of Miletus. Following centuries of relative inactivity, a renaissance of studies begins in the 9th century. This paper considers these later developments. Recently new texts, largely from the 9th to 12th centuries, have been added to those already known. These new texts contain practical examples of measuring land areas, which are presented as recipes and clearly sufficed for the layman who had no need to know the secrets of geometry. These texts contain numerous errors, and often the figures drawn to illustrate examples do not conform to the requirements of the text. If there is little mathematical pleasure to be found here, these documents are of interest for their use of technical language, their techniques for calculation, and for the various methods which they demonstrate.

Die Leistung von Byzanz auf dem Gebiet der Mathematik bestand darin, daß es die klassischen Werke, die zuerst dort allein bewahrt wurden, dem Abendland weitergab, einmal auf dem Weg über Sizilien und Italien und dann in Übersetzungen durch die Muslime [1]. Eigene, den Umfang

* Herrn Prof. Dr. Kurt-R. Biermann zum 60. Geburtstag gewidmet.

ISBN 0-12-204050-3

erweiternde Leistungen waren gering; doch lassen sich einige Perioden größerer Aktivität erkennen. Schon am Anfang, als Ost- und Westrom auseinanderlebten, hatte *Theodosius II.* im Jahre 425 eine hohe Schule aus der Zeit *Konstantins des Großen* erneuert, die in Beziehung stand zu den beiden anderen Reichsuniversitäten Athen und Alexandria. Zu jener Zeit des Übergangs von hellenistischer zu byzantinischer Wissenschaft waren es vor allem *Eutokios*, *Athenaios von Tralleis* und *Isidoros von Milet*, die der griechischen Mathematik in Byzanz Eingang verschafften. Nach einem durch äußere und innere Kämpfe verursachten wissenschaftlichen Vakuum sehen wir vom 9. Jahrhhundert an wieder ein Aufblühen der Studien. Am Anfang stand *Leon* [2], Rektor der im Jahre 863 neu gegründeten weltlichen Universität, der in seiner Jugendzeit lange nach einem Lehrer für Mathematik und nach Büchern gesucht hatte, war nicht nur als Lehrer und als Forscher tätig [3]; sein Hauptverdienst bestand in seinen Bemühungen um die Erhaltung der großen Klassiker der Mathematik. Es entstanden damals die Handschriften, auf denen fast die gesamte spätere Überlieferung beruht. Die Universität wurde im Jahre 1045 unter *Konstantin XX. Monomachos* erneuert. An der Spitze der Philosophischen Fakultät stand der ungemein vielseitige und geniale *Michael Psellos*; er trug auch über das Quadrivium vor, verfaßt eine Schrift über den Zweck der Geometrie [4] sowie ein für Vermessungsbeamte bestimmtes Gedicht über geometrische Inhaltsformeln [5].

Während des unrühmlichen lateinischen Kaiserreichs (1204–1261) ist von wissenschaftlichem Interesse und mathematischen Studien nichts zu berichten. Sie setzten erst nach der Restauration unter den Palaiologen wieder ein. Die Situation war ähnlich der bei *Leon.* So schildert *Theodoros Metochites* (ca. 1260–1332), daß er zuerst für die exakte Mathematik weder einen Lehrer noch einen Schüler gesehen habe [6]. Als Geometer in dieser Zeit ist der Polyhistor *Georgios Pachymeres* (ca. 1242–ca. 1310) zu nennen; der Abschnitt Geometrie in seinem allen anderen überlegenen Quadrivium schließt sich in vielem eng an *Euklid* an [7].

Sonst aber ist in Byzanz von einem allgemeiner verbreiteten Interesse an geometrischen Fragen nichts zu verspüren; hier waren sie echte Romaioi. Während im alten Griechenland—wie Cicero es sagt [8]—die Geometrie in hohem Ansehen stand, beschränkte sich bei den Römern diese Kunst lediglich auf das Messen. Interesse bestand für alles, was nützlich war dem Handwerker, dem Landwirt oder dem Vermessungsbeamten (μετρητής), der im Auftrag des kaiserlichen Sekretärs (νοτάριος) die Größe des Grundstückes feststellte, was ja neben der Qualität die Grundlage für die Steuererhebung bildete. Dazu genügten ihnen Näherungsverfahren. Man ging ebenso wie die römischen Agrimensoren von den Schriften *Herons* aus (ca. 75 n. Chr.), der in seinen "Metrika" die wissenschaftlichen Grundlagen dargelegt hatte. Unter seinem Namen erschienen auch in der Folgezeit zahlreiche Bearbeitungen und Auszüge, in denen auf jede Beweisführung verzichtet wurde und die nur die Berechnungsformeln anhand der Beispiele vorführten. Sie stehen in den unechten Schriften Herons (Geometrica, Stereometrica und "De mensuris"). Auch die Geodäsie eines *Heron von Byzanz* (ca. 940) gehört dazu [9]. Eine ausführliche geodätische Schrift stammt von *Pediasimos* (ca. 1330), der sich auch auf *Heron* beruft und ausdrücklich betont, daß die technische Vermessung, die Geodäsie mit Recht auch Geometrie genannt werden darf [10].

Neuerdings sind zu den bisher bekannten Texten zahlreiche weitere, meist aus der Zeit vom 9. bis 12. Jahrhundert stammende Texte kleineren oder größeren Umfangs dazu gekommen. *E. Schilbach* hat sie ohne Übersetzung, aber mit ausführlichem Kommentar als Quellenbuch [11] herausgegeben für die von ihm erstmals ausführlich dargelegte byzantinische Metrologie [12]. Diese neuen Texte enthalten praktische Beispiele zur Vermessung von Bodenflächen, was rezeptmäßig vorgerechnet wird und für den Laien, den nicht in die Geheimnisse der Geometrie Eingeweihten (ἀμύητος) offenbar genügte [13]. Sie enthalten zahlreiche Fehler [14], auch passen die beigegebenen Zeichnungen nicht immer zum Text. Wenn also hier auch kein mathematischer Genuß zu erwarten ist, so ist der Inhalt doch von einigem Interesse wegen der verwendeten Fachsprache und Rechentechnik sowie wegen mancherlei Methoden, die man auch bei den römischen Agrimensoren, bei "*Alkuin*" und anderen Autoren des Mittelalters findet [15].

Die beim Ausmessen der Bodenflächen verwendeten, aus antiken Maßen hervorgegangenen hauptsächlichen *Längenmaße* sind das Klafter (ὀργυιά zu ca. 2,11 m) und sein Zehnfaches, das "Meßseil" Schoinion. Durch sie sind die *Flächenmaße* Quadrat-Klafter und Quadrat-Schoinion bestimmt (ca. 4,44 bzw. 444 Quadratmeter). Doch wurden diese Maße zeitlich und örtlich verändert; zudem waren neben den offiziellen auch Lokalmaße in Gebrauch. Die Länge des Klafters wurde durch die Gnade, die der Kaiser *Michael IV* (1034–1042) den Bauern gewähren wollte, von 9 auf $9\frac{1}{4}$ Spannen (also von 2,11 auf 2,17 m) festgesetzt [16]. Als Flächenmaß dienten ferner die Menge sowie das Gewicht des benötigten Saatgutes; dies waren der "Scheffel" (μόδιος) (ca. 17,1 Liter) für 200 und die Litra, das Pfund (ca. 320 Gramm) für 5 Quadratklafter.

Demnach gilt die Relation: 1 Scheffel = 2 q-Schoinien = 40 Litren = 200 q-Klafter [17]. Ein Beispiel für die Umrechnung ist: $56\frac{1}{4}$ q-Klafter = $\frac{1}{2}$ Scheffel + 1 Litra + $1\frac{1}{4}$ q-Klafter [18]. Es gab aber eine Ausnahme: wenn es sich nämlich um Wiesen erster Qualität handelte. Hier gehen schon 100 q-Klafter auf einen Scheffel. So konnte also eine Fläche je nach der Bodenart verschiedene "Scheffel" groß sein; 15 q-Schoinien sind zum Beispiel im allgemeinen $7\frac{1}{2}$ Scheffel, sind es aber Wiesen erster Güte, dann sind es 15 Scheffel [19]. Die Fläche Scheffel zeigt sich hier weniger als Flächen—sondern als Verrechnungseinheit [20] bei der Festlegung von Preisen und Steuern [21]. Einer gerechteren Preisbildung diente auch die Vorschrift, daß minderwertiges Land wie steinige Felder und salzige Wiesen zu einem Schoinion (statt von 10 Klaftern) von 12 Klaftern vermessen werden soll. Beim Weinland werden neben den genannten meist andere Maße verwendet wie das Plethron zu 3 Scheffel oder der Meßstab Kalamos. Dieser hat die Länge von 1, 2 oder 3 Abständen zwischen den Weinstöcken (φυτά). Hat ein rechteckiges Weinland die Seitenlänge a und b Kalamoi, dann ist der Flächeninhalt ab, $4ab$ oder $9ab$ "Weinstöcke" [22].

Die Flächen von *Quadrat* und *Rechteck* werden nur selten als $F = a^2$ bzw. $F = ab$ berechnet. Ein Beispiel dafür ist das Quadrat mit der Seite $a = 48$ Klafter [23] so ist der Inhalt 2304 q-Klafter = $11\frac{1}{2}$ Schoinien + $\frac{1}{2} + \frac{1}{5} + \frac{1}{10}$ Litren. Eine andere Aufgabe berechnet die "Tafel" $10 \cdot 20$ q-Klafter = 200 q-Klafter = 1 Scheffel [24]. Sonst aber wird immer die bekannte Edfu-Formel verwendet [25]. Hat man ein Viereck mit den aufeinanderfolgenden Seiten a, b, c, und d, dann ist die Fläche $F = [(a + c)/2] \cdot [(b + d)/2]$. Dabei darf auch eine Seite krumm sein (ἴλιγγος) [26]. Daß man für die gegenüberliegenden Seiten, zum Beispiel für den "Fuß" und den "Kopf" den Mittelwert $(a + c)/2$ bildet, wird damit erklärt, daß der Fuß dem Kopf oder der Kopf dem Fuß etwas leihen soll (δανείζειν), damit beide gleich gemacht werden (ἰσάζειν) [27]. Auch bei Quadrat und Rechteck wird diese Formel genommen [28], sogar bei einem Quadrat mit der Seite 1 Schoinion. Es heißt da umständlich

$$\frac{1+1}{2} \cdot \frac{1+1}{2} = 1 \cdot 1 = 1 \text{ q-Schoinion} = \text{ein halber Scheffel [29].}$$

Auch alle Aufgaben, in denen es sich um das allgemeine Viereck (τραπέζιον) handelt, rechnen so. In einem Beispiel aus der Geodäsie des Georgios [30] sind die Seiten 45, 31, 42 und 28 Klafter. dann ist

$$F = \frac{45 + 42}{2} \cdot \frac{31 + 28}{2} = 43\tfrac{1}{2} \cdot 29\tfrac{1}{2} \text{ q-Klafter.}$$

Diese Multiplikation wird jetzt im fortlaufenden Text ohne weiteres Schema durchgeführt als:

$$40 \cdot 20 + 40 \cdot 9 + 40 \cdot \tfrac{1}{2} + 3 \cdot 20 + 3 \cdot 9 + 3 \cdot \tfrac{1}{2}$$
$$+ \tfrac{1}{2} \cdot 20 + \tfrac{1}{2} \cdot 9 + \tfrac{1}{2} \cdot \tfrac{1}{2} = 1283\tfrac{1}{4} \text{ q-Klafter.}$$

Da es sich hier offenbar um Wiesen erster Güte handelt, wird durch 100 dividiert. Dies gibt—mit einer Aufrundung von $1283\frac{1}{4}$ auf $1283\frac{1}{3}$—$12 + \frac{1}{2} + \frac{1}{3}$ Scheffel. Es wird dabei noch angegeben, daß $\frac{1}{12}$ q-Klafter fehlten.

Diese Formel für das allgemeine Viereck ist auch im Abendland bekannt, so zum Beispiel bei Alkuin (=Geometria incerti autoris Nr. 32) [31], bei dem Anonymus von "De iugeribus metiundis" [32] oder in der Geometrie II des Boethius [33].

Bei der Berechnung von *Dreiecken* fällt auf, daß—auch wenn es sich um nichtrechtwinklige handelt—eine beliebige Seite immer Hypotenuse heißt; Sie verbindet die Endpunkte von Kathete und Basis. So kommt es vor, daß in einem annähernd rechtwinkligen Dreieck mit den Seiten 10, 12, und 16 die Seite 12 als Hypotenuse bezeichnet wird [34]. Die Fläche des Dreiecks wird wieder nach der Edfu-Formel berechnet, also $F=[(a+0)/2]\cdot[(c+d)/2]$ [35]. So hat einmal ein gleichschenkliges Dreieck den Fuß 9 und die beiden anderen Seiten sind $5\frac{1}{2}$ Schoinien, es heißt da: "Man muß Fuß und Kopf (der aber = 0 ist!) addieren und sagen, Kopf und Fuß sind 9 Schoinien" [36].

Geradlinig begrenzte *Polygone* [37] werden in Parzellen (Rechtecke und Dreiecke) zerlegt, die einzeln berechnet und deren Flächen dann addiert werden, wie es schon die Babylonier gemacht hatten [38]. Andere Figuren haben die Form eines Winkelhakens, des Zimmermannswinkels [39], oder eines Gamma (τὸ γαμμάτιον), das schon die Chinesen kannten [40], oder auch eines Lambda [41].

Die Aufgabe über den Kreis weisen auf verschiedene Quellen hin. Einmal kommt der babylonische Wert mit $\pi = 3$ vor [42], sonst ist auch $\pi = 3\frac{1}{7}$ bekannt. Mit diesem Wert wird der Durchmesser aus dem Umfang berechnet oder umgekehrt, als

$$d = u\cdot\tfrac{7}{22} \qquad \text{und} \qquad u = d\cdot 3\tfrac{1}{7}\ [43].$$

Auch die Fläche von Kreis und Halbkreis $d^2\cdot 11/14$ bzw. $d^2\cdot 11/28$ wird richtig angegeben und dabei noch erwähnt, daß das Produkt $u\cdot d = 4$ Kreisflächen ist [44].

In den meisten Fällen [45] aber wird die Kreisfläche aus dem Umfang bestimmt, nach einem Verfahren, das "Ringsumbegrenzung" (περιορισμός) genannt wird. Man macht den vierten Teil des Umfangs $u/4$ zu einer Quadratseite und setzt $(u/4)^2$ gleich der Kreisfläche, was mit einem $\pi = 4$ nichts zu tun hat [46]. Dieselbe Methode verwendet auch Alkuin [47], der Anonymus von "De iugeribus metiundis" [48]; auch Franco von Lüttich erwähnt sie, macht sie sich aber nicht zu eigen, wie schon Proklos diese Inhaltsberechnung abgelehnt hatte [49]. Einmal aber wird ein Kreis mit $u = 61$ auf andere Weise berechnet, nämlich er wird zu einem Rechteck [50] gemacht von der Fläche $15\cdot 15\frac{1}{2}$, was vielleicht bequemer war als $(61/4)^2$.

Dieses Rundumverfahren findet Anwendung bei *unregelmäßigen Grundstücken* aller Art [51], deren Grenze auch mit Einbuchtungen oder Wölbungen nach außen verlaufen kann. Da hat zum Beispiel eine trockene Wiese [52] einen Umfang von $10 + 20 + 18 + 48 + 90 = 186$ Schoinien. Weil aber in ihr Wege, Klüfte, Trockenbäche und sumpfige Stellen den Wert vermindern, wird ein Abzug von einem Schoinion auf 10 Schoinien, also von 10% zugelassen, was aufgerundet 19 ausmacht. So ergibt sich eine Fläche von $(41 + \frac{1}{2} + \frac{1}{4})^2$ Schoinien. In der nächsten Aufgabe werden auch zuerst 10% vom Umfang abgezogen, weil aber hier die Wiese bewässert ist und auch Mühlen betrieben werden, fällt dieser Abzug wieder weg.

Noch ein weiteres bei der Bestimmung von Inhalten verwendetes Verfahren ist wohl von Byzanz ins Abendland gekommen. Es ist eine Mittelwertsbildung die einem nicht näher zu datierenden Patrikios zugeschrieben wird [53]. Bei einem Feld mit gegebener Länge a und unterschiedlichen, an verschiedenen Stellen gemessenen Breiten wird der Mittelwert m gebildet, dies gibt die Fläche $F = a \cdot m$. In einer Aufgabe sind drei [54], in anderen fünf [55] Breiten gemessen worden; diese sind 2, 3, 4, 2, und 4 Schoinien so ist $m = 3$ und der Inhalt wird (mit $u = 20$) $F = 60$ q-Schoinien = 30 Scheffel [56].

Aus dem *räumlichen* Bereich enthalten die Texte wenig. Einmal soll ein Feld als Mantel eines Kegels, eines "Hügels" (βουνίον) berechnet werden, dessen Grundkreis den Umfang $u = 80$ hat und dessen Seitenkante $s = 60$ Schoinien ist. Statt $F = u/2 \cdot s$ steht aber $F = (s/2) \cdot (u/2)$ da [57]. Ähnliche Aufgaben, allerdings mit einem Kegelstumpfmantel, stehen in den mittelalterlichen Texten. Sie lassen sich von *Epaphroditus* bis zu *Widmann von Eger* verfolgen [58].

Noch eine stereometrische Aufgabe ist zu erwähnen, nämlich die Volumenbestimmung eines Schiffes [59]. Ähnliche Beispiele stehen am Ende von Herons Stereometrica [60]. Auch die Schrift *De iugeribus metiundis* schließt mit der Berechnung einer "archa" ab [61]

Die *mathematische Fachsprache* der geodätischen Texte bedient sich vielfach ungewöhnlicher, volkstümlicher Wendungen. So ist das Halbieren ein Wegschlagen (κόπτειν δισσῶς), Wegwerfen (ῥίπτειν τὰ ἥμισα), Wegschicken (ἀφιέναι), Fortjagen (διώκειν τὸ ἥμισυ), Beseitigen (ἀπολύειν τὸ ἥμισυ) oder Abschneiden (τέμνειν μέσον εἰς δύο) der Hälfte. Ungewöhnlich sind auch die Ausdrücke für Halbieren (ἡμισάζειν) und Halbierung (μεσασμός).

Die Multiplikation wird auch "Vereinigung" genannt (ἑνώτησις für ἕνωσις). Merkwürdigerweise heißt Multiplizieren und Multiplikation meistens "Befragen" und "Befragung" (ἐρωτᾶν und ἐρώτησις bzw. ἐπερώτησις). Ein Schreibfehler [62] für ἑνώτησις kann es wohl nicht sein, wenn es an einer Stelle heißt [63]: "Es soll die Länge die Breite befragen." Sind die Faktoren als Personen gedacht, die sich gegenseitig befragen [64]?

Besonders bei den geometrischen Figuren werden statt der klassischen Fachwörter volkstümliche verwendet. Pediasimos sagt daß die nicht in die Mysterien Eingeweihten barbarisch und dumm (βαρβάρως καὶ ἀσόφως) daherreden, wenn sie gleichschenklige Dreiecke als Schwerter (ξίφη) oder ein aus Rechteck und Trapezion zusammengesetzte Figur als Sandalen (ὑποδήματα) [65] vergleichen. Die in den Beispielen vorkommenden Flächenstücke werden oft nur mit allgemeinen Ausdrücken eingeführt wie Fläche (ἐμβαδόν), Bodenfläche (γήδιον), Gegend (τόπος, τόπιον), Platz oder Feld (χωρίον, χωράφιον, κάμπος), Figur (σημάδιον) oder es wird nur gesagt, um was es sich handelt, zum Beispiel um einen Acker oder um eine Wiese.

Die Grundfigur, das Viereck, auch wenn es kein Rechteck ist, wird nach den vier Himmelsrichtungen (κλίματα) oder den vier Winden (ἀέρες) orientiert gedacht. Dabei entspricht die obere Seite, der Kopf (κεφαλή) dem Osten, die rechte Seite dem Süden, der Fuß (ποῦς, πόδωσις) dem Westen und die linke Seite dem Norden. Auch bei Chinesen und Ägyptern findet sich der gleiche Gedanke, nur ist bei den Chinesen der Norden oben [66] und bei den Ägypern der Westen [67].

Während bei Euklid klar definiert wird, nämlich das Quadrat als τετράγωνον, das Rechteck als τετράγωνον ἑτερόμηκες, die Raute als ῥόμβος, das Rhomboid als ῥομβοειδές und das allgemeine Viereck als Trapez (τραπέζιον) herrscht in den byzantinischen Texten ziemliche Willkür. Das allgemeine Viereck wird als τετράγωγογ ἑτερόμηκες, auch als Pfahlwerk (ἀλλεπάλληλον) bezeichnet, das Trapez als παραλληλόγραμμον; für Rechteck findet sich auch die Bezeichnung Brett oder Tafel (ταῦλα) oder Großviereck (μακροτετράγωνον). Eine Figur in der Form eines Winkelhakens wird Gamma (γαμμάτιον oder ὑπογαμματίζον χωράφιον) genannt [68].

Auch das Dreieck wird manchmal bildhaft bezeichnet als segelförmiges Dreieck (τρίγωνον ἄρμενον, ἀρμενοειδές) [69] oder als schildförmiges Dreieck (τρίγωνον σκουταροειδές) [70]. An einer Stelle findet sich der Hinweis, daß die alten das ungleichseitige Dreieck παράσκελον genannt hätten [71]. So steht es auch einmal bei "Heron" [72].

Der Kreis ist für den Landmesser ein rundes Feld (στρογγύλον χωράφιον) oder eine Sonnenscheibe (ἅλων, ἀλώνιον) [73]. Der Umfang ist ein Ring ὁ γῦρος, τὸ γυρομέτριν, der γύρωθεν gemessen werden soll [74]. Der Halbkreis bzw. das Segment wird Apsis genannt [75]. Der Radius heißt einmal Keil (σφηνά) "mitten durch den Kreis" [76]. Einen Kreis, in dem zwei aufeinanderstehende Durchmesser gezeichnet sind "kreuzartig" (σταυροειδές) [77] zu messen, geht natürlich nicht. An anderer Stelle [78] ist ja die kreuzförmige Figur durch die Zeichnung als Raute definiert. Zudem wird hier der Kreis wie üblich als $F = (u/4)^2$ gerechnet. Zu einer mondartigen Figur σχῆμα σεληνοειδές fehlt im Text die Ausrechnung [79].

Der Umfang einer von vielen Linien begrenzten unregelmäßigen Fläche πολύγραμμον hat nicht nur gerade Stücke, sondern auch gekrümmte (καμπυλοειδές), die von der geraden abweichen (ἀπὸ τῆς εὐθείας παρεκκλίνειν) [80]. Das Verfahren zur Berechnung der Fläche, die Rundumberechnung heißt περιορισμός [81] oder Berechnung nach dem ganzen

Umkreis (κατὰ το ολόγυρον). Der 10%-ige Abzug ist δεκατισμός [82] das Zeitwort dazu ἀποδεκατίζειν oder ἀποδεκατοῦν.

Vielleicht regt die in diesen neuen byzantinischen Texten dargebotene volkstümliche Geometrie den Jubilar, der ein Freund der Kombinatorik und Wahrscheinlichkeitsrechnung ist, dazu an, sich Gedanken zu machen unter welchen Bedingungen die Näherungsformeln zu einem hinreichend brauchbaren Ergebnis führen können.

ANMERKUNGEN

1. J. L. Heiberg, "Den graeske Mathematiks Overleveringshistorie," *Kongelige Danske Videnskabernes Selskabs Forhandlinger* (Kopenhagen, 1896), 77–93.
2. J. L. Heiberg, "Der byzantinische Mathematiker Leon," *Biblioteca Mathematica 1* (1887), 33–36.
3. K. Vogel, "Buchstabenrechnung und Indische Ziffern in Byzanz," *Akten des XI. Internationalen Byzantinistenkongresses 1958, Muünchen* (1960), 660–664.
4. ποῖον τὸ τέλος τῆς γεωμετρίας in J. F. Boissonade ψέλλος (Norimbergae, 1838), 159–163.
5. K. Krumbacher, *Geschichte der Byzantinischen Literatur* (München, 1897), 436.
6. I. L. Heiberg, *Euclidis Elementa V* (Lipsiae, 1888), xcv.
7. P. Tannery, *Quadrivium de Georges Pachymère* (Studi e testi 94; Città del Vaticano, 1940), 201–328.
8. Cicero, *Tusculanarum disputationum*, I, 5.
9. I. L. Heiberg, *Heronis Opera V* (Lipsiae, 1914), lxx-xcvii. Andere gehen unter dem Namen Isaac Argyros, Georgios oder sind annonym, ebenda xcviii-cxi.
10. G. Friedlein, *Die Geometrie des Pediasimos* (Programm Ansbach, 1866), 7.
11. E. Schilbach, *Byzantinische metrologische Quellen* (Düsseldorf, 1970), hier abgekürzt als BMQ.
12. E. Schilbach, *Byzantinische Metrologie* (= *Byzantinisches Handbuch*, vierter Teil, München, 1970), abgekürzt als BM. Abschnitt über das Vermessen von Bodenflächen, 244–248.
13. Friedlein, am angeführten Ort, Seite 7.
14. Siehe den Kommentar in BMQ, 145–174.
15. Die Hauptquellen sind folgende Texte:
    - Text 1. Epaphroditus (5. Jahrhundert?), ed. N. Bubnov, *Gerberti Opera Mathematica* (Berlin, 1899; Neudruck Hildesheim, 1963), 516–551.
    - Text 2. Anonymus (7. Jahrhundert?) von "De iugeribus metiundis"; ed. F. Blume, K. Lachmann und A. Rudorff, *Die Schriften der römischen Feldmesser I* (Berlin, 1848), 354–356.
    - Text 3. Alkuins "Propositiones" (8./9. Jahrhundert); ed. M. Folkerts, *Die Alkuin zugeschriebenen Propositiones ad acuendos juvenes Oesterreichische Akademie der Wissenschaften, mathematischenaturwissenschaftliche Klasse. Denkschrift 116* (6. Abhandlung) (1978).
    - Text 4. Anonymus (10. Jahrhundert) der "Geometria incerti autoris", ed. in Bubnov, am angeführten Ort, 310–365.
    - Text 5. Boethius (11. Jahrhundert), ed. M. Folkerts, *Boethius Geometrie II* (Wiesbaden, 1970), 113–171.
    - Text 6. Clm 13084, fol. 58ʳ–65ʳ (10. Jahrhundert), ed. V. Mortet in P. Tannery, *Mémoires scientifiques V* (1922), 42–78. Eine Variante dazu ist in Clm 14836.

Text 7. Clm 26639, fol. 1r–6v (15. Jahrhundert), ed. W. Kaunzner, *Über die Handschrift Clm 26639 der Bayerisches Staatsbibliothek München* (Hildesheim, 1978).

16. BMQ, 54.24 (=Seite 54, Zeile 24): ἀπεχαρίσατο ὁ βασιλεὺς κῦρ Μιχαὴλ τοῖς χωρίταις
17. Vergleiche die Tabellen in Heron IV, 196–201. Wenn es dort heißt: Breite und Länge zu 5 Klafter machen einen Liter, so ist gemeint: Wenn Breite und Länge eine Fläche von 5 Quadratklafter einschliessen, ist es l Litra.
18. BMQ, 90.23 ff; Text $1\frac{1}{2}$ statt $1\frac{1}{4}$.
19. BMQ, 75.16 f.
20. BM, 58.
21. Zu den Richtpreisen für Ackerland und Wiesen verschiedener Güte siehe BM, 249 ff.
22. Zur Vermessung von Weinland, siehe BM, 81–89.
23. BMQ, 73.26 ff.
24. BMQ, 60.15.
25. Siehe zum Beispiel M. Cantor, *Vorlesungen über Geschichte der Mathematik I* (1907^3), 110 f.
26. Wie in der Zeichnung BMQ, 118.
27. BMQ, 81.21 ff. Ähnlich in "De iugeribus metiundis" (siehe oben, Nr. 15 Text 2, Seite 355): aequas maiorem partem cum minore.
28. Zum Beispiel, BMQ 55.26 ff und 55.32 ff.
29. BMQ, 83.19 ff.
30. BMQ, 89.12 ff.
31. Siehe oben (Anmerkung 15) Text 3, Aufgabe Nr. 23 = Text 4, Nr. IV, 32.
32. Text 2, Seite 355.
33. Text 5, Liber II, XXII, Seite 161.
34. BMQ, 91.11 ff.
35. In BM, 245, muß es heißen $\frac{a(b + d)}{4}$.
36. BMQ, 84.21 ff. Daß der Kopf Null ist, wird im Text 2 (Anmerkung 15), Seite 355 betont, wenn es heißt: "... et in alio in punctum desierit." Siehe auch die Aufgaben von Text 3, Nr. 24 und 28 sowie Text 4, IV 33.
37. BMQ, 70, 72.
38. Siehe den Felderplan bei B. Meissner, *Babylonien und Assyrien II* (Heidelberg, 1925), 390.
39. BMQ, 118–126.
40. Siehe Lam Lay Yong, *A Critical Study of the Yang Hui Suan Fa* (Singapore, 1977), 103, 105.
41. BMQ, 119.11.
42. BMQ, 78.1.
43. BMQ, 91 f.
44. BMQ, 92.32.
45. BMQ, 56.25; 60.8 f; 76, 81, 85, 118.
46. Siehe hierzu A. J. E. M. Smeur, "On the Value Equivalent to π in Ancient Mathematical Texts," *Archive for History of Exact Sciences 6* (1969/70), 249–270; 249 f.
47. Text 3 (Anmerkung 15), Aufgabe Nr. 25.
48. Text 2, Seite 354.
49. Smeur, am angeführten Ort, Seite 250; Proklos Prop. XXXVII, Theorem XXVII (ed. Friedlein, Seite 403).
50. BMQ, 72.17 ff.
51. Zu den Grenzbegehungen in Byzanz siehe F. Dölger, *Aus den Schatzkammern des heiligen Berges*, (München, 1948), 143, 186. Die Frage nach der Methode (Seite 193) ist durch $(u/4)^2$ gelöst.
52. BMQ, 63 f.
53. Siehe Heron IV, 386. 25.

54. BMQ, 62.
55. BMQ 61.
56. Derartige Mittelbildungen auch bei Alkuin, Aufgabe Nr. 22 = *Geometria incerti auctoris* IV, 31.
57. BMQ, 76.26 ff.
58. So ist Epaphroditus II, 8 (Bubnov, Seite 526) = *Geometria incerti auctoris* IV, 26; (Bubnov, Seite 350) = Boethius II, Nr. XXXV (Folkerts, Seite 167) = Clm 26639 (Kaunzner Nr. 19, Seite 35) = Widman Rechenbuch 1489 [f. 222[r]]. Desgleichen: Epaphroditus II, 10 = Geometria incerti auctoris IV, 45 = Boethius XXXVII = Clm 26639 (Nr. 20, Seite 35) = Widman [f. 222[v]].
59. BMQ, 127–132.
60. Heron V.56, 128 ff.
61. Text 2 (Anmerkung 15), Seite 356.
62. F. Dölger, "*Beiträge zur byzantinischen Finanzverwaltung* = *Byzantanisches Archiv 9* (1927), 85.
63. BMQ, 50.1.
64. BMQ, 148.
65. Pediasimos, am angeführten Ort, Seite 7.
66. Die senkrecht stehende Länge ist ein Pflügen von Norden nach Süden, die Breite liegt in der Richtung Ost-West.
67. Siehe R. Parker, *Demotic mathematical papyri* (London, 1972), Problem 64 und 65.
68. BMQ, 118.26 f. Im Lexikon von Liddell-Scott ist ein γαμμοειδές εἶδος bei Paulos von Aigina erwähnt.
69. BMQ, 57, 58, 81.
70. BMQ, 57.10.
71. BMQ, 77.20. Das Dreieck ist nicht rechtwinklig (BMQ, 199), wie man es aus der beigegebenen Figur meinen könnte.
72. Heron V, Seite 206.22.
73. BMQ, 118.20, 58.25; 60, 28; 81, 13.
74. BMQ, 76.28; 81.13, und 85.3.
75. BMQ, 92.28.
76. BMQ, 78.14.
77. BMQ, 61.3.
78. BMQ, 80.21.
79. BMQ, 85.1.
80. BMQ, 71.25 f.
81. BMQ, 199.
82. BMQ, 193.

München, Bundesrepublik Deutschland

Deutsche Mathematiker—Auswärtige Mitglieder der Akademie der Wissenschaften der UdSSR*

A. P. JUSCHKEWITSCH

The historical value of the evaluations written on behalf of candidates for membership in scientific societies has been amply demonstrated by Professor Kurt-R. Biermann in his excellent monograph: "Vorschläge zur Wahl von Mathematikern in die Berliner Akademie," *Abhandlungen der Deutschen Akademie der Wissenschaften zu Berlin. Klasse für Mathematik, Physik und Technik* (Berlin: Akademie-Verlag, 1960). Following his example, this paper discusses the election of German mathematicians as foreign members of the Academy of Sciences of the Soviet Union. Often these members were also named honorary members (EM). In the middle of the 18th century the category of corresponding member (KM) was introduced as well. All chronological dates used here are taken from the recently published *Directory of Members of the Academy of Sciences of the USSR*.

Wie wertvoll die Angaben von den Wahlen der Mitglieder irgendeiner Akademie der Wissenschaften für die Wissenschaftsgeschichte sein können, hat uns Prof. K.-R. Biermann in seinem vortrefflichen Beitrag "Vorschläge zur Wahl von Mathematikern in die Berliner Akademie" (Berlin, 1960) gezeigt. Seinem Beispiel folgend, möchte ich hier über die Wahlen der deutschen Mathematiker zu Auswärtigen Mitgliedern der Akademie der

* Herrn Prof. Dr. Kurt-R. Biermann zum 60. Geburtstag gewidmet.

ISBN 0-12-204050-3

Wissenschaften der UdSSR[1] sehr knapp berichten. Es sei vorbemerkt, dass die Auswärtigen Mitglieder dieser Akademie oft auch Ehrenmitglieder (EM) genannt wurden; auch der Rang eines korrespondierenden Mitgliedes (KM) wurde in der Mitte des 18. Jahrhunderts eingeführt. Alle chronologischen Daten sind aus dem neulich erschienenen Verzeichnis der Mitglieder der Akademie der Wissenschaften der UdSSR entnommen [1]. Alle Daten für das 18, und 19. Jahrhundert sind zweifach—sowohl nach dem alten, als auch nach dem neuen Stil—angegeben; die neue Zeitrechnung in der UdSSR wurde anfangs 1918 offiziell eingeführt.

Im 18, Jahrhundert wurden insgesamt 11 deutsche Mathematiker zu EM oder KM der AdW ernannt, im 19. Jahrhundert 15 Gelehrte und in der ersten Hälfte des 20, Jahrhunderts—nämlich zwischen 1922 und 1932—noch 4 Personen. Die allgemein bekannten historischen Ereignisse haben in diesem letzten Fall eine wichtige und manchmal entscheidende Rolle gespielt.

Jetzt folgt zuerst die Liste der im 18, Jahrhundert ernannten Wissenschaftler; am Ende jeder Zeile ist das Ernennungsdatum angemerkt:

1. Chr. von Wolff (1679–1754); EM (30) 19.3.1725.
2. H. Kühn (1690–1769); EM (8.7) 27.6.1735.
3. G. W. Krafft (1701–1754); EM (12) 1.1.1745.
4. J. A. von Segner (1704–1777); EM (29) 18.8.1754.
5. A. G. Kästner (1719–1800); EM (3.11) 23.10.1786.
6. J. F. Pfaff (1765–1825); KM (6.9) 26.8.1793, EM (18) 7.5.1798.
7. K. F. Hindenburg (1741–1808); EM (8.8) 28.7.1794.
8. G. S. Klügel (1739–1813); EM (8.8) 28.7.1794.
9. A. Burja (1752–1816); EM (8.8) 28.7.1794.
10. M. von Prasse (1769–1814); KM (30) 19.9.1796.
11. Chr. F. Kausler (1760–1825); KM (6.3) 23.2.1797, EM (18) 7.5.1798.

Wie man sieht, wurde als erster der bekannte deutsche Philosoph und Polyhistor—damals Professor an der Universität Marburg—Christian von Wolff ernannt, der an den Besprechungen verschiedener Projekte der Organisation der Petersburger Akademie aktiv teilgenommen hat. Dank seiner Vermittlung waren einige Gelehrte, wie z.B. J. Hermann, D. Bernoulli u.a., zur Arbeit in der neugegründeten Akademie angestellt; man hatte auch Wolff selbst eingeladen, aber nach langen Verhandlungen blieb er schliesslich in Deutschland. Unter Wolffs Führung studierte 1736–1739 an der Universität Marburg der grosse M. W. Lomonossow (1711–1765), der bald nach seiner Rückkehr zuerst Adjunkt und dann ordentliches Mitglied der Petersburger Akademie wurde. Immer verehrte Lomonossow seinen alten

[1] Diese Benennung trägt die Akademie vom 27. Juli 1925 an. Bis zum Anfang des ersten Weltkrieges nannte man sie "Petersburger Akademie der Wissenschaften," und etwas später "Russische Akademie der Wissenschaften."

Lehrer, obgleich er dessen metaphysische und naturphilosophische Aussichten durchaus nicht teilte. Wolff war kein schöpferischer Mathematiker, jedoch fanden seine Lehrbücher eine grosse Verbreitung, auch in Russland. Ein Schüler L. Eulers, das Akademiemitglied S. K. Kotelnikow (1723–1806), veröffentlichte die zweibändige russische Bearbeitung des Wolff'schen "Auszuges aus den Anfangsgründen aller Mathematischen Wissenschaften" (Halle, 1713) [2].

Die Ernennung Wolffs als einer der führenden deutschen Universitätsprofessoren seiner Zeit zum EM war typisch für die Einstellung der Petersburger Akademie im 18. Jahrhundert. Auch die anderen oben genannten Mathematiker waren dank ihrer Lehrtätigkeit—und meistens auch dank ihrer Lehrbücher—wohlbekannte Professoren. Krafft war an der Universität in Tübingen tätig[2]; Segner in Göttingen und später in Halle, der Erfinder des nach ihm benannten Wasserrades; Kästner in Göttingen; Pfaff in Helmstadt und dann in Halle; Hindenburg und Prasse in Leipzig und Klügel in Halle. Auch Kühn war Gymnasialprofessor in Danzig (Gdansk), und Burja war Professor der Académie militaire in Berlin. Alle diese Leute haben auch einige mathematische Aufsätze oder Bücher veröffentlicht, die aber mehr Gelehrsamkeit als Begabung bezeugen[3]. Ausnahmen sind Hindenburg, ein hervorragender Vertreter der sogenannten Kombinatorischen Schule, und Pfaff, ebenfalls ein Anhänger dieser Schule, die die Mathematik wissenschaftlich bereicherten; übrigens die wichtigsten Forschungen Pfaffs, die zur Theorie der partiellen Differentialgleichungen gehören, wurden erst später, im Jahre 1815, veröffentlicht (die heute übliche Benennung "Pfaffsches Problem" stammt von C. G. Jacobi). Deutschland hatte in der Zwischenzeit nach Leibniz' Tod und vor dem Auftreten von Gauss keine eigenen grossen Mathematiker. Die so hohe Schätzung der pädagogischen Leistungen deutscher Professoren in dem Milieu der Petersburger Akademiker des 18. Jahrhunderts war ganz natürlich: die Ausbildung der inländischen Gelehrtenkader stellte damals in Russland ein dringendes und noch weitgehend ungelöstes Problem dar.

Wolff und Kästner übten einen grossen Einfluss auf die mathematische Bildung aus, und das nicht nur in deutschprachigen Ländern, sondern auch in Russland. Die reichhaltigen und auf einem ganz modernen Niveau verfassten "Anfangsgründe der Analysis der endlichen Grössen" (Göttingen, 1759) und "Anfangsgründe der Analysis des Unendlichen" Kästners (Göttingen, 1760–1761; danach viele Ausgaben) waren lange Zeit das beliebteste und das am

[2] G. W. Krafft war Mitglied der Petersburger Akademie von 1727 bis 1744; er verliess sie mit dem Titel eines EM und einer jährlichen Pension.

[3] Vgl. [3, nach dem Namenregister]. Was C. F. Kausler betrifft, so war er Professor der Französischen Sprache in Stuttgart. Über die mathematische Leistungen dieser Gelehrten vgl. [3, nach dem Namenregister].

meisten gebrauchte Universitätskompendium. Das erste Lehrbuch wurde von dem Astronomen und Petersburger Akademiker P. B. Inochodzew (1742–1806) ins Russische übersetzt und in zwei Teilen veröffentlicht [4]. Das zweite Buch Kästners, wie auch die Schriften L. Eulers, wurden von dem Moskauer Magister A. D. Barsow (gest. um 1800) benutzt bei der Verfassung seines Werkes über die Infinitesimal- und Variationsrechnung (Moskau, 1797); das war eine der ersten Darstellungen dieses Gegenstandes, von einem russischen Autor geschrieben [5]. Mehrere eminente Gelehrte waren Kästners Zuhörer, so z.B. die obengenannten Klügel, Hindenburg, Inochodzew und Pfaff, wie auch J. Ide (1775–1806), der später als Professor an der Moskauer Universität tätig war (1804–1806). Ferner M. Bartels (1769–1836), der künftige Professor an der Kasaner Universität (1808–1820), wo er unter seine Schüler den grossen N. I. Lobatschewski zählte; später arbeitete Bartels an der Universität Dorpat (jetzt Tartu). Auch F. Bolyai und C. F. Gauss waren in ihren Lehrjahren Schüler von Kästners; doch von einem Einfluss Kästners auf Gauss kann natürlich nicht die Rede sein.

Im 19. Jahrhundert, in neuen sozial-historischen und kulturellen Verhältnissen, beginnt in Russland wie auch in Deutschland ein neuer und immer wachsender Aufschwung der Mathematik. Schon die Namen Lobatschewski's, Ostrogradski's, Tschebyschew's und anderseits Gauss', Lejeune-Dirichlet's und C. G. Jacobi's bezeugen es. Dies zeigt auch die Liste der neuerwählten—mit wenigen Ausnahmen immer erstrangigen—deutschen KM oder EM der Petersburger Akademie[4]:

12. C. F. Gauss (1777–1855); KM (12.2) 31.1.1802, EM (5.4) 24.3.1824.
13. M. Ohm (1792–1872); KM (6.11) 25.10.1826.
14. C. G. J. Jacobi (1804–1851); KM (20) 8.12.1830, EM (25) 13.12.1833.
15. A. L. Crelle (1780–1855); KM (31) 19.12.1834.
16. J. P. G. Lejeune-Dirichlet (1805–1859); KM (3.1.1838) 22.12.1837.
17. F. E. Neumann (1798–1895); KM (2.1.1839) 21.12.1838.
18. E. E. Kummer (1810–1893); KM (19) 7.12.1862.
19. K. T. W. Weierstrass (1815–1897); KM (16) 4.12.1864, EM (14) 2.12.1895.
20. L. Kronecker (1823–1891); KM (20) 8.12.1872.
21. K. W. Borchardt (1817–1880); KM (19) 7.12.1879.
22. I. L. Fuchs (1833–1902); KM (14) 2.12.1895.
23. Chr. F. Klein (1849–1925); KM (14) 2.12.1895.
24. K. H. A. Schwarz (1843–1921); KM (25) 13.12.1897.
25. F. Engel (1861–1941); KM (16) 4.12.1899.
26. M. B. Cantor (1829–1920); KM (14) 2.12.1900.

[4] Diese Liste enthält nicht die Namen der Astronomen, die sich einigermassen auch in der Mathematik schöpferisch waren, wie z.B.
F. W. Bessel (1784–1846); EM (6.6) 25.5.1814,
J. F. Encke (1791–1865); EM (21) 9.12.1829,
P. A. Hansen (1795–1874); KM (25) 13.12.1833.

C. F. Gauss, der **princeps mathematicorum**, der die soeben angeführte Liste der deutschen KM oder EM der Petersburger Akademie eröffnet, stand am Anfang des 19. Jahrhunderts in regem Verkehr mit dieser Akademie. In der Leningrader Abteilung des Archivs der Akademie sind acht Briefe von Gauss an den ständigen Sekretär, den bekannten Mathematiker und Schüler Eulers, N. I. Fuss (1755–1825) während der Jahre 1801–1807 aufbewahrt. Es geht darin meistens um astronomische Forschungen Gauss', aber auch um den an ihm offiziell gemachten Antrag, nach Petersburg zu übersiedeln, wo er zum Ordentliches Mitglied der hiesigen Akademie ernannt worden sollte. Wie verlockend aber dieses Angebot für Gauss, der damals in ziemlich unbefriedigenden Bedingungen in Braunschweig wohnte auch war, so konnte er es nicht annehmen und kam im Herbst 1807 nach Göttingen, wo ihm eine ehrenvolle und gut bezahlte Stellung des Direktors der Sternwarte und des Universitätsprofessors angeboten wurde. In demselben Archiv befindet sich eine noch vor dem Sommer 1801 von Gauss selbst verfasste "Nachricht von zwei mathematischen Schriften für H. Collegiensrath Fuss"; diese zwei Schriften waren nämlich die noch nicht edierten "Disquisitiones arithmeticae" und die etwas frühere algebraische Doktordissertation von Gauss (1799). Alle diese Dokumente sind veröffentlicht [6, S.209–238].

Nach Gauss und Bessel wurde im Jahre 1826 der drittrangige Professor der Berliner Universität M. Ohm [vgl. 7, S.15–19] zum KM der Akademie erwählt. Man muss beachten, dass die Mathematik damals in der Petersburger Akademie nur durch E. D. Collins (1791–1841) vertreten war, einen Gelehrten mit ziemlich begrenzten Ansichten, und durch den ständigen Sekretär P. H. Fuss (1798–1855), der seinem Vater N. I. Fuss in diesem Amt nachfolgte, jedoch gar keine besondere mathematische Begabung bekundete. Mit der Wahl im Jahre 1828 zu Ordentlichen Mitgliedern der Akademie von M. W. Ostrogradski (1801–1862) und V. Ja. Buniakowski (1804–1889) und im Jahre 1853 von P. L. Tschebyschew (1821–1894), des Gründers und langjährigen Leiters der berühmten Petersburger mathematischen Schule, wurde die Ernennung solcher Gelehrten wie M. Ohm unmöglich. Freilich war auch A. L. Crelle kein bedeutender Mathematiker, aber ihm gehören sehr grosse Verdienste in der Förderung der Mathematik im allgemeinen und hochbegabter Mathematiker im besonderen; dazu war er der Gründer des weltbekannten "Crelle's Journal" ("Journal für die reine und angewandte Mathematik"), dessen Herausgeber er von Anfang an bis seinem Tode blieb [7, S.22–23].

Für den im Jahre 1862 ernannten deutschen KM der Akademie E. E. Kummer besitzen wir die erste kurze, aber gründliche schriftliche Beurteilung seiner wissenschaftlichen Verdienste. Der Vorschlag, datiert vom (29) 17.10.1862, wurde höchst wahrscheinlich von Buniakowski geschrieben; er ist mitunterzeichnet von Tschebyschew und dem Akademiemitglied O. I.

Somow (1815–1876), der sowohl als Mathematiker, auch als Mechaniker tätig war. Dort lesen wir (der Bericht ist französisch geschrieben):

A la Classe Physico-Mathématique

Présentation

Les soussignés ont l'honneur de présenter pour une des places vacantes de Membre Correspondant dans la Section Mathématique Mr E. Kummer, Professeur à l'Université de Berlin et Membre de l'Académie des Sciences de cette ville. Le nom de Mr Kummer est trop connu dans le monde savant pour qu'il soit necéssaire d'exposer en détail ses titres scientifiques: son mémoire sur le dernier théorême de Fermat, couronné par l'Académie de Paris, et ses belles recherches sur les nombres complexes suffisent seuls pour le placer au rang des géomètres les plus célèbres de l'époque. (LO AAN, f.2; op.2, N°6, 1.157)[5]

Die einflussreichsten Petersburger Mathematiker haben also Kummers Theorie der idealen komplexen Zahlen sehr hoch geschätzt. Der Einfluss Kummers über die weitere Entwicklung der ganzen Theorie der algebraischen Zahlen ist wohlbekannt. In Russland wurde diese Theorie vertieft und verallgemeinert von dem Akademiker E. I. Zolotarew (1847–1878), einem Schüler Tschebyschews, der während seines Aufenthaltes im Berlin im Jahre 1872 sich mit Kummer, Weierstrass und wahrscheinlich auch mit Kronecker persönlich bekannt machte. Eine ausführliche vergleichende Analyse der entsprechenden Theorien Kummers, Zolotarews, Dedekinds und auch Kroneckers hat Frau I. Baschmakowa gegeben [8, S.93–122].

Sehr lobend war auch der Wahlvorschlag für Weierstrass, datiert vom (12.12) 30.11.1864 und unterzeichnet von O. Somov, V. Buniakowski und dem Akademiemitglied A. N. Sawitsch (1811–1883), einem Astronomen (LO AAN, f.2, op.17, N°6, 1.179–179v). Weierstrass ist dabei als "einer der erstrangigen Mathematiker" charakterisiert, dessen "wichtige Vervollkommnungen in der Theorie der elliptischen und abelschen Funktionen mit den grossen Entdeckungen Jakobis und Abels gleichgestellt sein können". Besonders hervorgeh oben ist, dass Weierstrass die "Umkehrfunktionen" in der Form von Brüchen ausdrückte, deren Zähler und Nenner für alle reellen oder komplexen Werte des Arguments konvergente unendliche Potenzreihen sind. Dabei ist die folgende Äusserung von Ch. Hermite (1822–1891) zitiert: "Abel avait entrevue et rapidement indiqué la possibilité de ce nouveau mode d'expression des fonctions elliptiques, mais c'est à M. Weierstrass que revient l'honneur d'avoir mis dans la science, au lieu d'un simple aperçu, une théorie profonde qui conduit directement à ces

[5] LO AAN = Leningrader Abteilung des Archivs der Akademie der Wissenschaften der UdSSR; f = Fonds; op. = Inventar; l. = Blatt.

nouvelles fonctions, non seulement dans le cas des transcendantes elliptiques, mais pour les transcendantes abéliennes à un nombre quelconque de variables". Auch einige andere Forschungen von Weierstrass sind erwähnt, wie z.B. sein Artikel über die analytischen Fakultäten (1843). Man muss jedoch zugeben, dass im allgemeinen die Entdeckungen des eminenten deutschen Mathematiker hier nur flüchtig beschrieben sind.

Der angeführte Vorschlag ist nicht von Tschebyschew unterzeichnet, und das war, so glauben wir, nicht zufällig. Die Forschungsrichtungen und Auffassungen Tschebyschews und Weierstrass waren im grossen und ganzen ziemlich weit voneinander entfernt. So war es mindest in sechziger Jahren. Später scheint die Stellung Tschebyschews geändert. So schrieb die damals in Berlin weilende Weierstrass' Schülerin S. W. Kowalewskaja (1850–1891) am 21.11.1881 an G. M. Mittag-Leffler (1846–1927) folgendes:

> En passant par Pétersbourg, J'ai eu l'occasion de voir M. Tchebycheff et de m'entretenir longuement avec lui. A mon grand étonnement, et à ma grande satisfaction aussi, je l'ai trouvé changé sous beaucoup de rapports. Il parle avec respect de l'école de Berlin et pour Vous personnellement, cher Monsieur, il exprime une très grande admiration; il m'a même confié, qu'il va tâcher de Vous proposer pour la vacance à l'académie de St. Pétersbourg, mais qu'il craint de rencontrer beaucoup d'obstacles sur son chemin.[6] Je n'ai pas besoin de Vous dire que ceci aussi **est confidentiel.** Il y a ici, entre les Zuhörer de M. Weierstrass, un jeune russe qui a été aussi envoyé spécialement par M. Tchebycheff.[7] N'est ce pas que c'est là du changement![8]

Weierstrass wurde später noch zum EM der Akademie erwählt; dies geschah jedoch ein Jahr nach dem Tode Tschebyschews; dann wurden OM der Akademie A. A. Markow der ältere (1856–1922) und N. Ja. Sonin (1849–1915).

Wir besitzen auch den brieflichen Wahvorschlag für Kummers Schüler Kronecker; dieser Vorschlag betrifft auch Sylvester, und beide Mathematiker waren gleichzeitig zur KM erwählt. Der Vorschlag, am (6.11) 25.10.1872 eingereicht, war von Tschebyschew eigenhändig geschrieben und von den Akademikern D. M. Perewoschtschikow (Astronom, 1790–1880), V. Bouniakowski und Somow mitunterzeichnet. Wir zitieren den russischen

[6] Zum KM der Petersburger Akademie war Mittag-Leffler am (19) 7.12.1896 erwählt, also zwei Jahre nach dem Tod Tschebyschews; zum EM am 6.12.1926.

[7] Das war, wahrscheinlich, D. F. Seliwanow (1855–1932), später Professor an der Petersburger Universität.

[8] Das Original des Briefes gehört dem Institut Mittag-Leffler (Mittag-Lefflers Brefsamling) in Djursholm, Schweden. Für die Erlaubnis, seine Kopie zu benutzen, danke ich herzlich Frau Akademiker P. Ja. Kotschina. Die Photokopien dieser Briefsammlung sind im Archiv der Akademie der Wissenschften der UdSSR aufbewahrt (AAN, f.603, op. 1, N°15).

Text in deutscher Übersetzung (LO AAN, f.2, op.17, N°6, 1.280–281):

> Unter unseren ausländischen Geometern, die noch nicht korrespondierende Mitglieder unserer Akademie sind, verdienen besondere Rücksicht dank ihrer vortrefflichen Arbeiten das Mitglied der Berliner Akademie Kronecker und das jenige der Londoner Königlichen Gesellschaft Sylvester. Beide haben schon lange durch ihre Entdeckungen in verschiedenen Gebieten der reinen Analysis einen grossen Ruhm erworben und die mathematische Literatur durch zahlreiche klassische Werke bereichert. Wir finden es überflüssig alle Abhandlungen und Notizen der H. H. Kronecker und Sylvester aufzählen, die während mehr als einem Vierteljahrhundert in den Memoiren der Berliner Akademie der Wissenschaften, der Londoner Königlichen Gesellschaft und verschiedenen mathematischen Zeitschriften gedruckt wurden; wir begnügen uns mit dem Hinweis auf die Abhandlungen Kroneckers, betitelt: I) **De unitatibus complexis** (1843); 2) **Zwei Sätze über Gleichungen mit ganzzahligen Coefficienten** (1851); 3) **Über complexe Einheiten**; 4) **Sur la résolution de l'équation du 5e degré** sowie auf die Abhandlungen und Notizen Sylvesters: **über die Funktionen von Sturm**; **über die Elimination der Unbekannten**; **über die Invarianten**; **über die Lösung der unbestimmten Gleichungen**; **über lineare Approximationen**. Diese Arbeiten allein genügen, um die Namen ihrer Autoren zu verewigen ....[9]

Diese knappe Beurteilung der Forschungen Kroneckers und Sylvesters-sowie der Forschungen Kummers-ist besonders interessant: sie zeigt, dass der Kreis der mathematischen Interessen Tschebyschews viel breiter war als man gewöhnlich meint.

Für Burchardt, Fuchs, Klein, Schwarz und M. Cantor haben wir keine briefliche Wahlvorschläge gefunden. Bemerkenswert ist doch die Wahl M. Cantors, des hervorragenden Mathematikhistorikers. Zu dieser Zeit gab es, wie gesagt, in der Petersburger Akademie nur zwei OM-Mathematiker, nämlich A. A. Markow derältere und N. Ja. Sonin, und beide interessierten sich einigermassen für die Geschichte der Mathematik.

Nur wenige erstklassige deutsche Mathematiker des 19. Jahrhunderts wurden nicht zu EM oder KM der Petersburger Akademie erwählt. Es waren zunächst Geometer wie A. F. Möbius u. a.: die Geometrie, als solche, hatte in der Akademie im 19. Jahrhundert keinen Vertreter. Es waren auch B. Riemann, R. Dedekind und G. Cantor: die Forschungsrichtungen dieser grossen Männer waren in der alten Petersburger Schule nicht hoch geschätzt. Die Lage änderte sich mit dem Auftreten der jüngeren Generation. Die entscheidende Rolle spielte dabei der einflussreiche Akademiker W. A. Steklow (1863–1926), der in den Jahren 1919–1926 auch Vize-Präsident der Akademie wurde. Als Schüler des Akademikers A. M. Liapunow (1857–1918) befasste sich W. A. Steklow mit Problemen der mathematischen Physik und den daran anküpfenden analytischen Fragen. Er war kein Verehrer der neuesten Richtungen, die im ersten Viertel des 20. Jahrhunderts, besonders in Deutschland, Frankreich und Italien

[9] Hier und weiter haben wir keinen Platz für genauere bibliographische Angaben.

gepflegt wurden. Die Wichtigkeit der Ergebnisse der Göttinger, Pariser u.a. Schulen wusste er jedoch richtig einzuschätzen und dabei wurde er von seinen eigenen jungeren Schülern und Mitarbeitern unterstützt.

Zwischen dem ersten Weltkrieg und dem Jahre 1934 wurden vier deutsche Mathematiker zur KM oder EM der Akademie der Wissenschaften der UdSSR erwählt:

27. D. Hilbert (1862–1943); KM 2.12.1922, EM 12.2.1934.
28. A. Kneser (1864–1930), KM 6.12.1924.
29. E. G. H. Landau (1877–1938); KM 6.12.1924, EM 29.3.1932.
30. I. Schur (1875–1941); KM 31.1.1929.

Die Wahlvorschläge für Hilbert, Kneser und Landau sind von Steklow selbst geschrieben. Das briefliche Urteil über Schur's Arbeiten ist mir unbekannt geblieben.

"David Hilbert," schrieb Steklow (LO AAN, f.162, op.3, N°27, 1.2–4), "geboren im Jahre 1862, ist Professor an der Universität Göttingen und gehört zu den erstklassigen Geometern der Welt. Seine Forschungen in den verschiedenartigesten Gebieten der mathematischen Wissenschaften sind allgemein anerkannt und beeinflussten deren Entwicklung wesentlich". Weiter folgt eine Analyse der verschiendenen Richtungen und zugleich Etappen der wissenschaftlichen Tätigkeit Hilberts, beginnend mit seinen Arbeiten über Invariantentheorie, deren wichtigste "Über die Theorie der algebraischen Formen" im Jahre 1890 erschien. Zu den Entdeckungen Hilberts in der Theorie der algebraischen Zahlen und zum wohlbekannten "Zahlbericht" (1894) übergehend, sagt Steklow, Hilbert gebe in dieser Schrift "eine systematische Darlegung seiner eigenen sowie der Forschungen seiner Vorgänger. Dieses Werk, das viele neuen Ideen und wichtigste Ergebnisse enthält, beeinflusste die weitere Entwicklung dieser Theorie stark und dient heute als Handbuch für jeden, der in den entsprechenden Wissenschaftsbereichen arbeite".

Nach der Erwähnung noch einiger späterer Ergebnisse Hilberts in der Klassenkörpertheorie fasst Steklow diesen Abschnitt seines Gutachtens folgendermassen zusammen: "Nach der Wichtigkeit der Methode und der Ergebnisse kann man diese Forschungen Hilberts mit den klassischen Werken Kummers, Dedekinds und Kroneckers, des Begründers dieser Theorie, mit Recht gleichstellen". Auch der Hilbertschen geistreichen Lösung des Waringschen Problem (1909), die zu einem anderen Zweig der Zahlentheorie gehört, sind einige Zeilen gewidmet.

Dann werden die geometrischen Arbeiten Hilberts betrachtet und insbesondere seine "Grundlagen der Geometrie" (1899); dabei ist auch die berühmte Probevorlesung Riemanns "Uber die Hypothesen, welche der Geometrie zugrunde liegen" genannt. Auch der Hilbertsche Beweis, dass es eine singularitätenfreie Realisierung der ganzen Lobatschewskischen Ebene durch eine Fläche im dreidimensionalen Raum nicht geben könne, wurde nicht vergessen.[10]

Sehr eingehend sind die Hilbertschen "Grundzüge einer allgemeinen Theorie der linearen Integralgleichungen" (1904) analysiert. Zuerst ist die Vorgeschichte der Theorie kurz skizziert und die Verdienste von E. I. Fredholm (1866–1927), der zum ersten Male (1903) eine allgemeine Theorie der abelschen funktionellen Gleichungen (wie er es bezeichnete) schuf, besonders hervorgeholen. Dann fährt Steklow fort: "Obgleich die Frage somit nicht neu sei, sind die Forschungen Hilberts in dieser Richtung von sehr grosser Bedeutung, da er es von dem allgemeinsten Standpunkt ausarbeitete . . . Seine Theorie fundierte er auf ganz neuen Prinzipien, und als Grundstein seiner Forschungen legte er die Theorie der Funktionen von unendlich vielen Veränderlichen, deren Anfänge er in einer der obenerwähnten Abhandlungen entwickelte. Dann zeigte er die Grundzüge seiner Anwendung auf die gewöhnlichen Differentialgleichungen, partielle Differentialgleichungen, allgemeine Theorie der Funktionen einer komplexen Grösse, Geometrie, Hydrodynamik und an die Variationsrechnung, solcherweise mit einer Methode, die verschiedenartigsten Probleme der reinen und auch der angewandten Mathematik umfassend". Dazu werden noch einige Mathematiker erwähnt, die die Hilbert–Fredholmsche Theorie weiterentwickelten, wie A. Kneser, E. Schmidt (1876–1959), H. Weyl (1885–1955), M. Fréchet (1878–1973) u.a. Endlich fügt Steklow hinzu, dass diese Theorie auch auf die nicht linearen Integralgleichungen ausgedehnt geworden ist und eine Verallgemeinerung in der Theorie der Integrodifferentialgleichungen von Vito Volterra (1860–1940) findet.

Die vorwiegende Aufmerksamkeit Steklows auf die Theorie der Integralgleichungen war ganz natürlich: das Problem der Reihenentwicklungen der willkürlichen Funktionen nach irgendwelchen orthogonalen Funktionensystemen stand im Zentrum seiner langjährigen Forschungen und wurde von ihm mit Hilfe seiner eigenen Methoden gelöst. So sagte Akademiker W. I. Smirnow (1887–1974), ein Schüler Steklows, von seinem Lehrer:

> Ihm gehört die Schaffung der Abgeschlossenheitstheorie, eine ganze Reihe der allgemeinen Sätze über die Funktionsentwicklungen und eine ausführliche Untersuchung der Entwicklungen nach den speziellen orthonormierten Systemen. Die Beweismethoden dieser Sätze sind sehr vershiedenartig, originell und manchmal

[10] In den geschildenten Abschnitten seines Gutachtens hat Steklow die von dem damaligen Akademiemitglied Ja. W. Uspenski (1883–1947) hergestellte Notiz benutzt. Diese Notiz ist in demselben Dossier aufbewahrt wie der Wahlvorschlag von Steklow. Der Wahlvorschlag war offiziell von Steklow, Uspenski und Akademiker A. F. Ioffe (1880–1960), dem Physiker, unterzeichnet.

unerwartet, doch einfach, besonders in seinen späteren Arbeiten. In manchen Arbeiten findet man die Keime der zukünftigen funktionalanalytischen Verfahren. Bis zum Erscheinen von Knesers Arbeit, verbunden mit der Asymptotik der Sturm-Liouville'schen Funktionen (1903) und auch Hilberts Arbeiten über die Integralgleichungen (1904), waren die Arbeiten . . . [Steklows] in der ganzen mathematischen Literatur über die Entwicklungsprobleme die hauptsächlichen, [9, S.11–12].

Zum Schluss werden die Hilbertschen Forschungen über die Variationsrechnung zusammenfasst und die verschiedene Akademien und mathematische Gesellschaften, denen er schon angehörte, aufgezählt.

Sehr knapp doch sehr inhaltsreich war auch der Wahlvorschlag für Kneser, einen Schüler Kroneckers und Weierstrass', Professor an der Universität Dorpat (Tartu) von 1889–1900 und an der Universität Breslau (Wroclaw) ab 1905. Den Vorschlag unterzeichneten noch zwei Akademiker: A. A. Belopolski (1854–1934), ein Astronom, und P. P. Lazarew (1878–1942), ein Physiker. Das Dokument ist undatiert, es muss aber Ende November oder Anfang Dezember 1942 geschrieben sein, kurz vor dem Tag der Wahlen (LO AAN, f.162, op.3, N°36). "Die zahlreichen Forschungen Knesers," so schrieb Steklow, "gehören zu verschiedenen Bereichen der reinen Mathematik und der Mechanik: Theorie der Bewegungsstabilität, Reihentheorie, Funktionentheorie, Differential-und Integralgleichungen und Variationsrechnung"; dann sind die Ergebnisse Knesers in jedem der genannten Richtungen betrachtet.

Zuerst sind die mechanischen Arbeiten Knesers analysiert, darunter die Abhandlung "Bewegungsvorgänge in der Umgebung instabiler Gleichgewichtslagen" (Crelle's Journ., 118), wo Kneser die Ergebnisse seiner Vorgänger, wie z.B. S. Lie (1842–1899), O. Staude (1857–1928), Liapunow u.a. verallgemeinerte. Nebenbei sei bemerkt (in dem Vorschlag ist davon nicht, Rede), dass mit diesen Arbeiten von Knesser auch tiefe Forschungen des lettischen Mathematikers P. Bohl (1865-1921) im Zusammenhang standen und dass Kneser als einer der offiziellen Rezensenten der Doktordissertation von Bohl (Jurjew-Tartu, 1900) eingeladen wurde. Weiter lesen wir: "Eine grosse Zahl seiner [Knesers] Forschungen betreffen die Differentialgleichungen der Erkaltung des inhomogenen festen Stabes, d.h. das Sturm-Liouville'sche Problem, als auch mit diesem unmittelbar verbundenen Problem der Entwicklung willkürlicher Funktionen nach Sturm-Liouville'schen Funktionen. Dank seinen eigenen, wie auch den Forschungen anderer Mathematiker, kann man jetzt dieses Problem im allgemeinen als gelöst betrachten". Es geht hier um eine wichtige Abhandlung Knesers von 1903, die auch in die oben zitierten Worten Smirnows erwähnt wurde, -eine Abhandlung, die Steklow zu eigenen originellen Forschungen veranlasste (veröff.1907). Dasselbe gilt auch für die Hilbertschen Arbeiten über die Integralgleichungen. Eine kompetente Zusammenstellung der entsprechenden Ergebnisse Hilberts, Knesers und Steklows findet man in einem Artikel eines Schülers des letzteren, des korrespondierendes Mitglieds der Akademie der Wissenschaften der UdSSR N. M. Günter (1871–1951), der im Jahre 1927 veröffentlicht und später neugedruckt war [10, S.35–36].

Auch die Forschungen Knesers im Bereich der Integralgleichungen und sein bekanntes Lehrbuch (1911), sowie sein umfangreiches Lehrbuch der Variationsrechnung (1900), sind erwähnt. "Besonders bekannt sind," betonte Steklow, "Knesers Forschungen über die Variationsrechnung, wo er die klassische Richtung, die diesem Gebiet der Mathematik Eulers, Lagranges, Jacobis und Mayers[11] gaben, fortsetzte".

Der Vorschlag zur Wahl Landaus, verfasst um dieselbe Zeit wie der vorangehende (Landau und Kneser waren zur KM am selben Tag ernannt) ist auch von Belopolski und Lazarew mitunterzeichnet. Schon die ersten Zeilen bezeugen die Breite der Interessen und Kenntnisse Steklows. "Die Fragen, die mit der Theorie der Primzahlenverteilung verbunden sind, gehören zu den sehr wichtigen und schwierigsten der Zahlentheorie. Die Forschungen Tschebyschews, so kann man sagen, schliessen die arithmetische Periode dieser Lehre ab, und es ist kaum zu erwarten, dass irgendwelche neue wichtige Ergebnisse mit Hilfe dieser Verfahren erreichbar seien. Neue Wege zu derartigen Forschungen waren durch das berühmte Memoire Riemanns "Uber die Anzahl der Primzahlen unter einer gegebenen Grösse" eröffnet, obgleich es einige bisher unbewiesene Resultate enthält".[12] Dann finden wir eine Aufzählung der wichtigsten Ergebnisse Landaus (Theorie der ζ (s), Dirichlet'sche Reihen, Theorie der Zahlenfunktion μ (n), ideale Zahlen, Beweis einiger Sätze Tschebyschews, usw.). Weiter wird Landaus zweibändiges "Handbuch der Lehre von der Verteilung der Primzahlen" 1909 hochgepriesen und am Ende sagt Steklow folgendes: "E. Landaus ist jetzt Professor der Universität Göttingen und ein würdiger Nachfolger Minkowski's". Acht Jahre später, dem Vorschlag der Akademiker I. M. Winogradow und A. N. Krylow (1863–1945) folgend (LO AAN, f.1, op.1–1932, N°260, 1.1), wurde Landau zum EM ernannt[13].

Abschliessend möchte ich sagen, dass die oben angeführten Angaben—und die briefliche Wahlvorschläge ganz besonders—ein neues Licht nicht nur auf die wissenschaftlichen Kontakte zwischen deutschen und russischen Mathematikern, sondern auch auf die Geschichte der Mathematik beider Länder im allgemeinen werfen.

[11] Gemeint ist der leipziger Mathematiker Adolf Mayer (1839–1908).

[12] Als Steklow diese Meinung geäussert hatte, konnte man nicht vermuten, das später einige ganz neue wirksame "arithmetische" Methoden in der Primzahlenlehre eingeführt werden sollten. (P. Erdös und A. Selberg, 1949, u.a.).—Vermutlich hat bei der Ernennung Landaus auch Ja. W. Uspenski mitgewirkt.

[13] Die Wahlvorschläge für Hilbert, Landau und Kneser sind in [11] gedruckt.

LITERATURVERZEICHNIS

1. Академия наук СССР. Персональный состав. Кн.1-2.''Наука''. Москва,1974.
2. Хр.Вольф. Сокращения первых оснований математики. Т.1-2,С.-Петербург, 1770–1771.
3. M. Cantor. Vorlesungen über Geschichte der Mathematik. Bd.3,2 Aufl., Leipzig, 1900–1901; Bd.4, Leipzig, 1908.
4. А.Г.Кестнер. Начальные основания математики. Ч.1-2. С.-Петербург, 1792–1794.
5. А.Д.Барсов. Новая алгебра, содержащая в себе не только простую аналитику, но также дифференциальное, интегральное и вариационное исчисление. Москва,1797.
6. Архив истории науки и техники, III. Издат. Академии наук СССР. Ленинград, 1934.
7. K.-R.Biermann. Die Mathematik und ihre Dozenten in der Berliner Universität. 1810–1920. Akademie-Verlag. Berlin,1973.
8. Математика X1X века. Математическая логика, Алгебра. Теория чисел. Теория вероятностей. Под редакцией А.Н. Колмогорова и А.П. Юшкевича. ''Наука''. Москва,1978.
9. В.И. Смирнов. Памяти Владимира Андреевича Стеклова.—Труды Математического института АН СССР, 73,1964, с.11–12.
10. Н.М.Гюнтер.Труды В.А.Стеклова по математической физике.—Успехи математических наук,1/3/, 1946, с.35–36.
11. Известия Российской Академии наук, VI сер., XVI, 1922, с. 29–32; XVIII, 1924, с.451–453.

Institute of the History of Science and Technology
Moscow, USSR

Bibliographie

Veröffentlichungen von Prof. Dr.rer.nat.habil.
Kurt-R. Biermann, Berlin (DDR)

Die folgende chronologische Übersicht stellt insofern einen *Auszug* aus dem vollständigen Schriftenverzeichnis dar, als hier Rezensionen und Referate, Tagungsberichte und Übersetzungen sowie die Mehrzahl der populärwissenschaftlichen Artikel aus Platzgründen unberücksichtigt bleiben. Das Verzeichnis gibt den Stand vom *31.12.1979* wieder. Ein *Index* ist angefügt.

I. *Monographien*

1959 1. "Johann Peter Gustav Lejeune Dirichlet. Dokumente für sein Leben und Wirken," *Abhandlungen der Deutschen Akademie der Wissenschaften zu Berlin. Klasse für Mathematik, Physik und Technik* (Berlin: Akademie-Verlag, Nr. *2*, 87 S., 5 Abb. 4°).

1960 2. *Deutsche Akademie der Wissenschaften zu Berlin. Biographischer Index der Mitglieder* (Gemeinsam mit G. Dunken), (Berlin: Akademie-Verlag, XII und 148 S., 10 Tafeln. 8°).

3. "Vorschläge zur Wahl von Mathematikern in die Berliner Akademie," *Abhandlungen der Deutschen Akademie der Wissenschaften zu Berlin. Klasse für Mathematik, Physik und Technik* (Berlin: Akademie-Verlag, Nr. *3*, 75 S. 4°).

1968 4. *Alexander von Humboldt. Chronologische Übersicht über wichtige Daten seines Lebens* (Gemeinsam mit I. Jahn und F. G. Lange), (Berlin: Akademie-Verlag, XVII und 86 S. 8°; *Beiträge zur Alexander-von-Humboldt-Forschung*, Bd. *1*).

1973 5. *Die Mathematik und ihre Dozenten an der Berliner Universität 1810–1920. Stationen auf dem Wege eines mathematischen Zentrums von Weltgeltung* (Berlin: Akademie-Verlag, X und 265 S., 14 Tafeln. 4°).

1977 6. *Briefwechsel zwischen Alexander von Humboldt und Carl Friedrich Gauß* (Berlin: Akademie-Verlag, 202 S., 4 Tafel. 8°; *Beiträge zur Alexander-von-Humboldt-Forschung*, Bd. *4*).

1979 7. *Briefwechsel zwischen Alexander von Humboldt und Heinrich Christian Schumacher* (Berlin: Akademie-Verlag, 192 S., 4 Tafeln. 8°; *Beiträge zur Alexander-von-Humboldt-Forschung*, Bd. *6*).

1980 8. Alexander von Humboldt. Leipzig: BSB B. G. Teubner Verlagsgesellschaft 1980. 128 S., 12 Abb. 8° (Biographien hervorragender Naturwissenschaftler, Techniker und Mediziner, Bd. 47).

II. *Abhandlungen in Zeitschriften, Schriftenreihen und Sammelbänden*

1954 1. "Über die Untersuchung einer speziellen Frage der Kombinatorik durch G. W. Leibniz," *Forschungen und Fortschritte*, *28*, 357–361.

1955 2. "Über eine Studie von G. W. Leibniz zu Fragen der Wahrscheinlichkeitsrechnung," *Forschungen und Fortschritte*, *29*, 110–113.

3. "Eine Untersuchung von G. W. Leibniz über die jährliche Sterblichkeitsrate," *Forschungen und Fortschritte*, *29*, 205–208.

1956 4. "Aus der Geschichte der Wahrscheinlichkeitsrechnung," *Wissenschaftliche Annalen*, *5*, 542–548. Tschechische Übersetzung in: *Pokroky Matematiky, Fysiky a Astronomie*, *2* (1957), 31–35.

5. "Wissenschaftsgeschichtliche Notizen," *Wissenschaftliche Annalen*, *5*, Beiheft, 159–164.

6. "Spezielle Untersuchungen zur Kombinatorik durch G. W. Leibniz (2. Mitteilung)," *Forschungen und Fortschritte*, *30*, 169–172.

1957 7. "Eine Aufgabe aus den Anfängen der Wahrscheinlichkeitsrechnung," *Centaurus*, *5*, 142–150.

8. "Zadači genuéskogo loto v rabotach klassikov teorii verojatnostej," *Istoriko-matematičeskie Issledovanija*, *10*, 649–670.

9. "G. W. Leibniz' De incerti aestimatione," (Gemeinsam mit M. Faak), *Forschungen und Fortschritte*, *31*, 45–50.

10. "Leonhard Euler," *Wissen und Leben*, *2*, 282–284 u. 288. Tschechische Übersetzung in: *Věda a život* (1957). H. *4*, 200–201.

1958 11. "Überprüfung einer frühen Anwendung der Kombinatorik in der Logik," (Gemeinsam mit J. Mau), *The Journal of Symbolic Logic*, *23*, 129–132.

12. "Iteratorik bei Leonhard Euler," *L'Enseignement Mathématique*, *4*, 19–24.

13. "Alexander von Humboldt als Protektor Gotthold Eisensteins und dessen Wahl in die Berliner Akademie der Wissenschaften," *Forschungen und Fortschritte*, *32*, 78–81.

14. "Vom Glücksspiel zur Wahrscheinlichkeitstheorie," *Technische Rundschau*, *50*, Nr. *27*, 15.

15. "Eine Notiz N. H. Abels für A. L. Crelle auf einem Manuskript Otto Auberts," (Gemeinsam mit V. Brun), *Nordisk Matematisk Tidskrift*, *6*, 84–86.

16. "Zur Geschichte der Ehrenpromotion Gotthold Eisensteins," *Forschungen und Fortschritte*, *32*, 332–335.

17. "Aus den Anfängen des Determinantengebrauchs," *Technische Rundschau*, *50*, Nr. *46*, 15.

18. "Zum Verhältnis zwischen Alexander von Humboldt und Carl Friedrich Gauß," *Wissenschaftliche Zeitschrift der Humboldt-Universität Berlin*, *math.-nat. Reihe*, *8* (1958/59), 121–130.

1959 19. "Einige Euleriana aus dem Archiv der Deutschen Akademie der Wissenschaften zu Berlin," in *Leonhard-Euler-Sammelband* (Berlin: Akademie-Verlag), 21–34.

20. "F. G. M. Eisenstein. Bibliografija ego sočinenij," *Istoriko-matematičeskie Issledovanija*, *12*, 493–502.

21. "Zur Anwendung einer mathematischen Sprachcharakteristik auf Texte Alexander von Humboldts," *Sudhoffs Archiv*, *43*, 183–185.

22. "Über die Förderung deutscher Mathematiker durch Alexander von Humboldt," *Alexander-von-Humboldt-Gedenkschrift* (Berlin: Akademie-Verlag), 83–159.

23. "A. L. Crelles Verhältnis zu Gotthold Eisenstein," *Monatsberichte der Deutschen Akademie der Wissenschaften zu Berlin*, *1*, 67–72.

24. "Ein bisher unveröffentlichter wissenschaftlicher Brief von Carl Friedrich Gauß an Alexander von Humboldt," (Gemeinsam mit H.-G. Körber), *Forschungen und Fortschritte*, *33*, 136–140.

25. "G. W. Leibniz und die Berechnung der Sterbewahrscheinlichkeit bei J. de Witt," (Gemeinsam mit M. Faak), *Forschungen und Fortschritte*, *33*, 168–173.

26. "P. G. Lejeune Dirichlet. 1859–1959," *Monatsberichte der Deutschen Akademie der Wissenschaften zu Berlin*, *1*, 320–323.

27. "Gotthold Eisenstein," *Neue Deutsche Biographie*, *4*, 420–421.

28. "Aus der Geschichte der Dyadik," *Technische Rundschau*, *51*, Nr. *49*, 27 u. 29.

1960 29. "Ein Gemeinschaftsunternehmen der Berliner und der Petersburger Akademie," *Monatsberichte der Deutschen Akademie der Wissenschaften zu Berlin*, *2*, 125–129.

30. "Aus der Geschichte der Iteratorik," *Technische Rundschau*, *52*, Nr. *41*, 15.

31. "Aus den Anfängen der wissenschaftlichen Laufbahn Franz Neumanns, des Begründers der mathematischen Physik in Deutschland," *Forschungen und Fortschritte*, *34*, 97–101.

32. "F. Woepckes Beziehungen zur Berliner Akademie," *Monatsberichte der Deutschen Akademie der Wissenschaften zu Berlin*, *2*, 240–249.

33. "Die Begründung der spektralanalytischen Methode im Urteil der Berliner Akademie," *Monatsberichte der Deutschen Akademie der Wissenschaften zu Berlin*, *2*, 315–318.

34. "Urteile A. L. Crelles über seine Autoren," *Journal für die reine und angewandte Mathematik*, *203*, 315–318.

35. "Dirichletiana," *Monatsberichte der Deutschen Akademie der Wissenschaften zu Berlin*, *2*, 386–389.

36. "Aus der Geschichte der Zahl π," *Technische Rundschau*, *52*, Nr. *52*, 13.

1961 37. "Einige neue Ergebnisse der Eisenstein-Forschung," *Schriftenreihe für Geschichte der Naturwissenschaften, Technik und Medizin NTM*, *1*, H. *2*, 1–12.

38. "Leonhard Euler und die Technik," *Technische Rundschau*, *53*, Nr. *10*, 3.

39. "Der Mathematiker Ferdinand Minding und die Berliner Akademie," *Monatsberichte der Deutschen Akademie der Wissenschaften zu Berlin*, *3*, 128–133.

40. "Eine unveröffentlichte Jugendarbeit C. G. J. Jacobis über wiederholte Funktionen," *Journal für die reine und angewandte Mathematik*, *207*, 96–112.

41. "Zur Geschichte der Zerfällung natürlicher Zahlen in Summanden," *Forschungen und Fortschritte*, *35*, 71–74.

42. "Figurierte Zahlen in der "Arithmetischen Schatzkammer" des Lorenz Biermann," *Forschungen und Fortschritte*, *35*, 195–198.

43. "Ein Hilfsmittel für Maya-Kalenderrechnungen," *Monatsberichte der Deutschen Akademie der Wissenschaften zu Berlin*, *3*, 456–462.

44. "Neopublikovannoe piśmo N. I. Lobačevskogo Berlinskoj Akademii nauk," *Istoriko-matematičeskie Issledovanija*, *14*, 623–625.
45. "Die Anwendung der Gaußschen Theorien," *Technische Rundschau*, *53*, Nr. *45*, 29 u. 31.
46. "Nikolaus Fuss," *Neue Deutsche Biographie*, *5*, 742–743.

1962

47. "Zum wissenschaftlichen Briefwechsel zwischen Carl Friedrich Gauß und Alexander von Humboldt," (Gemeinsam mit H.-G. Körber), *Forschungen und Fortschritte*, *36*, 41–44.
48. "Die Alexander-von-Humboldt-Briefausgabe," (Gemeinsam mit F. G. Lange), *Forschungen und Fortschritte*, *36*, 225–230.
49. "A. v. Humboldts 'Kosmos'—Vorhaben in Briefen an Bessel," *Monatsberichte der Deutschen Akademie der Wissenschaften zu Berlin*, *4*, 318–324.
50. "My goal is to reach the sources of the Ganges. A. v. Humboldt's plans for an exploratory expedition," *Picture News* (Nov.), 26–27.

1963

51. "Aus der Vorgeschichte der Aufforderung A. v. Humboldts von 1836 an den Präsidenten der Royal Society zur Errichtung geomagnetischer Stationen. Dokumente zu den Beziehungen zwischen A. v. Humboldt und C. F. Gauß," *Wissenschaftliche Zeitschrift der Humboldt-Universität Berlin, math.-nat. Reihe*, *12*, 209–227.
52. "Vozmožnye metody grečeskoj kombinatoriki," *Voprosy Istorii Estestvoznanija i Techniki*, *15*, 103–105.
53. "N. H. Abel und Alexander von Humboldt," *Nordisk Matematisk Tidskrift*, *11*, 59–63.
54. "Ein Briefwechsel zwischen C. F. Gauß und der Berliner Akademie," *Monatsberichte der Deutschen Akademie der Wissenschaften zu Berlin*, *5*, 43–46.
55. "Zwei ungeklärte Schlüsselworte von C. F. Gauß," *Monatsberichte der Deutschen Akademie der Wissenschaften zu Berlin*, *5*, 241–244.
56. "Jakob Steiner, der Geometer," *Forschungen und Fortschritte*, *37*, 125–126.
57. "Jakob Steiner. Eine biographische Skizze," *Nova Acta Leopoldina. Neue Folge*, *27*, Nr. *167*, 31–45.
58. "Einige Ergänzungen zur Biographie Jakob Steiners," *Archives Internationales d'Histoire des Sciences*, *16*, 167–171.
59. "Lagrange im Urteil und in der Erinnerung A. v. Humboldts," *Monatsberichte der Deutschen Akademie der Wissenschaften zu Berlin*, *5*, 445–450.
60. "Neue Briefe Alexander von Humboldts," *Spektrum*, *9*, 404–405. Spanische Übersetzung: *Revista* (1964), H. *4*, 27.
61. "Der Versuch einer Leonhard-Euler-Ausgabe von 1903/07 und ihre Beurteilung durch Max Planck," *Forschungen und Fortschritte*, *37*, 236–239.

1964

62. "Christian Goldbach," *Neue Deutsche Biographie*, *6*, 602.
63. "Aus der Geschichte Berliner mathematischer Preisaufgaben," *Wissenschaftliche Zeitschrift der Humboldt-Universität Berlin, math.-nat. Reihe*, *13*, 185–198.
64. "Aus dem mathematischen Berlin des vorigen Jahrhunderts," *Schriftenreihe für Geschichte der Naturwissenschaften, Technik und Medizin NTM* (Beiheft), 11–20.
65. "David Hilbert und die Berliner Akademie," *Mathematische Nachrichten*, *27*, 377–384.
66. "Gotthold Eisenstein. Die wichtigsten Daten seines Lebens und Wirkens," *Journal für die reine und angewandte Mathematik*, *214/215*, 19–30. Nachdruck in: G. Eisenstein: *Mathematische Werke*, Bd. *2* (New York: Chelsea, 1975), 919–929.

67. "Einige Episoden aus den russischen Sprachstudien des Mathematikers C. F. Gauß," *Forschungen und Fortschritte*, *38*, 44–46.

68. "Thomas Clausen, Mathematiker und Astronom," *Journal für die reine und angewandte Mathematik*, *216*, 159–198.

69. "Alejandro de Humboldt y Simón Bolívar, el Libertador," *Revista*, H. *8*, 26–27.

70. "Aus der Geschichte eines Anordnungsproblems," *Monatsberichte der Deutschen Akademie der Wissenschaften zu Berlin*, *6*, 198–203.

71. "Ein Urteil Alexander von Humboldts über Peter Simon Pallas und die Zuwahlen zur Berliner Akademie," *Monatsberichte der Deutschen Akademie der Wissenschaften zu Berlin*, *6*, 859–863.

1965 72. "Alejandro de Humboldt y el Canal de Panamá," *Revista*, H. *3*, 24–25.

73. "Die Behandlung des 'Problème des dés' in den Anfängen der Wahrscheinlichkeitsrechnung," *Monatsberichte der Deutschen Akademie der Wissenschaften zu Berlin*, *7*, 70–76.

74. "Aus der Entstehung der Fachsprache der Wahrscheinlichkeitsrechnung," (Mit unveröffentlichten Ausführungen von C. F. Gauß), *Forschungen und Fortschritte*, *39*, 142–144.

75. "Alejandro de Humboldt y Cuba," *Revista ilustrada de la RDA*, *6*, H. *3*, 35–36.

76. "Der Zugang an Briefen Alexander von Humboldts hält an," *Spektrum*, *11*, 55–58.

77. "Primzahl- und Zufallsserien," *Praxis der Mathematik*, *7*, 181–182.

78. "Die Probleme der Schwereänderung und der Polhöhenschwankung sowie Fragen der Sterblichkeits- und Blitzstatistik in einem Brief von C. F. Gauß an Alexander von Humboldt," *Forschungen und Fortschritte*, *39*, 357–361.

79. "Dirichlet über Weierstraß," *Praxis der Mathematik*, *7*, 309–312.

1966 80. "Über die Beziehungen zwischen C. F. Gauß und F. W. Bessel," *Mitteilungen der Gauß-Gesellschaft Göttingen*, Nr. *3*, 7–20.

81. "Karl Weierstraß. Ausgewählte Aspekte seiner Biographie," *Journal für die reine und angewandte Mathematik*, *223*, 191–220. Bulgarische Übersetzung in: *Fiziko- matematičesko Spisanie*, *9* (1966), 39–43.

82. "Die Berufung von Weierstraß nach Berlin," *Festschrift zur Gedächtnisfeier für Karl Weierstraß 1815–1965*. Hrsg. v. H. Behnke und K. Kopfermann (Köln und Opladen: Westdeutscher Verlag), 41–52.

83. "Karl Weierstraß in seinen wissenschaftlichen Grundsätzen," *Sudhoffs Archiv*, *50*, 305–309.

84. "K. Weierstraß und A. v. Humboldt," *Monatsberichte der Deutschen Akademie der Wissenschaften zu Berlin*, *8*, 33–37.

85. "Ein Brief von Wolfgang Bolyai," *Mathematische Nachrichten*, *32*, 341–346.

86. "Gottfried Wilhelm Leibniz," *Wissenschaft und Fortschritt*, *16*, 482–487.

87. "Richard Dedekind im Urteil der Berliner Akademie," *Forschungen und Fortschritte*, *40*, 301–302.

88. "Aus den Arbeiten der A.-v.-Humboldt-Kommission," *Spektrum*, *12*, 356–359.

1967 89. "Ein unbekanntes Schreiben von N. H. Abel an A. L. Crelle," *Nordisk Matematisk Tidskrift*, *15*, 25–32.

90. "Auf den Spuren des mathematischen Glückritters Ferdinand von Sommer," *Forschungen und Fortschritte*, *41*, 235–238.

91. "Zur Einleitung" in: *G. Eisenstein. Mathematische Abhandlungen* (Hildesheim: Georg Olms Verlagsbuchhandlung) V–IX.

92. "Zur Geschichte mathematischer Einsendungen an die Berliner Akademie," *Monatsberichte der Deutschen Akademie der Wissenschaften zu Berlin*, *9*, 216–222.

93. "Beurteilung und Verwendung einer 'lebenden Rechenmaschine' [Z. Dase] durch C. F. Gauß und die Berliner Akademie," *Forschungen und Fortschritte*, *41*, 361–364.

94. "Überblick über die Studien von G. W. Leibniz zur Wahrscheinlichkeitsrechnung," *Sudhoffs Archiv*, *51*, 79–85.

95. "Carl Friedrich Gauß im Spiegel seiner Korrespondenz mit Alexander von Humboldt," *Mitteilungen der Gauß-Gesellschaft Göttingen*, Nr. *4*, 5–18.

1968 96. "Alexander von Humboldts wissenschaftsorganisatorisches Programm bei der Übersiedlung nach Berlin," *Monatsberichte der Deutschen Akademie der Wissenschaften zu Berlin*, *10*, 142–147. Nachdruck in: *Journal für die reine und angewandte Mathematik*, *250* (1971), 1–2.

97. "Die Datierung der Briefe Alexander von Humboldts—dargestellt als Modellfall für die editorische Bearbeitung naturwissenschaftlicher Briefe und Dokumente," *Monatsberichte der Deutschen Akademie der Wissenschaften zu Berlin*, *10*, 639–647.

98. "Attempt at a Classification of Unpublished Sources in the more recent History of Astronomy in German-speaking Countries," *Vistas in Astronomy*, *9*, 237–243.

99. "O nezaveršennom izdanii trudov K. Wejerstrassa," *Actes du XIe Congrès International d'Histoire des Sciences 1965*, Vol. *3* (1968), 235–239.

1969 100. "Der Briefwechsel zwischen Alexander von Humboldt und C. G. J. Jacobi über die Entdeckung des Neptun," *Schriftenreihe für Geschichte der Naturwissenschaften, Technik und Medizin NTM*, *6*, H. *1*, 61–67.

101. "A. Quetelet über seinen Besuch bei C. F. Gauß," *Mitteilungen der Gauß-Gesellschaft Göttingen*, Nr. *6*, 4–6.

102. "Did Husserl take his doctor's degree under Weierstrass' supervision?" *Organon*, *6*, 261–264.

103. "Les relations entre les mathématiciens français et Al. de Humboldt," *Actes du XII Congrés International d'Histoire des Sciences 1968*, T. *11* (1971), 17–21. Deutsch in: *Monatsberichte der Deutschen Akademie der Wissenschaften zu Berlin*, *11* (1969), 458–463.

104. "Versuch der Deutung einer Gaußschen Chiffre," *Monatsberichte der Deutschen Akademie der Wissenschaften zu Berlin*, *11*, 526–530.

105. "Alexander von Humboldts Weg zum Naturwissenschaftler und Forschungsreisenden," (Gemeinsam mit F. G. Lange), *A. v. Humboldt-Festschrift* (Berlin: Akademie-Verlag), 87–102. Spanische Übersetzung: S.103–117.

106. "A. v. Humboldt und die Deutsche Akademie der Wissenschaften zu Berlin," *A. v. Humboldt-Festschrift* (Berlin: Akademie-Verlag), 119–131. Spanische Übersetzung: S.133–144.

107. "Ausgewählte Illustrationen aus A. v. Humboldts amerikanischem Reisewerk," *A. v. Humboldt-Festschrift* (Berlin: Akademie-Verlag), 145–147. Spanische Übersetzung: S.149–151.

1970 108. "C. F. Gauß in Autographenkatalogen," *Schriftenreihe für Geschichte der Naturwissenschaften, Technik und Medizin NTM*, *7*, H. *1*, 60–65.

109. "Die Mathematik und ihre Dozenten an der Berliner Universität 1810–1920," (Autoreferat), *Monatsberichte der Deutschen Akademie der Wissenschaften zu Berlin*, *12*, 400–404.

110. "A. v. Humboldt in seinen Beziehungen zur Astronomie in Berlin," *Archenhold-Sternwarte. Vorträge und Schriften*, Nr. *37* (Berlin), 22 S.

111. "Alexander von Humboldts maritime Unternehmungen," *Spektrum*, [N. F.], *1*, H. *8*, 34–35.

112. "Alexander von Humboldt–ausgewählte Aspekte seines Lebens und Wirkens," *Schriftenreihe für Geschichte der Naturwissenschaften, Technik und Medizin NTM*, *7* H. *2*, 51–67.

113. "Heranziehung von Wasserzeichen zur Datierung von Briefen A. v. Humbolts," *Monatsberichte der Deutschen Akademie der Wissenschaften zu Berlin*, *12*, 540–544.

114. "Von Goethe zu Gauß. Stationen auf einer Reise Adolphe Quetelet's," *Archives Internationales d'Histoire des Sciences*, *23*, 207–213.

115. "Thomas Clausen als Astronom," *Janus*, *57*, 299–305.

1971 116. "Thomas Clausen," *Dictionary of Scientific Biography*, *3*, 302–303.

117. "Zu Dirichlets geplantem Nachruf auf Gauß," *Schriftenreihe für Geschichte der Naturwissenschaften, Technik und Medizin NTM*, *8*, H. *1*, S.9–12. Nachdruck in: *Mitteilungen der Gauß-Gesellschaft Göttingen*, Nr. *9* (1972), 47–50.

118. "Richard Dedekind," *Dictionary of Scientific Biography*, *4*, 1–5.

119. "Gotthold Eisenstein," *Dictionary of Scientific Biography*, *4*, 340–343.

120. "Streiflichter auf geophysikalische Aktivitäten Alexander von Humboldts," *Gerlands Beiträge zur Geophysik*, *80*, 277–291.

121. "Zum Gaußschen Kryptogramm von 1812," *Monatsberichte der Deutschen Akademie der Wissenschaften zu Berlin*, *13*, 152–157.

122. "Der Brief Alexander von Humboldts an Wilhelm Weber von Ende 1831—ein bedeutendes Dokument zur Geschichte der Erforschung des Geomagnetismus," *Monatsberichte der Deutschen Akademie der Wissenschaften zu Berlin*, *13*, 234–242.

123. "Die 'Memoiren Alexander von Humboldt's,'" *Monatsberichte der Deutschen Akademie der Wissenschaften zu Berlin*, *13*, 382–392.

1972 124. "Alexander von Humboldt über den Vorläufer des programmgesteuerten Rechenautomaten," *Schriftenreihe für Geschichte der Naturwissenschaften, Technik und Medizin NTM*, *9*, H. *1*, 21–24.

125. "Alexander von Humboldt," *Dictionary of Scientific Biography*, *6*, 549–555.

126. "Reinhold Hoppe," *Neue Deutsche Biographie*, *9*, 614–615.

1973 127. "Ferdinand Joachimsthal," *Dictionary of Scientific Biography*, *7*, 108–110.

128. "Leopold Kronecker," *Dictionary of Scientific Biography*, *7*, 505–509.

129. "Ernst Eduard Kummer," *Dictionary of Scientific Biography*, *7*, 521–524.

130. "Alexander von Humboldts Forschungsprogramm von 1812 und dessen Stellung in Humboldts indischen und sibirischen Reiseplänen," *Studia z dziejów geografii i kartografii* (Wroclaw, Warszawa, Kraków, Gdánsk), 471–483.

131. "Vorwort" in: *Die Jugendbriefe Alexander von Humboldts* 1787–1799. Hrsg. v. Ilse Jahn und Fritz G. Lange (Berlin: Akademie-Verlag), VII–XXII. (*Beiträge zur Alexander-von-Humboldt-Forschung*. Bd.*2*).

132. "Die Briefe von Martin Bartels an C. F. Gauß," *Schriftenreihe für Geschichte der Naturwissenschaften, Technik und Medizin NTM*, *10*, H. *1*, 5–22.

133. "Ob izbranii N. I. Lobačevskogo členom-korrespondentom Gettingskogo Naučnogo Obščestva," *Istoriko-matematičeskie Issledovanija*, *18*, 322–325.

134. "Aimé Bonpland im Urteil Alexander von Humboldts," *Wissenschaftliche Zeitschrift der Ernst-Moritz-Arndt-Universität Greifswald, math.-nat. Reihe*, *22*, 97–105.

1974 135. "Die Alexander-von-Humboldt-Forschung an der Akademie der Wissenschaften der DDR—Ergebnisse und Ziele," *Boston Studies in the Philosophy of Science*, *15*, 295–305.

136. "Alexander von Humboldt als Münzreformer," *Jahrbuch für Wirtschaftsgeschichte*, T. *2*, 201–220.

137. "Eugen Netto," *Dictionary of Scientific Biography*, *10*, 24.
138. "Übersiedlung eines deutschen Mathematikers von Braunschweig nach Kazań im Jahre 1807/08. Zur Biographie von M. Bartels," *Historia Mathematica*, *1*, 65–67.
139. "Über die statistischen Zahlenregister von C. F. Gauß," *Trudy XIII Meždunarodnogo Kongressa po Istorii nauki 1971*. Sekc. *5* (1974), 150–157.
140. "O pervych naučnych rabotach M. F. Bartel'sa," *Voprosy Istorii Estestvoznanija i Techniki*, *47/48*, Nr. *2*, 119–122.
141. "Alexander von Humboldt zu Newton in Beziehung gesetzt durch C. F. Gauß," *Mitteilungen der Mathematischen Gesellschaft der DDR*, H. *1/2*, 162–167.
142. "DDR-Schrifttum über C. F. Gauß," *Mitteilungen der Mathematischen Gesellschaft der DDR*, H. *4*, 73–80.
143. "Zur Einführung" in: *Alexander von Humboldt. Eine Bibliographie der in der DDR erschienenen Literatur*. Zusammengestellt von Fritz G. Lange (Berlin: Akademie-Verlag), 5–7. (*Beiträge zur Alexander-von-Humboldt-Forschung*, Bd. *3*)
144. "F. W. Bessels Projekt einer Populären Astronomie in seinem Briefwechsel mit Alexander von Humboldt," *Archenhold-Sternwarte. Veröffentlichungen*, Nr. *6* (Berlin), 35–43.
145. "Die Gauß-Briefe in Goethes Besitz," *Schriftenreihe für Geschichte der Naturwissenschaften, Technik und Medizin NTM*, *11*, H. *1*, 2–10.

1975 146. "Carl Gustav Jacob Jacobi," in *Biographien bedeutender Mathematiker*. Hrsg. v. H. Wußing und W. Arnold (Berlin: Volk und Wissen), 375–388; (Köln: Aulis Verlag, 1978), S.375–388.
147. "Zu den Beziehungen von C. F. Gauß und A. v. Humboldt zu A. F. Möbius," *Schriftenreihe für Geschichte der Naturwissenschaften, Technik und Medizin NTM*, *12*, H. *1*, 12–15.
148. "Alexander von Humboldts Interesse an Japan," *Schriftenreihe für Geschichte der Naturwissenschaften, Technik und Medizin NTM*, *12*, H. *2*, 70–75.
149. "Gauß und Goethe. Versuch einer Interpretation ausgebliebener Begegnung," *Goethe-Jahrbuch*, *92*, 195–219.
150. "Münzgeschichtliche Studien Alexander von Humboldts," *Numismatische Beiträge*, H. *2*, 17–21.
151. "A. v. Humboldts Einflußnahme auf Reformen der Berliner Akademie," (Gemeinsam mit W. Hartke), *Wissenschaft und Fortschritt*, *25*, 162–168.
152. "Ein Porträt Alexander von Humboldts von Emma Gaggiotti-Richards," *Acta Historica Leopoldina*, *9*, 51–57.

1976 153. "Karl Weierstraß," *Dictionary of Scientific Biography*, *14*, 219–224.
154. "Schlüsselworte bei C. F. Gauß," *Archives Internationales d'Histoire des Sciences*, *26*, 264–267.
155. "Historische Einführung" in: *C. F. Gauß. Mathematisches Tagebuch 1796–1814* (Leipzig: Akademische Verlagsgesellschaft Geest & Portig K.-G.), 7–20. 2. Aufl. 1979. (*Ostwalds Klassiker*, Bd. *256*).
156. "Aus der Vorgeschichte der Pläne Alexander von Humboldts für eine russisch-sibirische Forschungsreise," *Zeitschrift für geologische Wissenschaften*, *4*, 331–336.
157. "Eine Selbstbiographie von G. Eisenstein," *Mitteilungen der Mathematischen Gesellschaft der DDR*, H. *3/4*, 150–153.
158. "Einige Abkürzungen und Zeichen des historischen Münzwesens," *Jahrbuch für Wirtschaftsgeschichte*, T. *3*, 275–278.
159. "Alexander von Humboldts Stellung in der Geschichte der Fotografie," *Bild und Ton*, *29*, 121–122.

160. "Gauss i Goethe," *Istoriko-matematičeskie Issledovanija*, *21*, 261–272.

161. "Zum Einfluß von F. von Zach auf Alexander von Humboldt," *Die Sterne*, *52*, 166–171.

162. "Adol'f Pavlovič Juškevič. Zur Vollendung seines 70. Lebensjahres," *Schriftenreihe für Geschichte der Naturwissenschaften, Technik und Medizin NTM*, *13*, *13*, H. *2*, 101–104.

1977 163. "Carl Friedrich Gauß und Alexander von Humboldt in ihren Beziehungen zur Berliner Sternwarte," *Sternzeiten*, Bd. *2*, S.5–16. (*Veröffentlichungen des Forschungsbereichs Geo- und Kosmoswissenschaften der Akademie der Wissenschaften der DDR*, H. *7*).

164. "Aus unveröffentlichten Aufzeichnungen des jungen Gauß," *Wissenschaftliche Zeitschrift der Technischen Hochschule Ilmenau*, *23*, H. 4, 7–24.

165. C. F. Gauß in seinem Verhältnis zur britischen Wissenschaft und Literatur," *Schriftenreihe für Geschichte der Naturwissenschaften, Technik und Medizin NTM*, *14*, H. *1*, 7–15.

166. "Wie entziffert man Handschriften? *Wissenschaft und Fortschritt*, *27*, 348–351.

167. "Gauß und Heyne," (Gemeinsam mit W. Hartke), *Das Altertum*, *23*, 179–184.

168. "Wie Gauß zum Astronomen wurde," *Die Sterne*, *53*, 146–150.

169. "Zwei Briefe von Gauß [an F. R. Hassler] über die Berichtigung des Heliotrops und die Organisation erdmagnetischer Messungen," *Gerlands Beiträge zur Geophysik*, *86*, 1–10.

1978 170. "Alexander von Humboldt als Initiator und Organisator internationaler Zusammenarbeit auf geophysikalischem Gebiet," *Proceedings of the XVth International Congress of the History of Science 1977*, 126–138.

171. "Sonja Kowalewski. Stationen auf ihrem Lebensweg," *Für Dich*, H. *33*, 14–15.

172. "Martin Bartels—eine Schlüsselfigur in der Geschichte der nichteuklidischen Geometrie?" *Mitteilungen der Deutschen Akademie der Naturforscher Leopoldina*. Reihe 3, *21* (1975), 137–157.

173. "Gauß als Persönlichkeit—Ansätze für ein neues Verständnis," *Abhandlungen der Akademie der Wissenschaften der DDR*, Nr. *3N*, 39–49.

1979 174. "Chr. Huygens im Spiegel von Al. von Humboldts 'Kosmos'," *Janus*, *66*, 243–247.

175. "'Herr Eugen Dühring' und die Berliner Mathematiker," *Mitteilungen der Mathematischen Gesellschaft der DDR*, H. *4*, 59–64.

176. "Bernhard August von Lindenau, Weggefährte und 'Widersacher' Goethes," *Goethe-Jahrbuch*, *96*, 221–242.

177. "J.-H. Lambert und die Berliner Akademie der Wissenschaften," *Actes du Colloque International Jean-Henri Lambert* (*1728–1777*) (Paris: Editions Ophrys) 115–126.

1980 178. "Nekotorye resul'taty novych issledovanij o Gausse," *Istoriko-matematiceskie Issledovanija*, *25*, 266–280.

179. "Weierstraß uber Gauß," *Mitteilungen der Mathematischen Gesellschaft der DDR*, *1*, 76–80.

INDEX